AF247440

Modeling Nanowire and Double-Gate Junctionless Field-Effect Transistors

FARZAN JAZAERI

École Polytechnique Fédérale de Lausanne

JEAN-MICHEL SALLESE

École Polytechnique Fédérale de Lausanne

CAMBRIDGE
UNIVERSITY PRESS

University Printing House, Cambridge CB2 8BS, United Kingdom

One Liberty Plaza, 20th Floor, New York, NY 10006, USA

477 Williamstown Road, Port Melbourne, VIC 3207, Australia

314–321, 3rd Floor, Plot 3, Splendor Forum, Jasola District Centre, New Delhi – 110025, India

79 Anson Road, #06-04/06, Singapore 079906

Cambridge University Press is part of the University of Cambridge.

It furthers the University's mission by disseminating knowledge in the pursuit of education, learning, and research at the highest international levels of excellence.

www.cambridge.org
Information on this title: www.cambridge.org/9781107162044
DOI: 10.1017/9781316676899

© Cambridge University Press 2018

First published 2018

Printed in the United Kingdom by Clays, St Ives plc

A catalogue record for this publication is available from the British Library.

ISBN 978-1-107-16204-4 Hardback

Contents

Foreword

Since its first publication in 2009, the junctionless transistor [1] proved to be a very popular device amongst semiconductor research groups worldwide. Junctionless transistors are very simple to manufacture and the junctionless architecture has been successfully applied to many semiconductor materials including single-crystal and polycrystalline silicon and germanium, III-V compounds, ZnO, Indium-tin oxide (ITO), transition metal dichalcogenides, etc.

Figure 0.1 shows the number of publications and citations in Web of Knowledge corresponding to the search word "junctionless transistor". Well over a hundred papers were published on junctionless transistors each year during 2009–16, with more than a thousand corresponding citations.

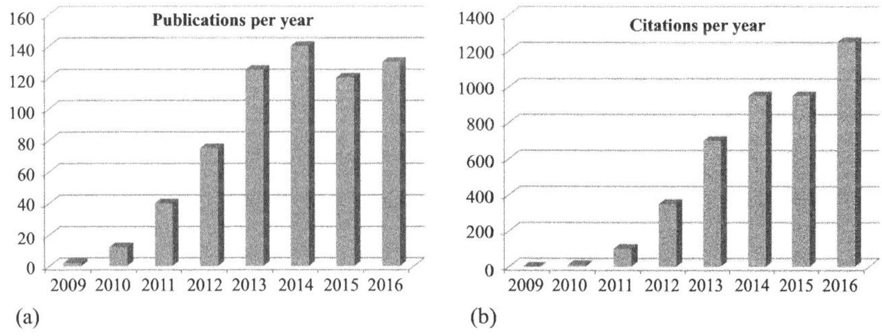

Figure 0.1 The number of publications (a) and citations (b) per year, corresponding to the search word "junctionless transistor" in Web of Knowledge.

Junctionless transistors are also increasingly being used as detectors such as gas, chemical, biochemical, and pH sensors [2, 3]. They are also being considered for the formation of low-thermal-budget active interconnects in future monolithic 3D integrated circuits [4].

Although the electrical characteristics of a junctionless transistor are quite similar to those of a conventional inversion-mode device, there are key differences in transport and capacitance properties: transport in a junctionless transistor is largely in the bulk of the semiconductor instead of in an inversion channel. As a result, surface scattering and trapping of carriers is reduced, which improves noise figure and allows one to make transistors in materials that have problematic interface properties, n-channel germanium MOSFETs being a good example.

Accurate modeling of the junctionless transistor is essential for comparing the performance of these devices with those of other types of transistors and for different types of applications. The publication of the book *Modeling Nanowire and Double-Gate Junctionless Field-Effect Transistors* by Doctors Farzan Jazaeri and Jean-Michel Sallese, two renown experts in the field, is thus timely and will bring valuable information to those interested in advanced device physics, simulation, and design.

Jean-Pierre Colinge
CEA-LETI, France

Preface

Metal-oxide semiconductor field-effect transistor scaling is following the prediction of the Moore's law enunciated in 1965 [5]. So far, this trend for miniaturization has never been invalided, enabling the industry of semiconductors to cope with the everlasting demand for higher performance at lower cost. However, this scaling becomes increasingly difficult to follow due to inherent process and device performance limitations for technology nodes beyond tents of nm. To stand the pace of downscaling, nonclassical device architectures have been continuously proposed in the ITRS roadmap.

The junctionless field-effect transistor is one of these nanoelectronics devices that is expected to withstand the downscaling of Complementary Metal-Oxide–Semiconductor (CMOS) technology by enabling easier fabrication processes, while allowing high performance. In addition, semiconductor nanowires largely used for label-free biosensing are essentially junctionless FETs without any gate.

Therefore, since the first implementation of junctionless FET nanowires by J. P. Colinge in 2009, growing interest in these devices in different fields of research motivated the authors to write a book dedicated to analytical modeling of double-gate and nanowire junctionless FETs. In contrast to the abundant literature on modeling and compact modeling of inversion-mode MOSFETs, analytical modeling of field-effect transistors without junctions is still following different strategies.

After discussing the advantages and limitations of junctionless field-effect transistors in the first introductory chapter, a thorough overview of published analytical models for double-gate and nanowires configurations is presented in Chapter 2, including the mains assumptions which are introduced.

In Chapter 3 and beyond, the analytical model of the so-called EPFL junctionless field-effect transistor is presented. After discussing the roots ending with a charge-based model valid in all the regions of operation, important features are introduced gradually in Chapters 4–13, each of which targets a specific feature. These topics include nanowire versus double-gate equivalence, technological design-space, junctionless FET performance, short-channel effects, transcapacitances, asymmetric operation, thermal noise, interface traps, and a revisited model for JFETs. In addition a general mobility extraction technique is proposed.

We suggest readers see Chapter 3 for a general introduction to the charge-based model before proceeding further. This is where the main ideas are introduced that will thus be used in the following chapters.

We hope this book will be useful to people interested in modeling these new types of field-effect transistors, which are likely to find diverse applications, not only in nanoelectronics, but also in a large variety of biosensors.

The authors would like to acknowledge Doctors Lucian Barbut and Didier Bouvet who actively participated in the research done at EPFL and also Doctors Maria-Anna Chalkiadaki, Wladek Grabinski, and Professor Christian Enz who provided invaluable support. Many thanks to Professors Christophe Lallement, Benjamin Iñiguez, Matthias Bucher and Ashkhen Yesayan for their involvement in this research. Finally, we would like to thank warmly Dr. Adil Koukab, Dr. Anurag Mangla, and Nikolaos Makris for the very constructive scientific discussions.

We are also very grateful to Elizabeth Horne, Heather Brolly, and Abirami Ulaganathan who continuously assisted us in the preparation of the book.

Last, we would like to warmly thank our families and our parents. Thanks in particular to Ferdos, Jamal, Farshid, Farzad, and Parnian and Baya, Yoanna, Marion, and Loris for their support and understanding during this seemingly endless task.

Farzan Jazaeri and Jean-Michel Sallese

Abbreviations

Abbreviation	Expansion
AC	Alternating current
Acc	Accumulation
ADS	Advanced design system
AM	Accumulation-mode
ASIC	Application specific integrated circuit
ASD	Asymmetric double-gate
BJT	Bipolar junction transistor
BOX	Buried oxide
CMOS	Complementary metal-oxide-semiconductor
DC	Direct current
Dep	Depletion
DFT	Density functional theory
DG	Double-gate
DIBL	Drain-induced barrier lowering
EDL	Electrical double layer
FB	Flat-band
FET	Field-effect transistor
FinFET	Fin-based field-effect transistor
GAA	Gate-all-around
GCA	Gradual-channel approximation
GIDL	Gate-induced drain leakage
Hyb	Hybrid
IC	Integrated circuits
IGFET	Insulated-gate semiconductor field-effect transistor
IGN	Induced-gate noise
IM	Inversion-mode
ISFET	Ion-sensitive field-effect transistor
ITRS	International technology roadmap for semiconductors
JFET	Junction FET
JL	Junctionless
LHS	Left-hand side
MOSFET	Metal-oxide semiconductor field-effect transistor
NEGF	Non-equilibrium green's function
NW FET	Nanowire field-effect transistor

Abbreviation	Expansion
PSD	Power spectral density
QCE	Quantum confinement effect
QEB	Quantum energy balance
QHO	Quantum harmonic oscillator
QME	Quantum mechanical effect
QW	Quantum well
RHS	Right-hand side
RF	Radio frequency
RSCE	Reverse short-channel effect
SC	Semiconductor channel
SCE	Short-channel effect
SOI	Silicon-on-insulator
SS	subthreshold swing
SW	Switch
TCAD	Technology computer-aided design
VeSFET	Vertical slit semiconductor field-effect transistor
VeSTIC	Integrated circuit vesfet-based

Symbols

Symbol	Description	Unit
q	Electron charge	$Coulomb$
T	Absolute temperature	Kelvin
h	Planck's constant	$J.s$
$\bar{h} = h/(2\pi)$	Reduced Planck's constant	$J.s$
k_B	Boltzmann's constant	J/K
ε_0	Permittivity of free space	F/m
ε_{si}	Permittivity of silicon	F/m
ε_{ox}	Permittivity of SiO_2	F/m
m_e^*	Electron effective mass	Kg
m_h^*	Hole effective mass	Kg
N_c	Effective density of states in conduction band	m^{-3}
N_v	Effective density of states in valence band	m^{-3}
$g_c(E)$	Density of states in conduction band	$m^{-3}J^{-1}$
$g_v(E)$	Density of states in valence band	$m^{-3}J^{-1}$
n_i	Intrinsic carrier concentration	m^{-3}
E_g	Silicon band-gap	eV
E_i	Intrinsic Fermi energy	eV
E_f	Fermi energy	eV
E_c	Bottom of conduction band energy level	eV
E_{cM}	Top of conduction band energy level	eV
E_{vm}	Bottom of valence band energy level	eV
E_v	Top of valence band energy level	eV
σ_n	Cross section of electrons	m^2
σ_p	Cross section of holes	m^2
e_n	Electron emission coefficient	s^{-1}
e_p	Hole emission coefficient	s^{-1}
v_n	Electron average velocity	ms^{-1}
v_p	Hole average velocity	ms^{-1}
τ_c	Average scattering time	s
σ	Average conductivity	$\Omega^{-1}m^{-1}$
$J_{n,drift} = qn\mu_n$	Drift electron current density	Acm^{-2}
$J_{p,drift} = qp\mu_p$	Drift hole current density	Acm^{-2}
$J_{n,diffusion}$	Diffusion electron current density	Acm^{-2}
$J_{p,diffusion}$	Diffusion hole current density	Acm^{-2}
J_{drift}	Total drift current density	Acm^{-2}
$J_{diffusion}$	Total diffusion current density	Acm^{-2}

Symbol	Description	Unit
$f(E)$	**Fermi–Dirac statistic**	–
$D_n = \mu_n k_B T$	**Electron diffusion constant (Einstein relation)**	$m^2 s^{-1}$
$D_p = \mu_p k_B T$	**Hole diffusion constant (Einstein relation)**	$m^2 s^{-1}$
W_m	**Metal work function**	V
W_s	**Semiconductor work function**	V
χ	**Electron affinity**	V
$W_{ms} = W_m - W_s$	**Metal-semiconductor work function difference**	V
$\Delta\phi_{ms}$	**Difference between metal work function and an intrinsic semiconductor reference** $\Delta\phi_{ms} = W_{ms} - (E_f - E_i)/q = W_{ms} - U_T \ln(N_D/n_i)$	V
μ	**Free-carrier mobility**	$cm^2/V.s$
μ_n	**Free electron mobility**	$cm^2/V.s$
μ_p	**Free hole mobility**	$cm^2/V.s$
μ_0	**Low-field surface mobility**	$cm^2/V.s$
μ_{eff}	**Effective carrier mobility**	$cm^2/V.s$
U_T	**Thermodynamic voltage**	V
t_{ox}	**Oxide thickness**	m
X_{dep}	**Depletion width**	m
$C_{ox} = \varepsilon_{ox}/t_{ox}$	**Gate oxide capacitance**	F/m^2
$C_{dep} = \varepsilon_{si}/X_{dep}$	**Depletion capacitance**	F/m^2
C_{eq}	**Series combination of C_{ox} and C_{dep}**	F/m^2
N_D	**Uniform donor concentration in channel**	cm^{-3}
N_A	**Uniform acceptor concentration in substrate**	cm^{-3}
N_s	**Trap density of states**	cm^{-2}
n	**Electron density**	cm^{-3}
p	**Hole density**	cm^{-3}
T_{sc}	**Silicon thickness**	m
$T_{sc,min}$	**Minimum silicon thickness in VeSFET**	m
$T_{sc,max}$	**Maximum silicon thickness in VeSFET**	m
$C_{si} = \varepsilon_{si}/T_{sc}$	**Channel capacitance**	F
W	**Channel width**	m
h	**Channel height**	m
Q_{fix}	**Fixed-charge density**	C/m^2
Q_m	**Mobile charge density**	C/m^2
$Q_{m,s}$	**Local mobile charge density at source**	C/m^2
$Q_{m,d}$	**Local mobile charge density at drain**	C/m^2
$Q_{m,FB}$	**Local mobile charge density at flat band**	C/m^2
$Q_{m1,2}$	**Internal mobile charge densities**	C/m^2
$\overline{Q}_{m,s}$	**Global source charge density**	C/m^2
$\overline{Q}_{m,d}$	**Global drain charge density**	C/m^2
$Q_G = \overline{Q}_{m,s} + \overline{Q}_{m,d}$	**Total gate charge density**	C/m^2
$Q_{sc} = Q_m + Q_{fix}$	**Total charge density in semiconductor**	$C/m2$
Q_{ss}	**Interface charged trap density**	$C/m2$
$V_{GS,FB}$	**Flat-band gate voltage at source**	V
$V_{ch}(y)$	**Channel potential**	V
x	**Coordinate across the gates**	m
y	**Coordinate along the channel**	m

Symbol	Description	Unit
V_{GS}	Gate to source voltage	V
V_{DS}	Drain-to-source voltage	V
V_{th}	Threshold voltage	V
$\Psi(x)$	Potential distribution	V
$\Psi_s = \Psi(x = \pm \frac{T_{sc}}{2})$	Surface potential	V
$\Psi_0 = \Psi(x = 0)$	Center potential	V
E_s	Electric field at surface	V/m
I_{DS}	Drain current	A
V_{th}	Threshold voltage	V
V_T	Charge threshold voltage	V
R	Radius	m
θ	Mobility reduction coefficient	$1/V$
$\Psi_{0,FB}$	Center potential at flat-band	V
$\Psi_{s,FB}$	Surface potential at flat-band	V
Ψ_{BCP}	Body center potential	V
$\Psi_{BCP,min}$	Minimum body center potential	V
$\Psi_{s,min}$	Minimum surface potential	V
Ψ_{0c}	Critical value for the center potential	V
Ψ_{sc}	Critical value for the surface potential	V
Ψ_{ox}	Oxide potential	V
Ψ_{ext}	Extremum potential across two gates	V
X_{ext}	Extremum position across two gates	m
E	Electric field	V/m
E_s	Surface electric field	V/m
E_0	Center electric field	V/m
E_{ox}	Electric field in the oxide	V/m
E_x	Longitudinal electric field	V/m
E_y	Lateral electric field	V/m
E_t	Trap-energy level	eV
$V_{GS,crit}$	Critical value for the gate potential	V
$I_{critical}$	Critical value for the drain-to-source current	A
g_{ds}	Drain trans conductance	Ω^{-1}
g_m	Gate transconductance	Ω^{-1}
V_{bi}	Built-in potential	V
g_{ch}	Local channel conductance	Ω^{-1}
$S_{\Delta I_{DS}}$	PSD of drain current thermal noise	A^2/Hz
$S_{\Delta I_G}$	PSD of induced gate noise	A^2/Hz
$S_{\Delta I_{DS}\Delta I_{G^*}}$	PSD of cross-correlation noise	A^2/Hz
$C_{GD} = \partial Q_G / \partial V_{DS}$	Gate–drain intrinsic capacitance	F/m^2
$C_{GG} = \partial Q_G / \partial V_{GS}$	Intrinsic gate capacitance	F/m^2
$C_{DD} = \partial \overline{Q}_{m,d} / \partial V_{DS}$	Drain–drain capacitance	F/m^2
$C_{SD} = \partial \overline{Q}_{m,s} / \partial V_{DS}$	Source–drain capacitance	F
$C_{DG} = \partial \overline{Q}_{m,d} / \partial V_{GS}$	Drain–gate capacitance	F/m^2
$C_{SG} = \partial \overline{Q}_{m,s} / \partial V_{GS}$	Source–gate capacitance	F
ω	Angular frequency	rad/s
f	Frequency	Hz

1   Introduction

1.1   The Birth of the Transistor

Since its invention in 1907 by Lee de Forest, the vacuum tube has been a major driver for electronics and communications technologies. However, vacuum tubes have conceptual limits. Besides being quite expensive, they have reliability issues, high energy consumption, and miniaturization limits. To address these limitations, AT&T, one of the main US phone companies, tasked its R&D unit – Bell Laboratories – with the development of an innovative device that would be cheaper, smaller, and more reliable – in short, a good substitute for the vacuum tube.

The initial idea for "transistors" came from Edgar Julius Lilienfeld in 1925 [6]. The scientist patented in Canada (1925) and successively in the United States (1928) the first field-effect semiconductor device designed to change the resistivity with applied voltages. However, due to the lack of dedicated technology, Lilienfeld never had the chance to validate his concept. It's interesting to note that his transistor idea was quite similar to today's accumulation-mode (AM) transistors.

The point-contact transistor built by John Bardeen and Walter Brattain at Bell Laboratories in the United States in December 1947 was the first solid-state transistor ever created. The two scientists, led by the physicist William Bradford Shockley, discovered the new amplification device while working on theory and experiments on solid-state materials (germanium in this case). This first proof of "controlled resistance" and "amplification" in a solid-state device is considered as the birth of the transistor, and its inventors were jointly awarded the Nobel Prize in physics in 1956 for this achievement.

After invention of the point-contact transistor, Bell Labs followed up with design and fabrication of solid-state amplifiers, gaining a deep understanding of transistors physics. A fully working junction transistor based on Shockley's design, consisting of adjacent semiconductors junctions, was fabricated in 1951 [7]. In 1954, at Texas Instruments Gordon Teal demonstrated the first silicon transistor using the knowledge gained by growing high purity crystals while working at Bell Labs.

1.2   The Metal-Oxide–Semiconductor Field-Effect Transistor

By the end of 1950s, all solid-state transistors were bipolar junction transistors (BJT). However, in 1959, John Atalla and Dawon Kahng at Bell Labs operated the field-effect transistor anticipated by Lilienfield and Shockley, called the insulated-gate semiconductor field-effect transistor (IGFET). The first version of IGFET was the metal-oxide–semiconductor field-effect transistor (MOSFET) and has been used since, although the metal was replaced for many years by doped poly-crystalline silicon (poly-Si) – from where the name metal insulator semiconductor field-effect transistor (MISFET). The idea is to modulate the conductivity of the channel (between source and drain) via voltage on the gate electrode. The conductivity in such MISFETs can be modulated in three different ways. One is by depleting the doped semiconductor from its majority carriers, which targets the well known junction FET (JFET). Another option is to enhance the majority carrier concentration, which is known as the accumulation-mode MOSFET. Last, the most popular way conductivity can be modulated is by using inversion-mode (IM) MOSFETs, where the gate voltage is used to invert the semiconductor in a region very close to the semiconductor–insulator interface. Among the technologies developed over time, CMOS technology has proven to be one of the most important achievements in engineering history.

1.3   Moore's Law, Limits of CMOS Scaling, and Alternative MOSFET Structures

1.3.1   Scaling in Bulk MOSFETs

With device scaling in MOSFETs, performance improvements in large-scale integrated circuits (LSI) have been impressive. In particular, reducing the device dimensions has greatly improved the performance of integrated circuits. For the past three decades, scaling of CMOS devices has been the key factor for improving device density, speed, and power reduction in integrated circuits. Scaling of CMOS technology is the main driving force in the semiconductor industry. However, scaling of conventional bulk MOSFETs is also reaching some intrinsic limits.

Scaling improves cost, speed, and power per function with every new technology generation. Gordon Moore made an empirical observation in the 1960s that the number of devices on a chip will double every 18 months [5]. The gate length of a MOSFET has been reduced from around $10\,\mu m$ in 1970 to $14\,nm$ in today's technology node. Maintaining the pace of downscaling is the driving force for further development of the semiconductor industry. However, improvement in performance through miniaturization must cope with severe short-channel effects [8, 9] and inherent limitations of silicon [10]. Decreasing the minimum feature size of the MOSFET brings forward some critical issues, which, in addition to the technology, are related to the device principle. For instance, higher gate capacitance is needed to drive more current while the supply voltage is decreasing. Similarly, to avoid destructive breakdown and leakage current in SiO_2 films thinner than $1.5\,nm$, high-k

dielectrics i.e., HfO_2 are used in place of SiO_2 [11–13]. The most critical issues that come along with miniaturization are as follows:

- Subthreshold swing degradation
- Drain-induced barrier lowering (DIBL)
- Random dopant fluctuations (RDF)
- Threshold voltage roll-off
- Hot carriers
- Reliability issues
- Gate leakage current.

The traditional approach to addressing these drawbacks is to increase the doping density in the channel and the gate capacitance. However, some semiconductor foundries have started investigating various solutions while proposing new device structures such as the fully depleted SOI MOSFET [14], the double-gate (DG) MOSFET [15], SiGe MOSFET [16], junctionless nanowire FETs [1], vertical slit FET [17–19], tunnel FETs [20], and even quantum dot devices [21].

As the density is further increased, extrinsic elements are becoming extremely important as well. These require innovative integration schemes such as vertical transistors and monolithic 3D integration [17, 18]. In the same spirit, a generic device concept known as the vertical slit field-effect transistor (VeSFET) was proposed by Wojciech Maly [17]. This 3D architecture aimed at being integrated in SOI technology was conceptualized in different configurations such as bipolar transistors and IM FETs for instance [22, 23].

1.3.2 Silicon-on-Insulator MOSFETs

Silicon-on-insulator (SOI) MOSFETs were the first category of new transistor architecture to disrupt the classical bulk CMOS evolution. In the 2000s, some companies (e.g., IBM, AMD) started using SOI substrates. Fully depleted SOI MOSFETs scale in a very similar way as bulk devices, but the SOI MOSFET has several advantages compared to its bulky counterpart, especially for short-channel effects. The buried oxide cuts most of the bulk leakage current that dominates in downscaled bulk MOSFETs due to the relatively large DIBL and subthreshold swing (SS), and the smaller source and drain junction areas minimize the junction capacitances, which helps increase the frequency and decrease the dynamic power consumption [24].

Another advantage of SOI technology is the use of multigate transistors with different topologies: double-gate, trigate, Ω-gate, π-gate, and gate-all-around (GAA), and so on. These devices have superior performance due to excellent control of short-channel effects giving rise to steeper subthreshold slope and lower DIBL [15].

The first commercially available CMOS technology making use of multigate architecture, the so-called trigate transistors, was introduced in 2012 by Intel for the $22\,nm$ node. The trend was followed successively by TSMC in late 2013. Samsung and Global Foundries announced their FinFET technology for the $12\,nm$ node in September 2017. In conclusion, fully depleted SOI technology is expected to be the short-term CMOS-compatible solution that combines low-power consumption and high CMOS circuit performance.

1.4   The Junctionless Concept

Although these devices based on ultrathin channels have been proposed and developed, the fabrication complexity and the related increased manufacturing costs explain why most of them are still far from being used in large-scale production.

The *pn* junction is a key element in MOSFET technology as it is used to isolate the device from the substrate and restrict the current flow in the inversion layer. Formation of source/channel and drain/channel junctions with high doping concentration gradients has been a critical issue in designing short-channel devices. In addition, controlling the source/drain junction depths at the nanometer scale is very difficult. For example, in a current CMOS technology, the doping must switch from a n-type doping higher than $1 \times 10^{19}\,cm^{-3}$ in the source/drain extensions, to a p-type doping of about $1 \times 10^{18}\,cm^{-3}$ in the channel, within a few nanometers [1]. An ultrafast annealing process and low thermal budged are needed to meet these requirements. From this point of view, having a device without a junction can be technologically and economically beneficial. Getting rid of this technological step was the main motivation for the integration of Schottky contacts for source and drain [25, 26].

Among these arrangements, one configuration of interest in this text is the so called junctionless field-effect transistor (JLFET). This type of device consists of a uniformly doped channel from source to drain [1, 27] i.e., they do not have a *pn* junction for source and drain, which helps get rid of abrupt doping profile issues.

The first experimental evidence came from Jean-Pierre Colinge and coworkers in 2010 [1] (see Figure 1.1) who proposed a planar triple-gate architecture implemented in ultrathin body SOI in 2009 [28]. It is worth noting that the principle of operation of junctionless FETs devices is similar to that of the JFETs invented by Shockley in 1952, in that they are voltage-controlled resistors based on majority carriers and without source and drain junctions. However, in JFET, some junctions are implemented for the gates reversed biased to control the channel (modeling of JFET will be revisited in Chapter 12).

1.4.1   Working Principle of Junctionless MOSFETs

The electrical characteristics of the junctionless transistor look very similar to the regular MOSFET i.e., for an n-type channel device the current increases when increasing the gate-to-source voltage. However, the principle of operation of junctionless double-gate MOSFETs is based on a current flowing in the volume of a heavily doped silicon layer and not at the Si–SiO_2 interface as in regular MOSFETs. In addition, the charge-voltage relationship is fundamentally different. Indeed, an

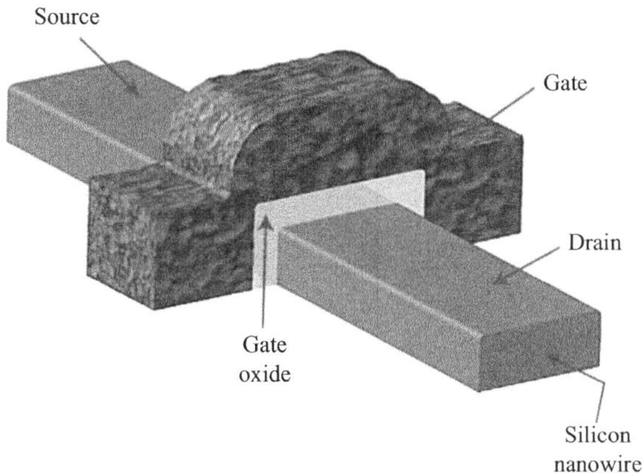

Figure 1.1 A junctionless double-gate MOSFET concept presented by Colinge et al. [1]. Reprinted from [1] with permission.

additional quadratic term comes into play in the junctionless double-gate MOSFET when operating below the flat-band.

The junctionless FET is made of a heavily doped semiconductor layer isolated from the gate electrodes by a thin dielectric, as for the IM MOSFET. However, besides this common feature, many differences exist between these two devices. The electrical conductivity in junctionless FETs is governed by majority carriers in the volume rather than by an inversion layer as in junction-based MOSFETs. Therefore, the channel conductance is modulated by depleting or enhancing majority carriers in the semiconductor layer. Whereas in inversion-mode MOSFETs, the flat-band voltage means that the device is turned off, in junctionless FETs flat-band means that the channel is neutral and therefore very conductive (the free-carrier density is almost equal to the doping density; see Figure 1.2). In fact, for n-type channels, the flat-band condition is situated above the threshold voltage ($V_{th} < V_{GS,FB}$). When the gate potential is below the threshold voltage in an n-type channel, the channel is depleted from majority carriers. As the gate voltage is further decreased, full depletion can be achieved and the device is turned off (see Figure 1.2a). On the other hand, when the gate potential is above the threshold voltage but still below the flat-band voltage, the channel region is partially depleted (see Figure 1.2c). Finally, increasing the gate voltage above the flat-band creates an accumulation channel that also "behaves" more or less as the inversion channel of an inversion-mode MOSFET.

Work function engineering is obviously critical in junctionless FETs since for the unbiased device to be in an *off*-state, the work function difference between the gate material and the doped semiconductor must be tuned so the channel is fully depleted at $V_{GS} = 0$ (see Figure 1.2a).

Figure 1.3a shows the drain current versus the gate voltage in n- and p-type devices having a width of 30 *nm* and a length of 1 *μm*. Figure 3b and c shows the experimental output characteristics of n- and p-channel gated resistors, respectively.

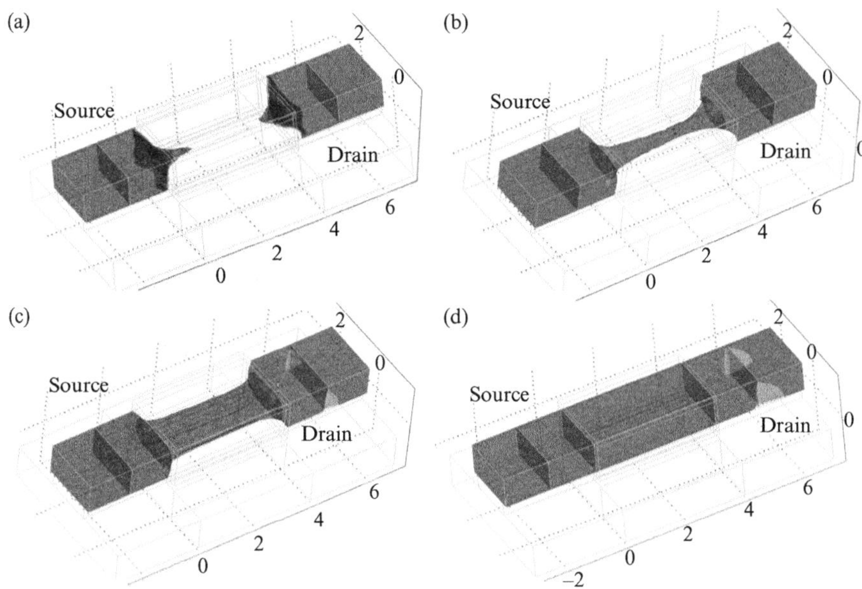

Figure 1.2 Electron concentration contour plots in an n-type junctionless gated resistor (a)–(d). Plots result from simulations carried out for a drain voltage of 50 mV and for different gate-voltage values: (a) below threshold, the channel region is depleted of electrons; (b) at threshold, a string-shaped channel of neutral n-type silicon connects source and drain; (c) above threshold, the channel neutral n-type silicon expands in width and thickness; (d) when a flat energy band situation is reached, the channel region has become a simple resistor. The plots were generated by solving the Poisson, drift-diffusion, and continuity equations. The device has a channel width, height, and length of 20, 10, and 40 nm, respectively. The n-type doping concentration is 10^{19} cm^{-3}. Reprinted from [1] with permission.

1.4.2   Diversity in Junctionless Architectures

The gate geometries of semiconductor field-effect devices have been improved to enhance their performance. Thus, junctionless architectures can have different shapes such as circular nanowires, multiple-gate nanowires [28], or vertical slit transistors (VeSFETs) [19]. The first working junctionless transistor was implemented as trigate silicon nanowire on SOI substrate [29].

Junctionless Double-Gate MOSFETs

Among the junctionless architectures that can sustain scalability is the double-gate topology (JL DG MOSFET). Despite its "simple" design, this is one of the most difficult architectures to implement because it requires precise alignment of the independent gates. Nevertheless, from a modeling point of view, the main advantage is to treat the device as a simple 1D system.

Junctionless Nanowire FETs

One-dimensional nanostructures such as nanowires and nanotubes are very attractive building blocks for nanoelectronics because of their relatively simple fabrication

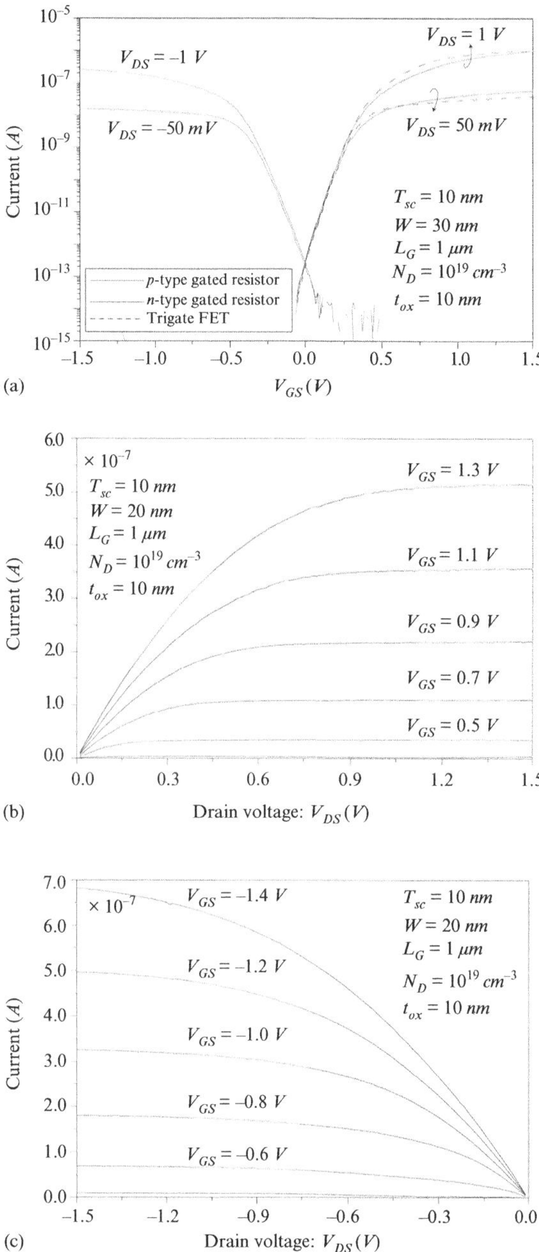

Figure 1.3 Current–voltage characteristics: Drain current versus gate voltage for drain voltages of +50 mV and +1 V. The width of the device is 30 nm. (a) The curve for a classical trigate FET is shown for comparison. Measured output characteristics of gated resistors: drain current versus drain voltage for different values of gate voltage for (b) an n-channel gated resistor and (c) a p-channel gated resistor. The width of the nanowires, W, is 20 nm and the gate length, L_G, is 1 μm. Reprinted from [1] with permission.

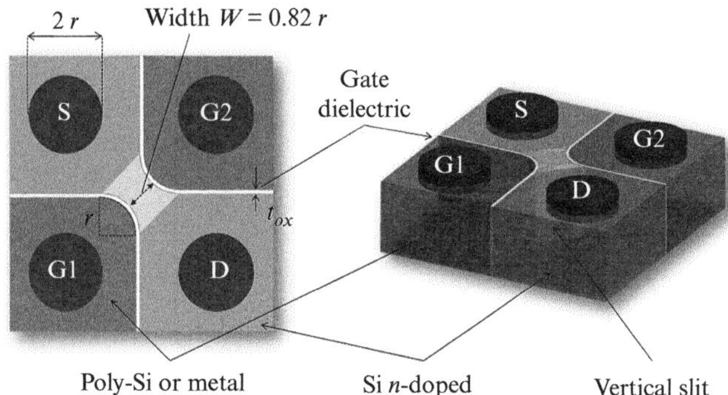

Figure 1.4 The architecture of a single-cell n-type VeSFET, as proposed by Maly [17], for high-density integration. Reprinted from [17] with permission.

processes combined with excellent electrostatic control of the channel. Indeed, junctionless nanowires can be obtained from different techniques (some of them relying on self-limited oxidation of silicon), which is obviously an advantage in academia for instance. Although electrical characteristics of double-gate (DG) and nanowire (NW) junctionless FETs (JL FETs) are quite similar, the way analytical models are derived is quite different and becomes more complex for nanowires due to analytical formulation of the Poisson–Boltzmann equation in cylindrical coordinates.

Junctionless Bulk FinFETs

Likewise Inversion mode bulk FinFETs which have been fabricated and used in real circuits (Intel), junctionless bulk FinFETs could be effectively implemented in bulk CMOS technology, getting rid of SOI technology constraints. Analysis by Han and coworkers [30, 31] suggested that short-channel junctionless FinFET transistors integrated in bulk silicon can perform as inversion-mode FinFETs and outperform SOI-based junctionless FinFETs, especially in regards to the off-current, DIBL, and subthreshold swing (SS). This means a cheaper bulk technology could be used to integrate junctionless FETs.

Junctionless Vertical Slit FETs

In addition to nanowire and double-gate shapes, the concept of junctionless FET was introduced by Wojciech Maly [17] in the name of VeSFETs. This new type of transistor is expected to relax lithography constraints by using highly regular arrays to achieve 3D integration at the nanometer scale (see Figure 1.4). All the devices in a VeSTIC can be arranged in regular arrays, evenly spaced, forming a "transistor canvas" (see Figure 1.5), with lower manufacturing costs and faster design capabilities [32]. Moreover, digital [33, 34] and analog [35] functions are expected to be easily integrated with a regular layout. Potentially, every transistor could be interconnected by upper and bottom terminals. In addition to the junctionless architecture, the "vertical slit" topology could also be applied to inversion-mode FETs and bipolar devices.

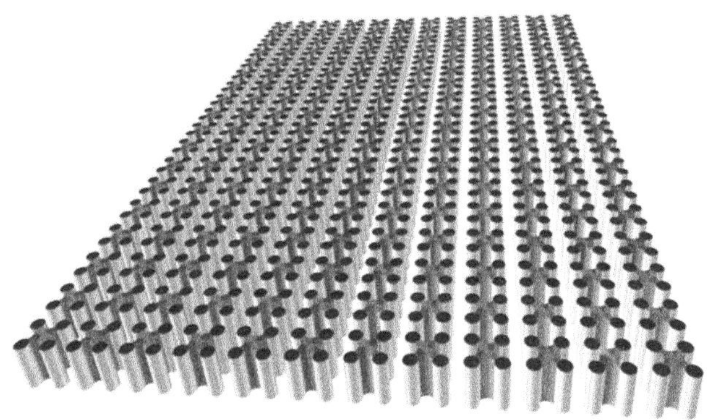

Figure 1.5 VeSFET in canvas allowing contacts both above and below the active device layer. Reprinted from [19] with permission.

Today, the principle operation has been proven [19] together with well controlled short-channel effect. It has been demonstrated [32, 33] that single independent gate VeSFETs can also act as AND and OR gates by tuning the slit width, providing a 60 percent area saving for a NAND logic gate.

1.5 Short-Channel Effects in Junctionless FETs

In addition to mitigating source and drain junctions implementation constraints, the principle of operation of junctionless FETs is expected to alleviate the short-channel drawbacks of MOSFETs. For instance, in the *off*-state, the charge depletion induced by the gate extends to the end of the gate electrode, inducing a longer resistive channel and lowering the *off*-state current. Conversely, in the *on*-state, the homojunction between the source/drain and the channel is expected to lower the access resistance. In addition, inversion-mode MOSFETs are sensitive to the semiconductor/gate insulator interface whereas for junctionless transistors the current flows in the "center" of the channel, meaning surface and roughness scattering should be less of a concern.

It has been reported that thin-film double-gate devices in AM have better short-channel effects and related DIBL than inverted channel multigate FETs [36], meaning that junctionless devices could also be more immune to short-channel effects.

The DIBL is another critical parameter that affects both static and dynamic operation in the off-state. Numerical simulations [37–41] and analytical solutions for DIBL in junction-based double-gate MOSFETs [42–48] have been done, some of them making use of the conformal mapping technique [48, 49]. A comparison of DIBL and subthreshold swing versus physical gate length for junctionless and IM multigate FETs is shown in Figure 1.6 [28]. As can be seen, the junctionless multigate FET has better short-channel characteristics than the inversion-mode device.

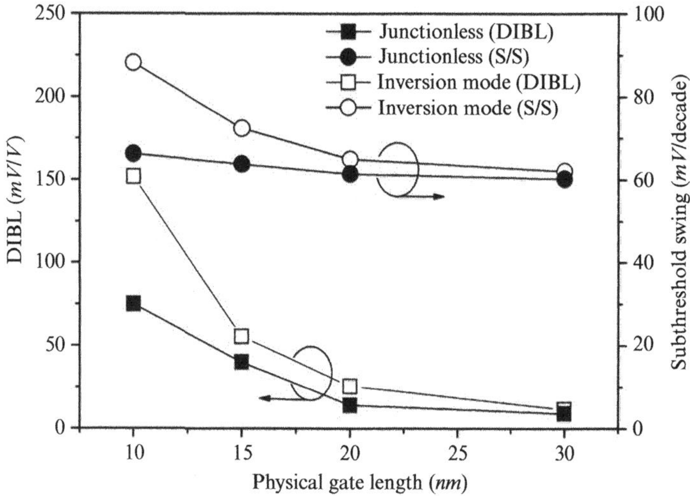

Figure 1.6 DIBL and subthreshold swing at $V_{DS} = 50\,mV$ in junctionless and regular IM devices with $T_{sc} = 5\,nm$. Reprinted from [28] with permission.

However, there are still some technological limitations that have to be quantitatively assessed in order to optimize the device performance while scaling down its dimensions. Increasing the doping to obtain a high *on*-current at flat-band must be mitigated in regard to other constraints. Indeed, one should be able to set the channel in full depletion to turn it off, which might be impossible if the channel thickness and/or the doping density are too high [50, 51]. In addition, the constraint on doping could also be a major hindrance to improve figures of merit such as time delay and power consumption, as discussed in [52].

1.6   Mobility in Junctionless FETs

The free-carrier mobility in junctionless devices was reported to be higher than in bulk semiconductor [53], still increasing with the gate bias above flat-band, in contrast to what is expected in highly doped semiconductors subjected to enhanced Coulomb scattering from ionized dopants. Upon measurement, a 40 to 110 percent increase in mobility compared to the values in bulk silicon for doping levels ranging from $1 \times 10^{19}\,cm^{-3}$ to $4 \times 10^{19}\,cm^{-3}$ has been reported. This unexpected higher mobility was attributed to lower Coulomb scattering due to a screening of ionized donors. Sore et al. [54] reported that mobility is also dependent on the channel width and that the surface roughness could be the dominant scattering mechanism, in addition to ionized impurities.

However, comparison of long-channel trigate inversion-mode FETs and junctionless FETs in [55] revealed a lower current in junctionless FETs, as for g_m and g_{ds}, which was attributed to lower mobility.

1.7   The Critical Aspect of Random Dopant Fluctuation

Another issue, and maybe the most severe one, is the fluctuation of the local doping density that may exhibit large variations at the nanometer scale. The so-called random dopant fluctuation (RDF) is inherent in very small volumes where the effective number of dopant impurities is so small that this parameter becomes meaningless, inducing significant threshold voltage variations from device to device. Similarly, the overall subthreshold behavior can be significantly impacted by RDF in junctionless FETs with channel doping levels greater than $10^{19}\ cm^{-3}$ [56].

Whatever the application envisioned, analog or digital, one of the most critical parameters to be controlled in field-effect transistors is the threshold voltage. In order to avoid excessive mismatch in transistors characteristics, this parameter in particular should be kept as stable as possible among the different devices. Despite clear advantages in downscaling field-effect transistors, reducing the size of the active region controlling the current comes along with critical issues dealing with statistical considerations. Therefore, the most common sources of variability are the line-edge roughness [57–59] and the fluctuations due to randomization of dopant atoms in the channel region.

The first observations of statistical variation in the threshold voltage due to RDFs were reported by Hagiwara et al. in 1982 [60]. Hagiwara developed a simple approach based on a global expression of the local impurity concentration dependence of the threshold voltage. However, the doping variation was still considered as a macroscopic parameter, which is questionable in nanometer-scaled geometries. The first evidence of RDF as a localized effect was reported by Washburn in 1988 in AlGaAs/GaAs wires [61] and then modeled at the atomic scale by Davies and Nixon in 1989 [62]. Davies demonstrated that the effect of potential fluctuations caused by the random positions of dopant impurities was expected to give rise to large aleatory threshold voltage in narrow and short submicrometer MOSFETs. Even though the dimensions of the devices were much larger than they are today, it was predicted that these fluctuations would limit the performance of submicrometer field-effect transistors.

In addition to RDF, the granularity of the metal gate could also be a concern for threshold-voltage variability with the introduction of high-k/metal gate technology. This was evidenced by Wang et al. [63] and the granularity effect will be added to the RDF contribution, both factors being uncorrelated. In some cases, this metal-gate granularity can dominate the threshold-voltage statistical distribution when the average grain size is comparable to the dimensions of the transistor.

Analytical approaches to model the effect of RDF on the threshold voltage in MOSFETs have been developed as well as 2D and 3D numerical simulations [60, 64–67]. However, the main issue in simulating these statistical effects at the atomic scale (in contrast to continuous doping) relates to the fine mesh discretization requirement, as initially addressed in [67]. Stolk et al. in [66] predicted that threshold-voltage variation due to dopant fluctuations could become a major concern in CMOS generations beyond $18\ nm$ and that a 3D approach was necessary to improve accuracy. Interestingly, the standard deviation of compact model parameters scales inversely with the square root of the active device area.

Asenov [58] proposed a systematic 3D numerical simulation analysis of threshold-voltage fluctuations on a very large set of deep submicrometer MOS-FETs. The Poisson equation with Boltzmann statistics was solved down to the individual dopant level to get the apparent threshold voltage. Microscopic arrangements of individual dopants were taken into account, which influences the threshold voltage. The variability increases faster than $\sqrt{N_A}$, which predicts a possible issue for junctionless FETs. It is worth noting that atomistic simulations predict a higher level of fluctuations than the continuous analytical models. Additionally, including quantum mechanical corrections has been shown to enhance further fluctuations and lower the threshold voltage [57].

With the continuous downscaling of CMOS technology, this topic has clearly become a real issue in modern CMOS, and a dedicated simulation tool is available [68]. For instance, the effect of a single dopant was shown to have a considerable impact in FET with gate length below $30\,nm$, affecting the subthreshold characteristics especially when the dopant is located in the middle of the channel.

1.7.1   Random Dopant Fluctuations in Junctionless FETs

Since RDF increases with the doping density [57], the threshold voltage variability is likely to be enhanced in junctionless FETs in regard to IM transistors with low doped channels. The first work investigating RFD in junctionless devices was published by Martinez et al. [69] who ran full 3D numerical simulations based on the nonequilibrium Green function (NEGF) in a junctionless gate-all-around silicon nanowire transistor with dopants randomly dispersed in a $4.2 \times 4.2\,nm^2$ cross-section with an average density of $10^{20}\,cm^{-3}$. The worst case in terms of threshold voltage deviation is when there is a lack of donors under the gate, with shifts of some hundreds of mV with respect to uniformly doped devices. Therefore, it is not the mean value of the doping concentration but the local distribution of dopants in a critical region that seems to be the most important for variability.

Analytical modeling of RDF and its impact on the threshold-voltage variance in long-channel junctionless transistors was developed for nanowire and planar double-gate structures by Gnudi et al. [70]. The model assumes that the full depletion approximation holds, which makes possible analytical evaluation of the different physical quantities including the surface potential. RDF is translated into a global fluctuation in terms of threshold voltage by calculating the variation in the gate voltage that leads to constant channel conductivity. Next, its mean square over all possible doping profiles based on Poisson statistics gives the variance of the threshold voltage, which is maximum when the device is biased in subthreshold.

Figure 1.7 illustrates the threshold-voltage standard deviation versus the gate bias for the cylindrical junctionless silicon nanowire (SiNW) n-FETs of different gate lengths ($L_G = 20$, 30, 50, 100, and $200\,nm$) and silicon radii ($R_s = 5$ and $10\,nm$). The threshold-voltage standard deviation versus the channel length for nanowire and

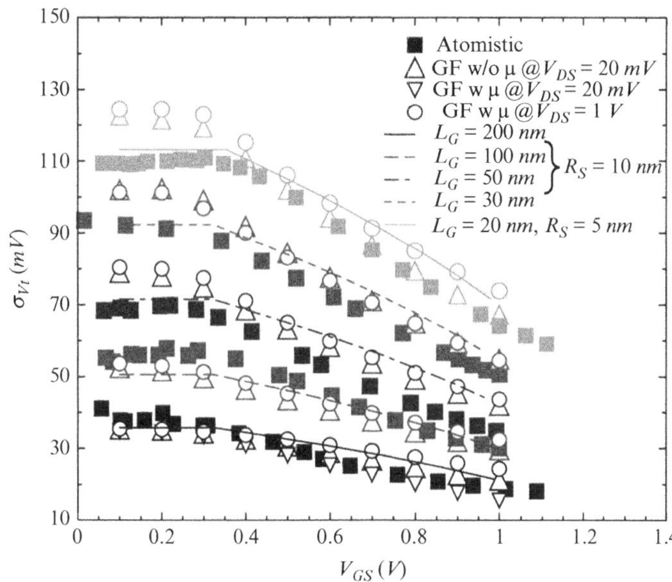

Figure 1.7 Threshold-voltage standard deviation versus gate bias for the cylindrical junctionless silicon nanowire n-FETs of different sizes computed with the proposed model in [70] (lines), with the atomistic method (Atomistic) and with the Green's functions method (GF). Reprinted from [70] with permission.

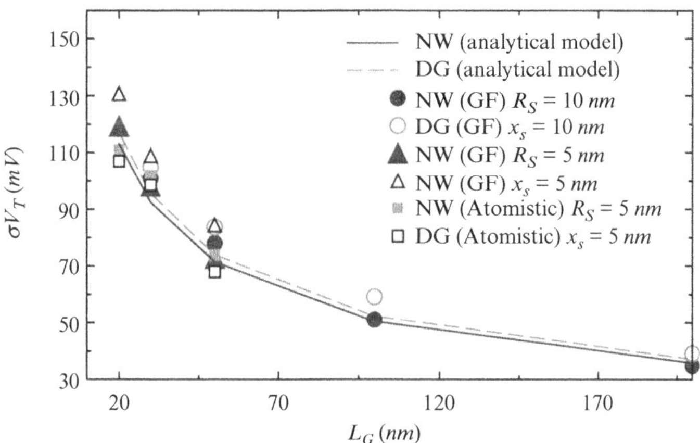

Figure 1.8 Threshold-voltage standard deviation versus gate length for the cylindrical nanowire and planar double-gate devices computed below threshold with the analytical models, the GF approach, and the atomistic approach. Reprinted from [70] with permission.

double-gate FETs is shown in Figure 1.8 [70]. When comparing the variation in threshold voltages between junctionless FET and inversion-mode FETs using 3D statistical device simulations, higher values are seen for junctionless devices in 45, 32, and 22 nm technology nodes whatever the architecture i.e., bulk, thin body SOI, and double.

In addition to RDF considerations, it is worth noting that recently full numerical modeling of the effect of doping on the mobility contrasts with the commonly accepted conclusions. Ueda et al. [71] investigated transport properties of heavily doped ultrascaled junctionless nanowire field-effect transistors by means of atomistic quantum transport simulations based on a tight binding model including electron-phonon scattering. It was found that increasing the doping above some $10^{19}\,cm^{-3}$ in nanowires increases the free-carrier mobility since screening enhances the impurity limited mobility contribution, which is known to be the dominant term in junctionless FETs [27].

Experimental evidence for this unexpected trend was reported in [72] at moderate levels of carrier density (but still high doping). This means high doping is not always synonym of a low carrier mobility, a fact that has always been considered to be the main limitation for junctionless FETs.

1.8   Summary

Since 2009 when the first junctionless FET were fabricated, many research teams from academia and industry have explored the junctionless technology. Various research groups have compared junctionless with inversion-mode FETs, mainly Fin-FETs, double-gate, and gate-all-around (GAA) nanowires, and highlighted that junctionless nanowire could be successfully used in place of inversion-mode FETs, as for instance in high-density NAND flash memory [73, 74]. Although a fully objective comparison is sometimes difficult, the conclusion is that junctionless transistors seem better than inversion-mode counterparts when considering low-power applications, while for high performance, inversion-mode FETs still offer the best option [75, 76]. Today, the variability induced by RDFs is not yet fully understood and could slow further technology developments, preventing this device from being used in nanoscale CMOS technology. Still, these breakthroughs in technology and compact modeling activities are useful for biosensing where core concepts developed in junctionless FETs can be used to simulate a new class of nanowire biosensors.

2 Review on Modeling Junctionless FETs

Along with the proofs of concept of junctionless FETs integrated in standard CMOS technology, significant modeling activities were initiated almost at the same time by different teams. These theoretical developments aimed at predicting electrical characteristics of junctionless FETs and bridged this new technology with circuit applications. A short review of the most relevant theoretical developments targeting modeling of junctionless FETs is given in this chapter.

The principle of operation of junctionless FETs is quite different from inversion-mode (IM) MOSFETs. Whereas in IM MOSFETs an inversion layer is created at the semiconductor–insulator interface, in junctionless FETs the role of the gate is to deplete, and in some cases accumulate, the majority carriers inside a doped semiconductor channel. Accordingly, in a junctionless FET the current flows essentially in the volume of the semiconductor.

Junctionless FETs are like the well-known JFET [1] except that these modulate the channel conductivity by means of reverse biased *pn* junctions. However, as it will be made obvious in this chapter, the way electrostatics and related current characteristics have been derived in junctionless FETs is quite different from conventional JFET models [77–79].

The purpose of this chapter is not to present an exhaustive overview of all what has been published in modeling activities in relation with junctionless FETs, but rather to illustrate some peculiar analytical approaches published at the time being. Due to different analytical treatments of the Poisson–Boltzmann equation, double gate and nanowires architectures will be discussed separately.

2.1 Modeling Junctionless Double-Gate MOSFETs

2.1.1 Full Depletion Approximation

The simplest approach to derive an analytical model including some exponential dependence of the current in deep depletion is developed in [80] (Figure 2.1). The authors propose an analytical expression of the Poisson's equation where the full depletion approximation is used to calculate the channel potential (as for accumulation-mode transistors [81]), resulting in simple expressions for the

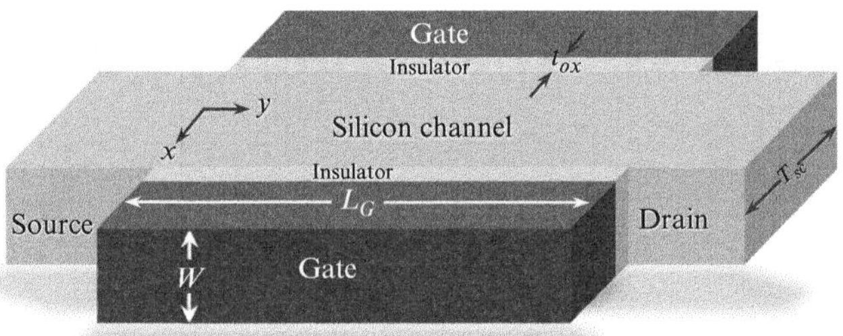

Figure 2.1 Schematic view of the *n*-type junctionless symmetric double-gate MOSFET.

subthreshold drain current. In a first step, the depletion width X_{dep} is evaluated as a function of the gate and Fermi potentials:

$$X_{dep} = \left(\frac{\varepsilon_{si}}{C_{ox}}\right)\left\{-1 + \sqrt{1 + \frac{2C_{ox}^2}{\varepsilon_{si}qN_D}\left[V_{GS} - V_{GS,FB} - V_{ch}(y)\right]}\right\}. \tag{2.1}$$

A threshold voltage is then defined by setting $X_{dep} = T_{sc}/2$ in (2.1). Next, using a Taylor expansion around the threshold voltage and neglecting higher-order terms while assuming a thin silicon channel, (2.1) can be simplified into

$$X_{dep} \approx \left(\frac{C_{eq}}{qN_D}\right) \times \left[V_{GS} - V_{GS,FB} - V_{ch}(y)\right], \tag{2.2}$$

where C_{eq} is the series combination of oxide and depletion capacitances i.e., C_{ox} and $C_{dep} = 2\varepsilon_{si}/T_{sc}$. In addition, the bulk current through the channel must satisfy Ohm's law i.e., $dV = IdR$, where dR is the differential channel resistance obtained by

$$dR = \frac{dy}{2W\mu qN_D\left(T_{sc}/2 - X_{dep}\right)}. \tag{2.3}$$

After introducing Ohm's law, the bulk current is readily obtained from (2.2) [80].

$$I_{DS} = 2\mu qN_D\left(W/L_G\right)\left[\left(\frac{T_{sc}}{2} + \frac{C_{eq}}{qN_D}\left(V_{GS} - V_{GS,FB}\right)\right)V_{DS} - \frac{C_{eq}}{qN_D}\frac{V_{DS}^2}{2}\right]. \tag{2.4}$$

Regarding the subthreshold regime i.e., deep depletion, a parabolic approximation across the gates is introduced and the current in subthreshold is expressed in terms of an error function. Since the error function tends to be "1" for a heavily doped channel, this gives back an explicit expression for the current in the subthreshold:

$$I_{DS} = qD_n\left(\frac{W}{L_G}\right)\frac{\alpha}{2}\sqrt{\frac{\pi}{\beta}}\left[1 - \exp\left(-\frac{V_{DS}}{U_T}\right)\right], \tag{2.5}$$

where α and β are respectively given by

$$\alpha = 2N_D \exp\left[\frac{1}{U_T}\left(\frac{qN_D T_{sc}^2}{8\varepsilon_{si}} + V_{GS} - V_{GS,FB} + \frac{qN_D T_{sc}}{2C_{ox}}\right)\right], \tag{2.6}$$

$$\beta = \frac{qN_D}{2U_T\varepsilon_{si}}. \tag{2.7}$$

2.1.2 Enhanced Depletion Approximation

Gnani et al. [82] derived an accurate solution of the Poisson–Boltzmann equation based on an enhanced depletion approximation for n-type and p-type channels. However, the Boltzmann statistics are still used to calculate the mobile charge density in the band. Imposing the boundary condition at the substrate/BOX (buried oxide) interface links the surface potential to the charge densities from different sources, namely Q_{SS}, the charge density at the upper surface of the substrate; Q_{bxe}, a weighted average of the BOX charge density per unit volume; and Q_{it} for charges trapped on interface states. In that study, two different situations relying on different initial conditions are considered to solve a normalized Poisson's equation within the channel.

- One is when the surface potentials at each channel interface are either accumulated or weakly depleted. This is when the space charge regions at each interface are electrically isolated by a neutral region at the center. This situation enables decoupling the surface potentials and solving the two equations independently.
- The other case is when the gate potential is close to some threshold voltage. Then, the neutral region becomes negligible and an extrema in the potential distribution is introduced as an additional unknown parameter.

In between, some approximations are introduced i.e., linearization of potential and depletion approximation, leading to a simplified solution.

Finally, the calculation of the drain current is carried out using gradual-channel approximation (GCA), which provides the subthreshold swing and threshold voltage variation versus source substrate bias.

2.1.3 Surface Potential-based Approach

A model based on the surface potential was proposed by Chen et al. in [83]. The authors calculated the electric field at the semiconductor interface (E_s) as a function of center and surface potentials and linked it to the space charge density.

When the channel is partially depleted, the channel is assumed to be fully depleted from the silicon-gate insulator interface down to a coordinate X_{dep}. Imposing a parabolic potential across the gates gives:

$$\Psi(x) = \Psi_0 - \frac{qN_D}{2\varepsilon_{si}}\left[x - \left(\frac{T_{sc}}{2} - X_{dep}\right)\right]^2 \qquad for \quad |X_{dep}| \le \frac{T_{sc}}{2}. \tag{2.8}$$

Replacing $\Psi_0 = V_{ch}$ the surface potential is obtained from

$$\Psi_s = V_{ch} - \frac{qN_D}{2\varepsilon_{si}} x_{dep}^2 . \tag{2.9}$$

The total space charge density in this limit is given by

$$Q_{sc} = 2qN_D X_{dep} = -2C_{ox}(V_{GS} - V_{GS,FB} - \Psi_s). \tag{2.10}$$

Combining Q_{sc} with (2.9) provides another link between the surface potential and the gate voltage:

$$\Psi_s = V_{ch} - \frac{C_{ox}^2}{2\varepsilon_{si}qN_D}(V_{GS} - V_{GS,FB} - \Psi_s)^2 , \tag{2.11}$$

where $\Psi_s = \Psi(T_{sc}/2)$ and $V_{GS,FB}$ is the gate voltage corresponding to the flat-band i.e., to a neutral channel.

On the other hand, expressing the gate voltage as the sum of the flat-band voltage, the voltage drop in the semiconductor and the potential drop across the oxide layer gives:

$$V_{GS} = V_{GS,FB} - \frac{qN_D}{2\varepsilon_{si}} X_{dep}^2 - \frac{qN_D}{\varepsilon_{ox}} X_{dep} t_{ox}. \tag{2.12}$$

This provides an analytical expression to calculate the depletion depth X_{dep} for a given gate voltage V_{GS}. In addition, when full depletion occurs i.e., when $X_{dep} = T_{sc}/2$, relation (2.12) defines a kind of threshold voltage (note that the meaning is different from the one in inversion-mode MOSFETs). This is where the full depletion approximation is the most accurate since the channel is totally depleted of mobile carriers.

When the device enters in accumulation, the depletion approximation becomes meaningless. Then, the electric field quickly drops down to zero, and the electric potential is $\Psi(x) \approx V_{ch}$. Thus, the electric field is obtained from the surface potential by using Taylor expansion:

$$E_s \approx \sqrt{\frac{2qN_D U_T}{\varepsilon_{si}} \left\{ \left[\exp\left(\frac{\Psi_s - V_{ch}}{U_T}\right) - 1 \right] - \frac{\Psi_s - V_{ch}}{U_T} \right\}}. \tag{2.13}$$

The authors propose to omit the term $(\Psi_s - V_{ch})/U_T$ in the square root, and the surface electric field in accumulation is expressed as

$$E_s \approx \sqrt{\frac{2qN_D U_T}{\varepsilon_{si}} \left[\exp\left(\frac{\Psi_s - V_{ch}}{U_T}\right) - 1 \right]}. \tag{2.14}$$

The continuity of the (surface) displacement vector at the semiconductor–insulator interface links the gate and surface potentials:

$$(V_{GS} - V_{GS,FB} - \Psi_s)^2 = \beta U_T \left[\exp\left(\frac{\Psi_s - V_{ch}}{U_T}\right) - 1 \right], \tag{2.15}$$

where $\beta = 2\varepsilon_{si}qN_D/C_{ox}^2$.

Whatever the channel state, accumulation ($V_{GS} > V_{GS,FB} + V_{DS}$) or depletion ($V_{GS} < V_{GS,FB}$), the drain current is obtained by the integration of the mobile charge with respect to the channel potential:

$$I_{DS} \frac{L_G}{W\mu} \overset{acc}{\approx} Q_{fix} V_{ch} - 4 C_{ox} U_T (V_{GS} - V_{GS,FB} - \Psi_s) - C_{ox}(V_{GS} - V_{GS,FB} - \Psi_s)^2$$

$$+ 4 C_{ox} U_T \sqrt{\beta U_T} \arctan\left(\frac{V_{GS} - V_{GS,FB} - \Psi_s}{\sqrt{\beta U_T}} \right) \Bigg|_S^D , \tag{2.16}$$

similarly, the drain current in depletion becomes

$$I_{DS} \frac{L_G}{W\mu} \overset{dep}{\approx} Q_{fix} V_{ch} - C_{ox}(V_{GS} - V_{GS,FB} - \Psi_s)^2 + \frac{4 C_{ox}}{3\beta} (V_{GS} - V_{GS,FB} - \Psi_s)^3 \Bigg|_S^D . \tag{2.17}$$

Finally when $V_{GS,FB} < V_{GS} < V_{GS,FB} + V_{DS}$, the channel is biased in accumulation from the source to the flat-band coordinate, and in depletion from the flat-band point up to the drain (called the hybrid state and discussed in Chapter 3). In this particular situation, the drain current is obtained by splitting the integral in two parts (see Chapter 3).

2.1.4 Simplified Current Model Involving Pinch-Off

Another approach was proposed by Sahu et al. [84] who derived the drain current using the depletion approximation, a common method for modeling junctionless FETs. When the whole channel is depleted, the authors assume that no more current flows between source and drain: the corresponding gate voltage is then defined as a pinch-off voltage, which is in fact the definition for the threshold voltage in some papers (see [83] for instance):

$$V_P = V_{GS,FB} - \frac{q N_D T_{sc}^2}{8\varepsilon_{si}} - \frac{q N_D T_{sc} t_{ox}}{2\varepsilon_{ox}} . \tag{2.18}$$

Next, the drain current is obtained by integrating the mobile charge density from source to drain:

$$I_{DS} = \frac{2\mu C_{GG} W}{L_G} \left[V_{DS}(V_{GS} - V_P) - \frac{V_{DS}^2}{2} \right] , \tag{2.19}$$

where $C_{GG}^{-1} = C_{ox}^{-1} + C_{dep}^{-1}$ and C_{dep} is the depletion region capacitance given by

$$C_{dep} = \frac{2\varepsilon_{si}}{\left(\dfrac{T_{sc}}{2} + X_{dep} \right)} . \tag{2.20}$$

Before reaching deep depletion, the current is expressed in terms of the gate capacitance and is approximated by

$$I_{DS} = \frac{\mu C_{GG} W}{L_G} (V_{GS} - V_P)^2 . \tag{2.21}$$

In deep depletion, the authors introduce an exponential law:

$$I_{DS} = \frac{U_T \mu N_D W T_{sc}}{L_G} \left(1 - \exp\left(-\frac{V_{DS}}{k_B T}\right)\right) \exp\left(-\frac{V_{GS} - V_P}{k_B T}\right). \qquad (2.22)$$

2.1.5   Semiempirical Charge-based Approach

A semicharge-based approach was derived in [85] by Cerdeira et al. The authors considered a specific link between surface and center potentials similar to what was proposed in [86] (see Chapter 3), resulting in a dependence between the applied gate voltage and the surface potential, which could only be solved numerically. It should be noticed that the effect of series resistances in the drain current is also included.

2.1.6   Analytical Approach based on Conventional Inversion-Mode MOSFETs

Jin and colleagues discuss an analytical approach for symmetric and asymmetric junctionless double-gate MOSET which relies on conventional inversion-mode MOSFET modeling [87, 88]. The authors assume that the fixed-charge density is zero, likewise in undoped channels. Then, the Poisson–Boltzmann equation reverts to the inversion-mode FET for which the analytical solution is known [89]. Using the solution in [89] as an initial guess, and substituting back into Poisson's equation, the authors derive an approximate solution, which can be expressed as

$$\Psi(x) = \frac{q}{\varepsilon_{si}} \left\{ \frac{N_D x^2}{2} - \frac{2\varepsilon_{si} KT}{q^2} \ln\left[\cos\left(\frac{2\beta x}{T_{sc}}\right)\right] \right\} + C_1 x + C_2, \qquad (2.23)$$

where N_D is the doping density, β is an integration constant, and C_1 and C_2 are obtained by imposing the boundary conditions at the interfaces. Even though the model assumes undoped channel, the analytical solution for the potential distribution is able to reflect a wide range of doping densities, including high values. Knowing the potential expression and integrating once, the drain current is obtained.

2.1.7   Parabolic Approximation and Full-Range Drain Current

Relying on parabolic approximation across two gates for the potential distribution, a full-range drain current was derived by Duarte et al. [90]. Adopting the Boltzmann statistics for mobile charge density, basic steps such as imposing the boundary conditions at surfaces link the applied gate voltage and the surface and center potentials. Thus, it is possible to obtain a closed form for the mobile charge in the channel:

$$-\frac{2\varepsilon_{ox}}{\beta t_{ox}}(V_{GS} - V_T - V_{ch}) = Q_m - \frac{2\varepsilon_{ox} U_T}{\beta t_{ox}} \ln\left(-Q_m \sqrt{\frac{qN_D T_{sc} + Q_m}{2\varepsilon_{si} \pi U_T q^2 N_D^2 T_{sc}}}\right), \qquad (2.24)$$

where $\beta = 1 + \varepsilon_{ox} T_{sc}/4\varepsilon_{si} t_{ox}$ and $V_T = V_{GS,FB} - qN_D T_{sc} t_{ox}/2\varepsilon_{ox} - qN_D T_{sc}^2/8\varepsilon_{si}$. This approach gives the following Pao–Sah integral, which includes the diffusion and drift components

$$I_{DS} = -\mu \frac{W}{L_G} \left(\frac{\beta t_{ox}}{4\varepsilon_{ox}} Q_m^2 - U_T Q_m \right)\Big|_S^D. \tag{2.25}$$

2.1.8　Gaussian Distribution of Mobile Charge Density

The model derived by Lime et al. [91] presents a simple approach. It also assumes a parabolic shape across two gates and integrating Boltzmann's distribution gives the mobile charge density. Noting that in full depletion the tail of the Gaussian distribution for mobile charges does not overlap with the gate insulator, an explicit link of mobile charge density in subthreshold and center potential is obtained. The mobile charge density in this particular regime can then be obtained integrating the Boltzmann's distribution as:

$$Q_m = \int_0^{T_{sc}/2} qN_D \exp\left(\frac{\Psi(x) - V_{ch}}{U_T} \right) dx \approx \frac{1}{2} \sqrt{\pi Q_{fix} Q_{cp}} \exp\left(\frac{\Psi_0 - V_{ch}}{U_T} \right), \tag{2.26}$$

where $Q_{cp} = 2C_{sc} U_T$ and $C_{sc} = \varepsilon_{si}/T_{sc}$. From the conditions at the boundaries, the difference between surface and center potential is a function of the mobile charge density:

$$\frac{\Psi - \Psi_0}{U_T} \approx \frac{\dfrac{1}{2}\sqrt{\pi Q_{fix} Q_{cp}} \exp\left(\dfrac{\Psi_0 - V_{ch}}{U_T} - \dfrac{Q_m}{2Q_{cp}} \right) + Q_m - \dfrac{Q_{fix}}{2}}{2Q_{cp}}. \tag{2.27}$$

Next, after integrating Poisson's equation, the surface electric field becomes dependent on the surface potentials. Last, combining these two relationships links the mobile charge density to the gate potential:

$$V = \ln\left(\frac{2Q_m}{Q_{fix}} \right) + \frac{Q_m}{C_{ox} U_T} - \ln\left(\frac{2Q_{cp}\left[1 - B\exp\left(-\dfrac{Q_m - \dfrac{Q_{fix}}{2}}{2Q_{cp}} \right) \right]}{Q_m - \dfrac{Q_{fix}}{2} - 2Q_{cp}\ln(B)} \right), \tag{2.28}$$

where

$$V = \frac{V_{GS} - \Delta\phi_{ms} - U_T \log\left(\dfrac{N_D}{n_i} \right) + \dfrac{Q_{fix}}{2C_{ox}} - V_{ch}}{U_T}, \tag{2.29}$$

and

$$B = 1 + \frac{1}{2}\sqrt{\pi \frac{Q_{fix}}{4Q_{cp}}} \exp\left(-\frac{Q_m}{2Q_{cp}} \right). \tag{2.30}$$

However, the solution is not fully analytic and needs some numerical integration. Further, depending on the operation mode (depletion and accumulation) approximate analytical expressions for the mobile charge density are introduced:

$$V + \frac{Q_{fix}}{4Q_{cp}} - \ln(B) \stackrel{dep}{\approx} \ln\left(\frac{Q_m}{2Q_{cp}}\right) + \frac{Q_m}{Q_{eq}}, \tag{2.31}$$

and

$$V \stackrel{acc}{\approx} \ln\left[\frac{Q_m}{2Q_{cp}}\left(\frac{2Q_m}{Q_{fix}} + 1\right)\right] + \frac{Q_m}{C_{ox}U_T}, \tag{2.32}$$

where

$$Q_{eq} = \frac{1}{(C_{ox}U_T)^{-1} + (2Q_{cp})^{-1}}. \tag{2.33}$$

Finally, because of discontinuity issues, the drain current is obtained by introducing a bridging function which has been found by trial and error:

$$I_{DS} = 2\frac{W}{L_G}\mu U_T \left[f(Q)|_{V_{ch}=0} - f(Q)|_{V_{ch}=V_{DS}}\right] = 2\frac{W}{L_G}\mu U_T \left[f(Q_s) - f(Q_d)\right], \tag{2.34}$$

with

$$f(Q) = \frac{Q^2}{2Q_{eq}} + 2Q - Q\ln\left[1 + \exp\left(\frac{Q - \frac{Q_{fix}}{2}}{4Q_{cp}}\right)\right]$$

$$+ \frac{Q_{fix}}{2}\ln\left[\frac{\frac{Q_{fix}}{2} - Q}{4Q_{cp}\left(1 - \exp\left(\frac{Q - \frac{Q_{fix}}{2}}{4Q_{cp}}\right)\right)}\right]. \tag{2.35}$$

2.1.9   Simple Model to Estimate Junctionless FET Performances

Koukab et al. [52] proposed a simple analytical expression of the mobile charge density above the threshold obtained without introducing Boltzmann statistics i.e., using electrostatic only with the depletion approximation. The model makes use of simple relationships where the concept of threshold voltage, defined as the potential where the channel is fully depleted, plays a central role and results in a simple expression for the drain current:

$$I_{DS} \approx 2\mu\frac{W}{L_G}\left(\frac{1}{C_{ox}} + \frac{1}{4C_{si}}\right)^{-1}V_{DS}(V_{on} - V_{off}), \tag{2.36}$$

where C_{si} is the silicon capacitance, ε_{si}/T_{sc}.

The authors demonstrated that junctionless FETs do scale as inversion-mode FETs, and exhibit almost the same CV/I delay and AC power consumption, in contrasts with earlier publication [1]:

$$\tau \approx \frac{C_{eq} V_{DD}}{I_{DS}} \approx \frac{\left(C_{ox}^{-1} + (4 C_{si})^{-1} \right)^{-1} 2 W L_G (V_{on} - V_{off})}{2 \mu \dfrac{W}{L_G} \left(C_{ox}^{-1} + (4 C_{si})^{-1} \right)^{-1} V_{DS}(V_{on} - V_{off})} = \frac{L_G^2}{\mu V_{DD}}. \tag{2.37}$$

2.1.10 Explicit Drain Current Model Relying on Charge-based Approach

Relying on an analytical charge-based approach from [86] (see Chapter 3), a fully explicit model for the long-channel junctionless double-gate MOSFET was proposed by Yesayan et al. [92], which is suitable for circuit simulation purposes and simple hand calculations. The accuracy of the original implicit formulation is preserved and compact equations are derived to compute explicitly the charge density and drain current in depletion and accumulation modes. This approach was tested with the EPFL junctionless model.

2.1.11 Modeling of Quantum Mechanical Effects

As device dimensions are scaled down targeting a technology node of $10\,nm$ before 2026 [93], quantum confinement effects (QCEs) will have to be included in coming, and possibly already in present, generations of compact models. In this context, a compact model including quantum mechanical effects in the subthreshold region was proposed by Duarte et al. [94]. Neglecting the mobile charge density in subthreshold, the Schrödinger is combined with Poisson's equations in the depletion approximation to explicit the potential distribution across the channel as a function of threshold voltage [80, 86, 90].

The Schrödinger equation is solved analytically for two cases [94]. For a thick channel, the situation is alike the harmonic oscillator and the sub-band energy levels are almost independent of the channel thickness, but depend on the doping N_D given as

$$V_{QM} = \frac{q^2 N_D}{2 \varepsilon_{si}} x^2. \tag{2.38}$$

Using the effective-mass approximation, the discrete sub-band energy levels for this case are expressed by

$$E_{QHO,k,n} = \left(n + \frac{1}{2} \right) \frac{h}{2\pi} \sqrt{\frac{q^2 N_D}{m_{c,k}^* \varepsilon_{si}}}, \tag{2.39}$$

where n is a quantum number ($n = 0, 1, 2, 3, ...$), and h is the Planck's constant.

For thin channels, the confined energy levels are obtained from the perturbation theory. The system is like a quantum well (QW) combined with a perturbation potential (PP). The energy levels are given by (2.38):

$$E_{QW+PP,k,n} = \frac{h^2 n^2}{8 m^*_{c,k} T^2_{sc}} + \frac{q^2 T^2_{sc} N_D}{24 \varepsilon_{si}} \left(1 - \frac{6}{(n\pi)^2} \right). \tag{2.40}$$

Further, the quantum electron density is given by [95].

The transition between these two cases takes place at a given silicon thickness $T_{sc,transition}$, where the QHO model switches to the QW+PP model. This can be found by solving $dE_{QW+PP,n}/dT_{sc} = 0$, for a given n and N_D:

$$T_{sc,transition} = \left[\frac{\dfrac{h^2 n^2}{m^*_{c,k}}}{\left(\dfrac{q^2 N_D}{3 \epsilon_{si}} \left(1 - \dfrac{6}{(n\pi)^2} \right) \right)} \right]^{\left(\frac{1}{4} \right)}. \tag{2.41}$$

The classical electron density calculated by ignoring quantum corrections can be fairly well predicted merely by shifting the threshold voltage [80, 86].

A numerical solution was proposed by Huang et al. [96] to simulate the junctionless FET adopting the nonequilibrium Green function (NEGF) method [97]. The approach was also applied to different geometries of junctionless FETs such as nanowires, trigate, FinFETs. The shift in the threshold voltage, the drain-induced-barrier-lowering and the subthreshold swing are expressed as a function of the device geometry and compared with inversion-mode transistors [96].

2.1.12   Short-Channel Effects in Subthreshold

Short-channel effects (SCE) in deep depletion consist essentially in calculating the drain-induced barrier lowering and subthreshold swing which requires estimating (at least) the 2D profile of the potential distribution (see [37–41] for numerical simulation studies).

In addition to full numerical computing, analytical approaches targeting implementation of SCE in compact models have been developed. Since at low current densities, the junctionless FET operates with an almost fully depleted channel, the full depletion approximation is always adopted to simulate the potential profile in two dimensions using approximative solutions for the 2D Poisson's equation.

Analytical expressions for the short-channel threshold voltage and subthreshold swing are discussed in [98, 99].

In other derivations [100–103], applying the principle of superposition proposed in [104] and using the variable separation technique, the threshold voltage and potential profile are obtained, but without explicit analytical expressions for the subthreshold behavior. In [105–108], the Poisson's equation is split in two parts, a 1D solution, which is a function of y, and a 2D solution, which is a function of (x, y). Next, the conformal mapping technique is used [105–108].

Assuming full depletion of the channel and a parabolic potential along the y-direction as proposed by Young [109], and imposing the boundary conditions (symmetry and continuity of the displacement vector at the silicon–insulator interface) results in explicit relationship for the subthreshold swing, subthreshold current, DIBL, and threshold voltage in [110, 111]. It should be noted that [110] is very similar to [111] which bases the analysis on the center potential profile along the channel. According to [111], the sensitivity of the minimum potential shift along the channel with applied drain voltage leads to a new definition of DIBL in [111]. This is discussed and compared to the standard definition of a threshold voltage shift [112]. Following [110, 111], the drain current in subthreshold is obtained analytically, as for the subthreshold swing and DIBL [111].

2.1.13　Transcapacitance Modeling

While there have been various efforts to model the DC characteristics of junctionless devices, there has been less interest in transcapacitance modeling even though transcapacitance matrix elements are mandatory for circuits simulation. In [113], the gate capacitance of junctionless FETs was evaluated by 3D numerical simulations. Performance assessment with respect to bulk (IM) and SOI junctionless FETs was analyzed in [31, 41, 114–117]. A junctionless FET operating as a single-transistor amplifier was also considered in [28], which showed that junctionless transistors exhibit larger voltage gain than inversion-mode devices of similar dimensions [28] and should perform better than inversion-mode transistors in the low-moderate frequency range.

A complete treatment of the transcapacitance matrix in double-gate junctionless FETs relying on the charge-based model of Chapter 3 [86] is detailed in Chapter 8 [118].

2.1.14　Modeling Asymmetry in Junctionless Double-Gate MOSFETs

Most modeling efforts have been devoted to simulate symmetric operation in junctionless double-gate MOSFETs, and only a few papers have addressed non-symmetric situations [82, 87, 102, 119, 120].

In [87], an analytical solution for the potential profile, the current, and the mobile charge density was proposed. The authors assumed an initial guess function with different boundary conditions for each of the gates. In [102], the variable separation technique was used to simulate the 2D potential and to predict the subthreshold characteristics including gate asymmetry.

Relying on the analysis developed in Chapter 3 [86] for symmetric operation, asymmetry is viewed as a combination of two virtual symmetric devices [119] in Chapter 9. This approach is basically model-independent and therefore generic.

2.2　Modeling Junctionless Nanowire MOSFETs

In [121], the drain current in long-channel junctionless nanowires is derived considering a parabolic potential distribution in the radial direction. Next, Gauss law links

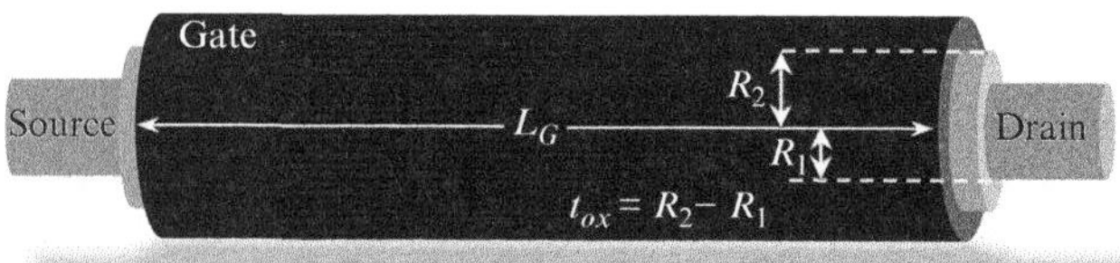

Figure 2.2 Schematic view of an *n*-type junctionless nanowire FET investigated in this chapter.

the surface and gate potentials by means of the gate capacitance. Finally, integrating the charge density along the channel links the surface to the center potentials. The current is obtained using the Pao–Sah integral.

Another analytical solution is proposed in [122]. Using a regional approach and suggesting simplifying assumptions in accumulation and depletion modes leads to an approximate solution of the Poisson's equation in cylindrical coordinates. Considering the junctionless nanowire as a planar device when operating in weak depletion, or using the linearization of the exponential term for the mobile charge density in other situations (in deep depletion), the space charge density and the potential profile are obtained and introduced in the current expression.

For relatively low doped channels (2×10^{18} cm^{-3} to 1×10^{19} cm^{-3}), an approximate solution of Poisson's equation in cylindrical coordinates valid in accumulation mode is proposed in [123]. The current is obtained by numerical integration.

As discussed in the following chapters, another study carried out in [124] proposed to map junctionless nanowire FETs on junctionless double-gate MOSFETs by introducing a trapezoidal integration of the Poisson–Boltzmann equation across the nanowire (Figure 2.2). This means that the double-gate MOSFET proposed in [86] could be the core model for junctionless nanowire FETs as well. To do so, equivalent parameters are defined. This approach is suitable to predict charges and currents.

Lime et al. [125] evaluated the mobile charge density in nanowire junctionless FET given by

$$Q_m = Q_{cp}Q_{fix}\frac{1 - \exp\left(\dfrac{-Q_{sc}}{Q_{cp}}\right)}{Q_{sc}}\exp\left(\frac{\Psi_s - V_{ch}}{U_T}\right), \tag{2.42}$$

where $Q_{sc} = Q_m - Q_{fix}$ is the algebraic sum of fixed (Q_{fix}) and mobile (Q_m) charges densities (per unit area) in the channel, Q_{cp} is given by $2\varepsilon_{si}U_T/R$. After some mathematical manipulation, the mobile charge density versus the gate potential is explicit:

$$Q_m\exp\left(\frac{Q_m}{C_{ox}U_T}\right) = \frac{Q_{cp}Q_{fix}}{Q_{sc}}\frac{1 - \exp\left(\dfrac{-Q_{sc}}{Q_{cp}}\right)}{\exp\left[-\dfrac{V_{GS} - \Delta\phi_{ms} - V_{ch} - U_T\ln(N_D/n_i) - Q_{fix}/C_{ox}}{U_T}\right]}, \tag{2.43}$$

or equivalently:

$$V_{GS} - \Delta\phi_{ms} - U_T \ln(N_D/n_i) - V_{ch} = -\frac{Q_{fix}}{C_{ox}} + U_T \ln\left(\frac{Q_m}{Q_{fix}}\right) + \frac{Q_m}{C_{ox}}$$

$$- U_T \ln\left(\frac{Q_{cp}\left(1 - \exp\left(-\frac{Q_m - Q_{fix}}{Q_{cp}}\right)\right)}{Q_m - Q_{fix}}\right). \qquad (2.44)$$

Relying on charge-based expressions, the drain current is then obtained numerically. A simple analytical model of drain current applying approximate solutions for the mobile charge density in depletion and accumulation modes is also available [125].

Finally, nonideal nanowires such as triple-gate junctionless nanowires were investigated by Trevisoli et al. [126]. The authors do not derive a model for the current, instead they propose to calculate a threshold voltage they define as the gate potential when drift and diffusion current become equal.

2.2.1 Short-Channel Effects in the Subthreshold

Solving the 2D Poisson's equation, Li et al. [127] developed an analytical subthreshold model for cylindrical junctionless nanowire. The mobile charge density is neglected in the Poisson's equation expressed in cylindrical coordinates. Next, the superposition technique is introduced. Introducing boundary conditions at the surface and center of the channel, the electrostatic potential is expressed in terms of Fourier–Bessel series. The subthreshold drain current and subthreshold swing are obtained analytically.

Similarly, in [103], SCE in cylindrical junctionless nanowire are modeled by neglecting the mobile charge density. Approximated solutions of the potential distribution, subthreshold drain current and SCE are obtained.

In the same spirit, a parabolic approximation, neglecting the free-carrier density in depletion mode of junctionless nanowire provides another approximate solution of potential distribution through the channel in [128]. According to minimum center potential obtained in [128], the short-channel threshold voltage and DIBL are obtained.

2.2.2 Transcapacitance Modeling in Junctionless Nanowire FETs

As for double-gate architectures, modeling transcapacitances in junctionless nanowires has been somehow a marginal activity in regards to $I\text{--}V$ and transconductances [129–133].

The approach in [134] adopts the analytical model developed for junctionless double-gate MOSFETs to calculate the transcapacitances in junctionless nanowire FETs while using the parameters transformation scheme proposed in [124]. Taking advantage of this "mapping" between planar and cylindrical junctionless FETs, explicit solutions for transcapacitance matrix elements are obtained (the Ward–Dutton charge partitioning method is used). The distortion due to the nonlinear relationship between charges and potentials is also a feature built into the model.

In the development from Moldovan et al., the authors neglect the mobile charge density in depletion while considering $Q_m \gg Q_{fix}$ in accumulation (and therefore the total charge density is approximated to Q_m). Using the Ward–Dutton charge partitioning method, all capacitances in depletion and accumulation modes are derived. In each domain, analytical expressions are obtained, but the relationship between local charge densities and potentials is essentially linear. When the channel is not in depletion/accumulation, these approximations become less valid and an hyperbolic tangent function with some fitting parameters is used to bridge deep depletion with accumulation regimes [135].

2.2.3   Quantum Mechanical Effects in Junctionless Nanowire FETs

There have been various simulation-based studies on the influence of quantum mechanical effects in junctionless nanowire FETs [136, 137]. One study developed a quantum-corrected transport simulator that includes quantum drift diffusion and quantum energy balance (QEB) (see [138]).

In [73], a multiscale quantum mechanics/electromagnetic (QM/EM) method was applied to simulate junctionless transistors with dimensions comparable to experimental devices. Adopting a multiscale modeling concept [139, 140], an effective way to include QM effect, transfer characteristics of a junctionless transistor, was simulated in [141]. Using the NEGF and density functional theory (DFT) gives the spectral density of occupied states and the Hamiltonian, respectively. The drift-diffusion current is obtained through a transmission coefficient.

2.3   Summary

Analytical modeling of junctionless FET targeting electrical simulations of circuit based on this new kind of device has been addressed in different ways. Except for numerical solutions, all approaches are seeking for a closed form solution of the current density and capacitances with respect to the applied voltages. Today, downscaling of CMOS technology is shifting modeling toward quantum computation and nonequilibrium transport. However, more classical approaches such as those discussed in this chapter can estimate the performance and sensitivity of the technological parameters.

3   The EPFL Charge-based Model of Junctionless Field-Effect Transistors

This chapter presents a charge-based model for junctionless field-effect transistors primarily developed at the Ecole Polytechnique Fédérale de Lausanne (EPFL) and details the derivation of the core relationships used to calculate charges and current in symmetric double-gate and nanowire topologies. These core relations will be reused in upcoming chapters when discussed more advanced aspects of modeling.

From a design and architecture point of view, inversion-mode (IM) and junctionless field-effect transistors look very similar. Each consists of a semiconductor channel and an insulator gate stack that aims to monitor the mobile carrier density in the semiconductor. However, in junctionless FETs the gate is used to deplete or accumulate **majority** carriers from a highly doped silicon layer, and not to create a **minority** carriers layer at the semiconductor–insulator interface as in IM FETs. In addition, in junctionless FETs the current spreads in the volume and is not confined at the semiconductor–insulator interfaces, as in IM FETs.

This explains why charge-potential relationships in JL FETs cannot be mapped on inversion-mode FETs, for which many models exist. For instance, as will be evidenced, junctionless FETs exhibit a quadratic charge-potential dependence that cannot be predicted with IM FET models.

3.1   Charge-based Modeling of Junctionless Double-Gate Field-Effect Transistors

3.1.1   Recalling Basics of Semiconductor Statistics

The carrier density in a semiconductor is obtained by integrating over the energy the density of states times the probability for that state to be occupied. For electrons in the conduction band, the integral is evaluated from the bottom to the top of the conduction band:

$$n = \int_{E_c}^{E_{cM}} g_c(E) f(E) dE, \tag{3.1}$$

where $g_c(E)$ is the density of states in the conduction band and $f(E)$ is the Fermi–Dirac distribution that determines the probability that a state of energy E is occupied with an electron

$$f(E) = \cfrac{1}{1 + \exp\left(\cfrac{E - E_f}{k_B T}\right)}, \tag{3.2}$$

where E_f is the Fermi energy, k_B is Boltzmann's constant, and T is the absolute temperature. Similarly for holes one obtains:

$$p = \int_{E_{vm}}^{E_v} g_v(E) \left[1 - f(E)\right] dE, \tag{3.3}$$

where $g_v(E)$ is the density of states of the valence band. Since holes are empty states in the valence band, the probability of having a hole is $[1 - f(E)]$.

The densities of states represent the number of states per unit volume and energy, for a given energy. They result from the solution of the Schrödinger's equation with appropriate boundary conditions. In a 3D semiconductor, these are given by

$$g_c(E) = \frac{8\pi \sqrt{2}}{h^3} (m_e^*)^{3/2} \sqrt{E - E_c} \quad \text{and} \quad E \geq E_c, \tag{3.4}$$

$$g_v(E) = \frac{8\pi \sqrt{2}}{h^3} \left(m_h^*\right)^{3/2} \sqrt{E_v - E} \quad \text{and} \quad E \leq E_v, \tag{3.5}$$

where m_e^* and m_h^* are the electron and hole effective masses, respectively. Using (3.4) and (3.5), the carrier densities for electrons and holes are:

$$n = \int_{E_c}^{E_{cM}} \frac{8\pi \sqrt{2}}{h^3} (m_e^*)^{3/2} \sqrt{E - E_c} \cfrac{1}{1 + \exp\left(\cfrac{E - E_f}{k_B T}\right)} dE, \tag{3.6}$$

$$p = \int_{E_{vm}}^{E_v} \frac{8\pi \sqrt{2}}{h^3} \left(m_h^*\right)^{3/2} \sqrt{E_v - E} \left[1 - \cfrac{1}{1 + \exp\left(\cfrac{E - E_f}{k_B T}\right)}\right] dE. \tag{3.7}$$

These integrals cannot be solved analytically. However, when the Fermi level lays inside of the band-gap i.e., $E - E_f \gg k_B T$ the semiconductor is nondegenerate and the Fermi–Dirac function can be approximated by the Maxwell–Boltzmann statistics:

$$f(E) = \exp\left(\frac{E - E_f}{k_B T}\right). \tag{3.8}$$

Adopting this expression in the integrals (3.6) and (3.7), the electrons and holes concentrations at equilibrium take simple forms:

$$n = N_c \exp\left(\frac{E_f - E_c}{k_B T}\right), \tag{3.9}$$

$$p = N_v \exp\left(\frac{E_v - E_f}{k_B T}\right), \tag{3.10}$$

where N_c and N_v are the effective densities of states for the conduction and valence bands, respectively.

$$N_c = 2\left(\frac{2\pi m_e^* k_B T}{h^2}\right)^{3/2},\qquad(3.11)$$

$$N_v = 2\left(\frac{2\pi m_p^* k_B T}{h^2}\right)^{3/2}.\qquad(3.12)$$

It is useful to introduce the Fermi energy and carrier density for the intrinsic semiconductor (undoped silicon) i.e., E_i and n_i. These are obtained by imposing equal concentration of electrons and holes ($n = p$):

$$E_f = E_i \Rightarrow n = n_i = N_c \exp\left(\frac{E_i - E_c}{k_B T}\right),\qquad(3.13)$$

$$E_f = E_i \Rightarrow p = n_i = N_v \exp\left(\frac{E_v - E_i}{k_B T}\right).\qquad(3.14)$$

Combining these two equations gives the intrinsic carrier density and the intrinsic Fermi energy:

$$n_i = \sqrt{N_v N_c}\,\exp\left(-\frac{E_c - E_v}{2k_B T}\right) = \sqrt{N_v N_c}\,\exp\left(-\frac{E_g}{2k_B T}\right),\qquad(3.15)$$

$$E_i = \frac{E_c + E_v}{2} + \frac{k_B T}{2}\ln\left(\frac{N_v}{N_c}\right).\qquad(3.16)$$

Conversely, the effective densities of states can be expressed as a function of intrinsic carrier density:

$$N_c = n_i \exp\left(-\frac{E_i - E_c}{k_B T}\right),\qquad(3.17)$$

$$N_v = n_i \exp\left(-\frac{E_v - E_i}{k_B T}\right).\qquad(3.18)$$

Alternatively, electron and hole densities are obtained as a function of the intrinsic Fermi energy and carrier density:

$$n(x) = n_i \exp\left(\frac{E_f - E_i}{k_B T}\right),\qquad(3.19)$$

$$p(x) = n_i \exp\left(\frac{E_i - E_f}{k_B T}\right).\qquad(3.20)$$

These relationships will be used to model electrostatics in junctionless double-gate FETs.

3.1.2   Approximate Solution of the Poisson–Boltzmann Equation in Junctionless Double-Gate MOSFET

As proposed in [86], we consider a n-type junctionless double-gate MOSFET where the channel is a semiconductor layer of thickness T_{sc} with a donor density N_D. A cross-section and energy representation are shown in Figures 3.1 and 3.2.

Defining the electrostatic potential $\Psi(x)$ as $\Psi(x) = -(E_i(x) - E_{fs})/q$, where E_{fs} is the Fermi energy at the source (see Figure 3.2) and making use of (3.19) and (3.20), electron and hole concentrations are given by

$$n(x) = n_i \exp\left[\frac{\Psi(x)}{U_T}\right], \tag{3.21}$$

$$p(x) = n_i \exp\left[-\frac{\Psi(x)}{U_T}\right], \tag{3.22}$$

where $U_T = k_B T/q$ is the thermal voltage. When a potential is applied on the drain, it shifts the quasi-Fermi energy at the drain by $-qV_{DS}$ and along the channel by $-qV_{ch}$.

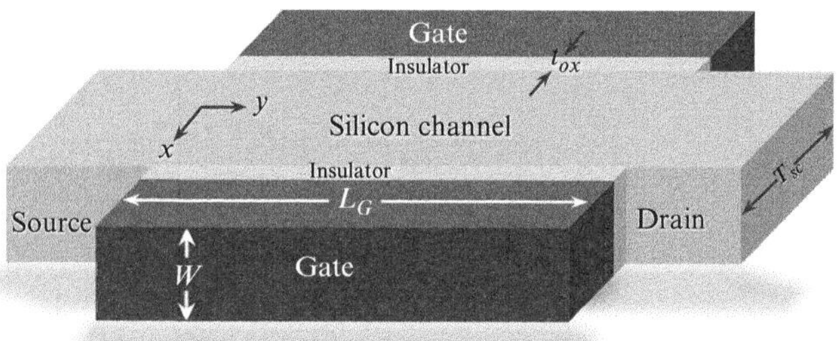

Figure 3.1 Schematic view of the n-type junctionless symmetric double-gate MOSFET.

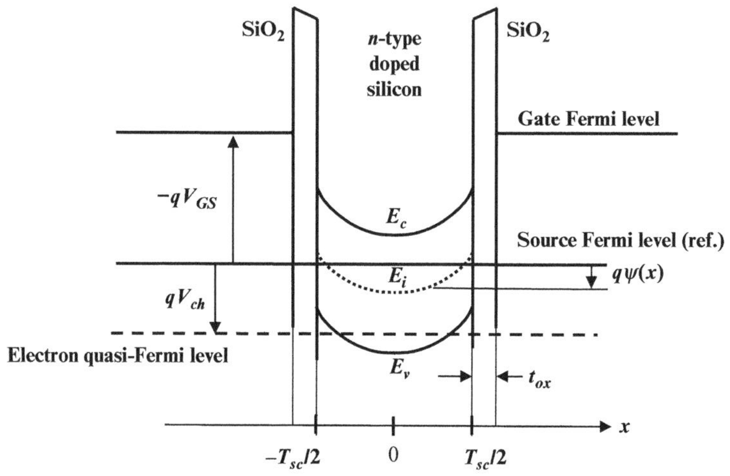

Figure 3.2 Sketch of the band-energy diagram for the n-type junctionless double-gate MOSFET. Reprinted from [86] with permission.

Then, the free electron and hole concentrations at "y" are obtained with a Fermi potential shifted by $-qV_{ch}(y)$ for electrons and $qV_{ch}(y)$ for holes:

$$n(x) = n_i \exp\left[\frac{\Psi(x) - V_{ch}(y)}{U_T}\right], \tag{3.23}$$

$$p(x) = n_i \exp\left[\frac{-\Psi(x) + V_{ch}(y)}{U_T}\right]. \tag{3.24}$$

A relevant potential is the flat-band voltage obtained from (3.21) when assuming the concentrations of mobile charges and donors (N_D) are equal:

$$\Psi_{FB}(x) = -\frac{E_i - E_f}{q} = U_T \ln\left(\frac{N_D}{n_i}\right). \tag{3.25}$$

Note that whereas in junctionless FETs, the flat-band corresponds to a mobile carrier density as large as the doping, i.e., highly conductive channel; in inversion-mode FETs the mobile carriers density is negligible at flat-band and so the channel is highly resistive.

Considering a nondegenerate semiconductor, merging the Poisson equation with the Boltzmann statistics leads to the partial differential equation:

$$\frac{\partial^2 \Psi(x,y)}{\partial x^2} + \frac{\partial^2 \Psi(x,y)}{\partial y^2} = \frac{q}{\varepsilon_{si}}\left[n_i \exp\left(\frac{\Psi(x,y) - V_{ch}(y)}{U_T}\right) - n_i \exp\left(-\frac{\Psi(x,y)}{U_T}\right) - N_D\right], \tag{3.26}$$

where ε_{si} is the silicon permittivity and $V_{ch}(y)$ is the shift in the electron quasi-Fermi potential at coordinate y along the channel. Neglecting the contribution of the electric field gradient along y (gradual channel approximation) and assuming that no current flows in the x-direction between the gates i.e., the Fermi potential does not vary along the x-direction for a given channel coordinate y gives:

$$\frac{\partial^2 \Psi(x,y)}{\partial x^2} = \frac{q}{\varepsilon_{si}}\left[n_i \exp\left(\frac{\Psi(x,y) - V_{ch}(y)}{U_T}\right) - n_i \exp\left(-\frac{\Psi(x,y)}{U_T}\right) - N_D\right], \tag{3.27}$$

Finally, neglecting holes in the n-type channel (3.27) can be simplified:

$$\frac{\partial^2 \Psi(x)}{\partial x^2} = \frac{q}{\varepsilon_{si}}\left[n_i \exp\left(\frac{\Psi(x) - V_{ch}}{U_T}\right) - N_D\right]. \tag{3.28}$$

Note that Boltzmann statistic used in (3.28) remains valid if the free-carrier density does not exceed the effective density of states. Concerning electrons in silicon, this would correspond to about 3×10^{19} cm^{-3} at room temperature, a condition fulfilled in real devices.

First, integration of (3.28) links the electric field to the potential evaluated at the surface (omitting the channel coordinate y to avoid lengthy expressions).

$$E_s^2 = \left(\frac{Q_{sc}}{2\varepsilon_{si}}\right)^2 = \frac{2qn_i U_T}{\varepsilon_{si}}\left[\exp\left(\frac{\Psi_s - V_{ch}}{U_T}\right) - \exp\left(\frac{\Psi_0 - V_{ch}}{U_T}\right) - \frac{N_D}{n_i U_T}(\Psi_s - \Psi_0)\right], \tag{3.29}$$

where Q_{sc} is the total charge density in the semiconductor, V_{ch} and $\Psi_s = \Psi(x = \pm T_{sc}/2)$ are the channel potential and the surface potential, respectively, and $\Psi_0 = \Psi(x = 0)$ is the potential at the center of the semiconductor.

However, Ψ_S and Ψ_0 are not independent. An additional link between the surface and center potentials exists (note that (3.28) imposes only one condition on the second-order differential equation (3.29)). The Laplacian in (3.28) can be approximated by the first four terms of a Maclaurin series around $x = 0$. Noting that the potential is symmetric with respect to the center and odd-order derivatives of $\Psi(x)$ versus x vanish, the potential distribution across the gates is approximated by

$$\Psi(x) \approx \Psi_0 + \frac{qn_i}{\varepsilon_{si}} \left[\exp\left(\frac{\Psi_0 - V_{ch}}{U_T} \right) - \frac{N_D}{n_i} \right] \frac{x^2}{2}. \tag{3.30}$$

Evaluating $\Psi(x)$ at the channel interfaces ($x = \pm T_{sc}/2$) links the surface and center potentials:

$$\Psi_s \approx \Psi_0 + \frac{qn_i T_{sc}^2}{8\varepsilon_{si}} \left[\exp\left(\frac{\Psi_0 - V_{ch}}{U_T} \right) - \frac{N_D}{n_i} \right]. \tag{3.31}$$

Relation (3.31) can be considered as a generalization of the well-known full depletion approximation, a strong assumption that has also been used in [142]. Indeed, when the mobile charge density is negligible with respect to the fixed-charge density, (3.31) reverts to the exact solution of the Poisson equation in a totally depleted semiconductor layer. However, in (3.31), the Boltzmann term introduces a carrier-density term that balances the staircase-like distribution of ionized donors in a full depletion scenario. Such a correction is fundamental when predicting below threshold operation.

Therefore, for a given center potential Ψ_0, the surface potential Ψ_s takes a well-defined value. According to (3.31), the charge density in the semiconductor Q_{sc} is uniquely determined as well:

$$Q_{sc} = 2\,\mathrm{sign}(\Psi_0 - \Psi_s) \sqrt{2qn_i U_T \varepsilon_{si}}$$
$$\times \sqrt{\exp\left(\frac{\Psi_s - V_{ch}}{U_T} \right) - \exp\left(\frac{\Psi_0 - V_{ch}}{U_T} \right) - \frac{N_D}{n_i U_T}(\Psi_s - \Psi_0)}, \tag{3.32}$$

where the sign function is consistent with the charge in the semiconductor. In contrast to the inversion-mode FET, Q_{sc} can be either positive or negative, depending on whether the potential at the surface Ψ_s is lower or higher than the one at the center Ψ_0. This transition happens at flat-band. From (3.28), semiconductor neutrality is achieved when $\partial^2\Psi/\partial x^2 = 0$, implying:

$$\Psi_{0,FB} = \Psi_{s,FB} = V_{ch} + U_T \ln\left(\frac{N_D}{n_i} \right). \tag{3.33}$$

Here, $\Psi_{0,FB}$ and $\Psi_{s,FB}$ refer to the flat-band center and surface potentials. In the next section, the mobile charge density will be calculated based on the former developments.

3.1.3 Introduction of Symmetric Gate Capacitances

The total charge density consists of mobile Q_m and fixed-charge densities Q_{fix} from the ionized donor atoms:

$$Q_{sc} = Q_m + Q_{fixs} = Q_m + qN_D T_{sc}. \tag{3.34}$$

On the other hand, the semiconductor charge density is also related to the potential drop across the gate capacitor:

$$Q_{sc} = -2C_{ox}(V_{GS} - \Delta\phi_{ms} - \Psi_s), \tag{3.35}$$

where V_{GS} is the gate-to-source voltage, $\Delta\phi_{ms}$ is the difference between the metal work function and an intrinsic reference in the semiconductor, and C_{ox} is the gate capacitance. It should be noted that $\Delta\phi_{ms}$ does not have the meaning of the metal-semiconductor work function difference, a parameter usually given by

$$W_{ms} = W_m - W_s = W_m - \left[\chi + \frac{E_g}{2q} - U_T \ln\left(\frac{N_D}{n_i}\right)\right], \tag{3.36}$$

with W_m being the work function of the metal, χ the electron affinity for the semiconductor, and E_g the semiconductor band-gap. As illustrated in Figure 3.3, $\Delta\phi_{ms}$ is defined as:

$$\Delta\phi_{ms} = W_m - \frac{E_i}{q} = W_m - \left(\chi + \frac{E_g}{2q}\right) = W_{ms} - U_T \ln\left(\frac{N_D}{n_i}\right). \tag{3.37}$$

A discussion of the definition of the work function difference used in this derivation is now in order.

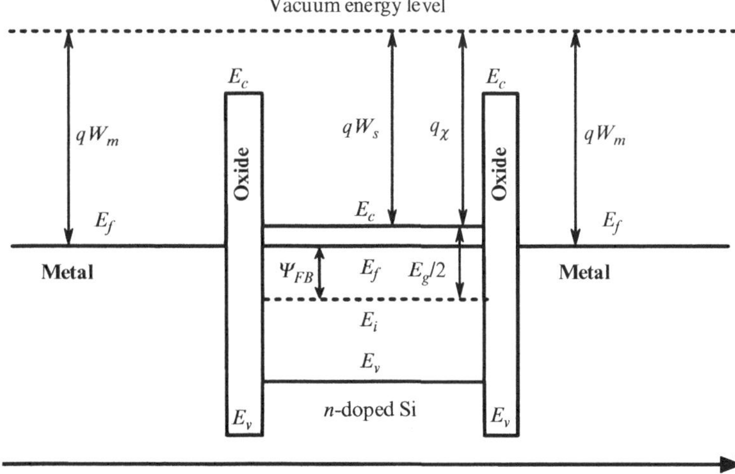

Figure 3.3 Sketch of the energy band diagram for the n-type double-gate MOSFET at flat-band condition.

According to (3.32), relationship (3.35) can be written as:

$$2C_{ox}(V_{GS} - \Delta\phi_{ms} - \Psi_s) = -2\,\mathrm{sign}(\Psi_0 - \Psi_s)\sqrt{2qn_i U_T \varepsilon_{si}}$$

$$\times \sqrt{\exp\left(\frac{\Psi_s - V_{ch}}{U_T}\right) - \exp\left(\frac{\Psi_0 - V_{ch}}{U_T}\right) - \frac{N_D}{n_i U_T}(\Psi_s - \Psi_0)}. \tag{3.38}$$

Combining (3.38) with (3.31) and defining $C = \exp\left(\dfrac{\Psi_0 - V_{ch}}{U_T}\right)$ and $K = \dfrac{qn_i T_{sc}^2}{8 U_T \varepsilon_{si}}$, the mobile charge density is obtained through the evaluation of "C" after solving the general equation:

$$V_{GS} - \Delta\phi_{ms} - V_{ch} = U_T \ln(C) + K U_T\left(C - \frac{N_D}{n_i}\right) - \mathrm{sign}(\Psi_0 - \Psi_s)\frac{\sqrt{2qn_i U_T \varepsilon_{si}}}{C_{ox}}$$

$$\times \sqrt{C\left\{\exp\left[K\left(C - \frac{N_D}{n_i}\right)\right] - \frac{N_D}{n_i}K - 1\right\} + K\left(\frac{N_D}{n_i}\right)^2}. \tag{3.39}$$

Once the unknown C is determined, the charge density is readily obtained from (3.31) and (3.35):

$$Q_{sc} = -2C_{ox}\left[V_{GS} - V_{ch} - \Delta\phi_{ms} + K U_T\frac{N_D}{n_i} - K U_T C - U_T \ln(C)\right]. \tag{3.40}$$

Even though we are not yet at the point where we get an explicit expression, such a formulation avoids solving the nonlinear second-order differential equation including the constraint from the gate bias. As such, the charge density in the semiconductor can already be computed with respect to the applied voltages.

However, it is possible to express charges and current in a more intuitive formulation that relies explicitly on device parameters and voltages. These developments are obtained through an analysis of (3.39) and are presented hereafter.

3.1.4   Derivation of Explicit Voltage–Charge Relationships

In this section, the derivation of the core model equations (3.39) and (3.40) is given for the two regimes of operation inherent to junctionless FETs, namely accumulation and depletion modes.

The Junctionless Double-Gate MOSFET Biased in Accumulation

The results from the numerical solution of the differential equation (3.28) are plotted in Figure 3.4. These reveal that the center potential remains very close to the value it gets at flat-band condition i.e., $\Psi_0 \sim \Psi_{0,FB}$ as soon as the channel is set in accumulation. For instance, considering a silicon layer of $20\,nm$ doped up to $5 \times 10^{18}\,cm^{-3}$ the upper limit for Ψ_0 is about $0.508\,V$, and the asymptote is even more pronounced when the doping concentration is increased. This "observation" is helpful when seeking an approximate expression featuring the "above flat-band" operation.

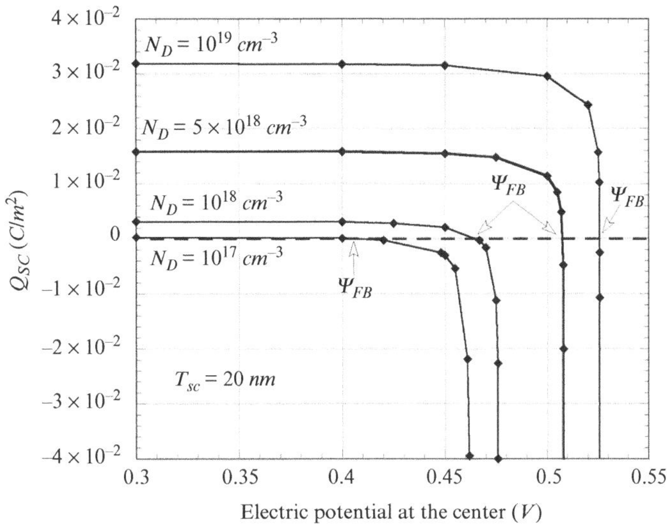

Figure 3.4 Semiconductor charge density as a function of the potential at the center in a 20 nm silicon layer of a junctionless double-gate MOSFET with various doping concentrations. The surface potential corresponding to flat-band conditions is shown by the arrows. Reprinted from [86] with permission.

After replacing Ψ_0 with $\Psi_{0,FB}$ in (3.32), we obtain:

$$Q_{sc} \stackrel{acc}{\approx} -2\sqrt{2qn_i U_T \varepsilon_{si}}\sqrt{\exp\left(\frac{\Psi_s - V_{ch}}{U_T}\right) - \frac{N_D}{n_i} - \left\{\frac{N_D}{n_i}\left[\frac{\Psi_s - V_{ch}}{U_T} - \ln\left(\frac{N_D}{n_i}\right)\right]\right\}}. \tag{3.41}$$

The ratio between the LHS and RHS terms in the square root is always larger than unity above flat-band (and becomes unity at flat-band), which justifies that the RHS term in the square root can be omitted in accumulation. Then, relation (3.41) simplifies to:

$$Q_{sc} \stackrel{acc}{\approx} -2\sqrt{2qn_i U_T \varepsilon_{si}}\sqrt{\exp\left(\frac{\Psi_s - V_{ch}}{U_T}\right) - \frac{N_D}{n_i}}. \tag{3.42}$$

Replacing $\Psi_s - V_{ch}$ obtained from (3.42) into (3.35) leads to an explicit expression of the potentials versus the channel charge density (which includes the fixed-charge density):

$$V_{GS} - V_{ch} - \Delta\phi_{ms} - U_T \ln\left(\frac{N_D}{n_i}\right) \stackrel{acc}{\approx} -\frac{Q_{sc}}{2C_{ox}} + U_T \ln\left(1 + \frac{Q_{sc}^2}{8qN_D\varepsilon_{si}U_T}\right). \tag{3.43}$$

From (3.43), the particular flat-band condition ($Q_{sc} = 0$) corresponds to $V_{FB} - \Delta\phi_{ms} = U_T \ln(N_D/n_i)$, which is the expected result.

The Junctionless Double-Gate MOSFET Biased in Depletion

Conversely, under depletion mode, the potential at the center is higher than the surface potential and the semiconductor is charged positively i.e., $Q_{sc} \geq 0$. This implies that the exponential in (3.39) is lower than unity. Again, some useful simplifications are possible. For instance, accurate results are obtained when assuming $C = N_D/n_i$ in the exponential term of (3.39), which is a kind of flat-band condition leading to the following relationship:

$$Q_{sc} \overset{dep}{\approx} 2 \sqrt{2qn_i U_T \varepsilon_{si}} \sqrt{K\frac{N_D}{n_i}\left(\frac{N_D}{n_i} - C\right)}. \tag{3.44}$$

Expressing the parameter C as a function of the total charge in the semiconductor (in depletion mode) gives:

$$C \overset{dep}{\approx} \frac{N_D}{n_i}\left[1 - \left(\frac{Q_{sc}}{qN_D T_{sc}}\right)^2\right]. \tag{3.45}$$

Next, introducing (3.45) in (3.39) yields a charge-based relationship in depletion:

$$V_{GS} - V_{ch} - \Delta\phi_{ms} - U_T \ln\left(\frac{N_D}{n_i}\right) \overset{dep}{\approx} -\frac{Q_{sc}^2}{8qN_D\varepsilon_{si}} - \frac{Q_{sc}}{2C_{ox}} + U_T \ln\left[1 - \left(\frac{Q_{sc}}{qN_D T_{sc}}\right)^2\right]. \tag{3.46}$$

Likewise for accumulation the flat-band voltage satisfies $V_{FB} - \Delta\phi_{ms} = U_T \ln(N_D/n_i)$ when $Q_{sc} = 0$. This ensures that the continuity of the charge density is satisfied when crossing depletion/accumulation modes. These approximations are evaluated with respect to the numerical solution obtained by combining (3.39) with (3.40) and represented in Figure 3.5. Here, the circles indicate the transition between accumulation and depletion modes where $Q_{sc} = 0$ according to (3.43) or (3.46).

3.1.5   Analytical versus Numerical Simulations

The charge–voltage relationships developed in (3.43) and (3.46) are compared with numerical Technology Computer-Aided Design (TCAD) simulations using a constant mobility for electrons of $(\mu = 0.1\,m^2/Vs)$, a channel length and width of $1\,\mu m$ to avoid short-narrow channel effects, and an oxide thickness of $1.5\,nm$. Channel quantization is ignored since silicon channels are more than $10\,nm$ thick.

Figures 3.5 to 3.8 compare the mobile charge density evaluated at source ($V_{ch} = 0\,V$) for three silicon thicknesses i.e., 10, 20, and $40\,nm$, while varying the doping concentration. Linear and semilogarithmic scales confirm the good agreement between the analytical expressions and the TCAD simulations from depletion to accumulation. The slope of the mobile charge versus potential does not give back the gate oxide capacitance C_{ox}, even though the device is biased well above the threshold. The nonlinear dependence show below the flat-band is inherent to the junctionless architecture and cannot be simulated with inversion-mode FET models. However, since this nonlinearity becomes less pronounced in thinner and lower doped channels, inversion-mode double-gate FET models could still fit junctionless double-gate FETs, but the technological parameters would have to be considered as fitting parameters.

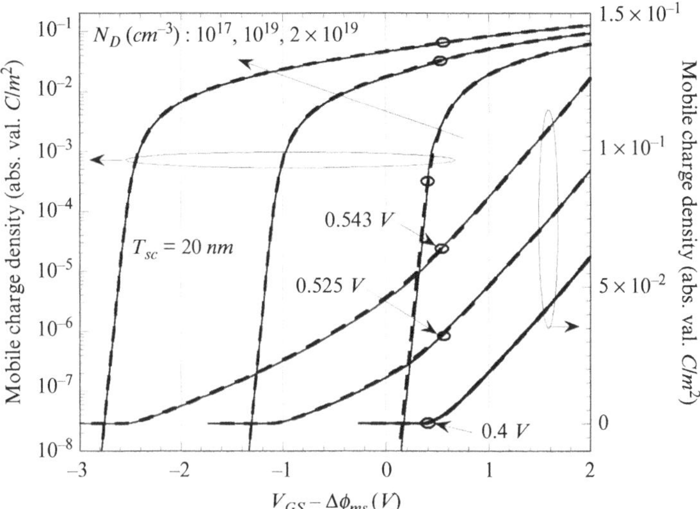

Figure 3.5 Comparison of the mobile charge density in a $20\,nm$ junctionless double-gate MOSFET obtained from (3.39) and (3.40) (dashed lines) and from "regional" approximations given by (3.43) and (3.46) (solid lines). The open circles highlight flat-band conditions that set the limit of validity for the approximated relationships. The numbers indicate the values of $V_{GS,FB} - \Delta\phi_{ms}$. Reprinted from [86] with permission.

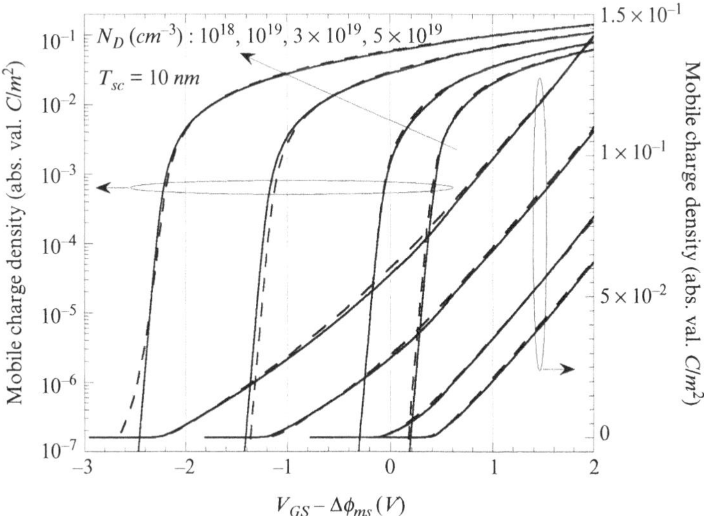

Figure 3.6 Mobile charge density versus the gate voltage for a $10\,nm$ silicon thickness junctionless double-gate MOSFET for different doping concentrations. Solid lines: model and dashed lines: TCAD simulations. Note that $\Delta\phi_{ms}$ is the difference between the metal work function and an intrinsic reference semiconductor. Reprinted from [86] with permission.

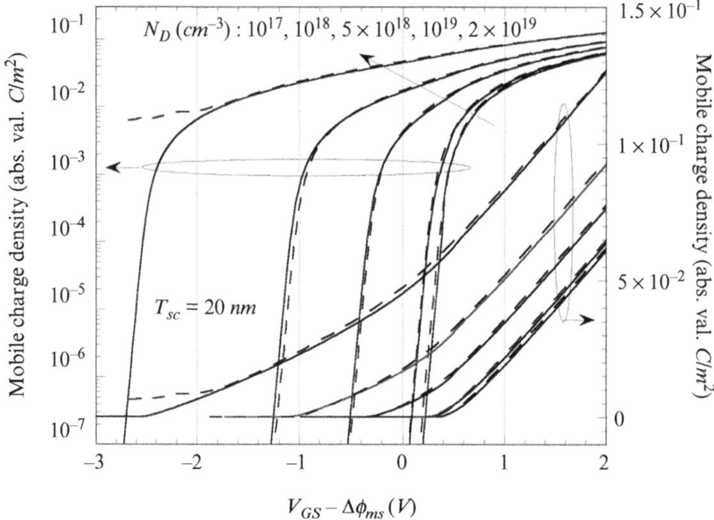

Figure 3.7 Mobile charge density versus the gate voltage for a 20 nm silicon thickness junctionless double-gate MOSFET for different doping concentrations. Solid lines: model and dashed lines: TCAD simulations. Note that $\Delta\phi_{ms}$ is difference between the metal work function and an intrinsic reference semiconductor. Reprinted from [86] with permission.

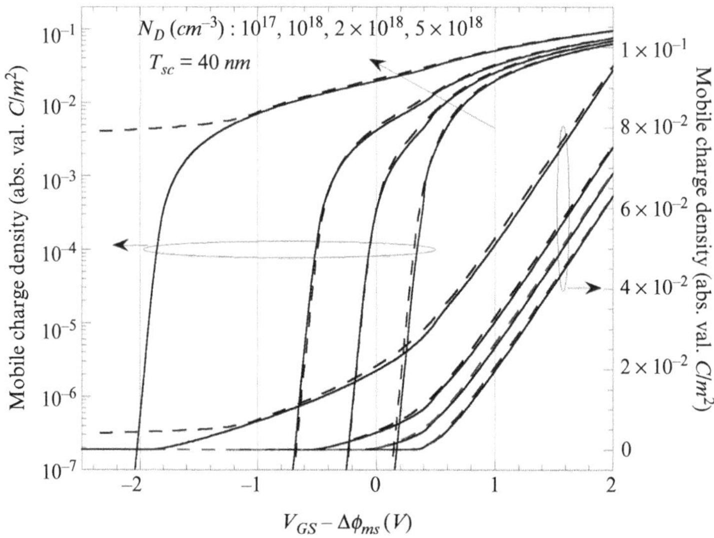

Figure 3.8 Mobile charge density versus the gate voltage for a 40 nm silicon thickness junctionless double-gate MOSFET for different doping concentrations. Solid lines: model and dashed lines: TCAD simulations. Note that $\Delta\phi_{ms}$ is difference between the metal work function and an intrinsic reference semiconductor. Reprinted from [86] with permission.

Well below flat-band, the mobile charge density varies exponentially with the gate voltage and reaches the ideal slope of $60\,mV$/decade (at room temperature, $300\,°K$). In this sense, the double-gate junctionless FET behaves as the inversion-mode double-gate FET biased in weak inversion, a feature that cannot be captured by only full depletion approximation.

3.1.6 Threshold Voltage in Junctionless FETs

The notion of threshold voltage in junctionless devices is questionable. Indeed, extrapolating the mobile charge density from the almost linear strong inversion asymptote is no longer valid when high doping concentration and/or thick layers are concerned. For instance, when considering a $20\,nm$ junctionless double-gate MOSFET, Figure 3.5 evidences "two" slopes in the Q–V dependence when the device is set in depletion (open circles indicate the flat-band condition). This unusual characteristic for junctionless double-gate MOSFETs comes from the square term in (3.46) when the device operates in depletion (which must happen before switching off the channel). But a threshold voltage can still be defined as the gate potential that cancels the mobile charge density when the logarithmic term is neglected. It is worth noting that this definition is consistent with the one used when considering double-gate MOSFET [143]. By adopting this definition and setting $Q_m = 0$ in (3.34), relation (3.46) leads to:

$$V_{th} = \Delta\phi_{ms} + U_T \ln\left(\frac{N_D}{n_i}\right) - qN_D T_{sc}\left(\frac{1}{2C_{ox}} + \frac{1}{8C_{si}}\right). \tag{3.47}$$

Intrinsic Limitations

There is an upper limit for the doping concentration above which the silicon channel cannot be depleted from majority carriers, thus preventing the device from switching off. This is illustrated in Figures 3.7 and 3.8 for 20 and $40\,nm$ silicon thickness. When the gate voltage is pushed to negative values, the charge density remains high and almost independent of the gate voltage. This is attributed to an inversion layer (holes in our case) that builds up at the Si/SiO_2 interfaces. For instance, this happens around ($V_{GS} - \Delta\phi_{ms} \sim -2\,V$) in the $20\,nm$ devices doped at $2 \times 10^{19}\,cm^{-3}$ and close to ($V_{GS} - \Delta\phi_{ms} \sim -1\,V$) in the $40\,nm$ doped at $5 \times 10^{18}\,cm^{-3}$. This behavior, which is not predicted by the model that suggests that full depletion can be reached even for lower gate voltages, represents an intrinsic limit that must be kept in mind when moving to large/highly doped junctionless devices. This will be addressed in detail in Chapters 4 and 5.

3.1.7 Derivation of the Channel Current

Drift Current

The total drift current density consists of electron and hole drift currents that can be calculated from the concentrations of electrons and holes with the corresponding velocities:

$$J_{drift} = J_{n,drift} + J_{p,drift} = qnv_n + qpv_p. \tag{3.48}$$

Here, n and p are the electron and hole densities and v_n and v_p are the electron and hole average velocities. Introducing the carrier mobility:

$$\mu = \frac{q\tau_c}{m^*}, \tag{3.49}$$

where τ_c is the average time between scattering events and m^* is the effective mass of a particle with charge q, thus giving:

$$J_{drift} = J_{n,drift} + J_{p,drift} = q(n\mu_n + p\mu_p)E = \sigma E, \tag{3.50}$$

where μ_n and μ_p are the electron and hole carrier mobilities, respectively, and σ is the average conductivity.

Diffusion Current

In addition to the drift currents governed by an electric field, a gradient in carrier concentration results in a current arising from diffusive events that satisfy Fick's law:

$$J_{n,diffusion} = qD_n \frac{\partial n}{\partial y}, \tag{3.51}$$

$$J_{p,diffusion} = -qD_p \frac{\partial p}{\partial y}, \tag{3.52}$$

where D_n and D_p are the diffusion constants linked to the mobilities through the Einstein relation:

$$D_n = \mu_n \frac{k_B T}{q} = \mu_n U_T, \tag{3.53}$$

$$D_p = \mu_p \frac{k_B T}{q} = \mu_p U_T. \tag{3.54}$$

The total diffusion current density for both electrons and holes is given by

$$J_{diffusion} = J_{p,diffusion} + J_{n,diffusion} = -q\left(\mu_p U_T\right)\frac{\partial p}{\partial y} + q\left(\mu_n U_T\right)\frac{\partial n}{\partial y}. \tag{3.55}$$

Alternative Expression of the Total Drift-Diffusion Current in Nondegenerate Semiconductors

The total electron current density is obtained by adding the diffusion and drift contributions:

$$J_n = J_{n,diffusion} + J_{n,drift} = q\mu_n nE + q\left(\mu_n U_T\right)\frac{\partial n}{\partial y}. \tag{3.56}$$

Similarly, the total hole current density is obtained from

$$J_p = J_{p,diffusion} + J_{p,drift} = q\mu_p pE + q\left(\mu_p U_T\right)\frac{\partial p}{\partial y}. \tag{3.57}$$

Then, the total current density is

$$J_{total} = J_p + J_n. \tag{3.58}$$

Neglecting holes in the n-type channel and combining (3.56) with the free electron concentration $n(x)$ given by (3.23) leads to:

$$J_n = q\mu_n n \left(-\frac{\partial \Psi}{\partial y}\right) + q\mu_n U_T \frac{\partial}{\partial y}(n). \tag{3.59}$$

In nondegenerate semiconductors, the concentrations of electrons and holes follow the Boltzmann statistics:

$$n = qn_i \exp\left(\frac{\Psi - V_{ch}}{U_T}\right). \tag{3.60}$$

Differentiating "n" with respect to y and replacing $(-\partial \Psi/\partial y)$ in (3.59), the current densities can be expressed in terms of the Fermi potential gradient:

$$J_n = -\mu_n q_m \frac{dV_{ch}}{dy}, \tag{3.61}$$

where q_m is the mobile charge density expressed per unit volume. The current is obtained by integrating the current density across the semiconductor thickness i.e., normal to the surface. Assuming that there is no Fermi potential gradient in this direction i.e., no current is expected to flow normal to the surface gives:

$$I_n = -\mu_n Q_m \frac{dV_{ch}}{dy}, \tag{3.62}$$

where Q_m is the electron density *per unit surface*. This is where the current continuity is introduced. Under steady state, the current should be constant from source to drain, which translates into the following relationship:

$$\int_S^D I dy = I_{DS} L_G = -W \int_S^D \mu_n Q_m dV_{ch}. \tag{3.63}$$

Therefore, the current satisfies:

$$I_{DS} = -\frac{W}{L_G} \int_S^D \mu_n Q_m dV_{ch}. \tag{3.64}$$

Current Continuity along the Channel

Introducing the silicon charge density given by (3.34) in (3.64) while assuming that the *mobility is constant along the channel*, we obtain the general relationship:

$$I_{DS} = \frac{W}{L_G}\mu_n \int_S^D (qN_D T_{sc} - Q_{sc})dV_{ch} = \frac{W}{L_G}\mu_n qN_D T_{sc} V_{DS} - \frac{W}{L_G}\mu \int_S^D Q_{sc} dV_{ch}. \tag{3.65}$$

This expression cannot be integrated analytically when using relation (3.39). However, using the approximate expressions of the charge density derived for accumulation and depletion regimes gives analytical expressions for current that are valid in all the regions of operation.

3.1.8 Evaluation of the Charge Integral

This section evaluates analytically the integral present in (3.65).

The Current in Accumulation

Here, relation (3.43) is used to calculate the current in accumulation. Differentiating (3.43) with respect to Q_{sc} gives the differential term inside the integral of (3.65):

$$Q_{sc}dV_{ch} \overset{acc}{\approx} \frac{Q_{sc}}{2C_{ox}}dQ_{sc} - \frac{\dfrac{Q_{sc}^2}{4qN_D\varepsilon_{si}}}{1 + \dfrac{Q_{sc}^2}{8qN_D\varepsilon_{si}U_T}}dQ_{sc}, \tag{3.66}$$

which gives the integral of (3.65):

$$\int_S^D Q_{sc}dV_{ch}\bigg|_{acc} = \frac{1}{4C_{ox}}Q_{sc}^2\bigg|_S^D - 2U_T\left[Q_{sc} - \sqrt{8qN_D\varepsilon_{si}U_T}\arctan\left(\frac{Q_{sc}}{\sqrt{8qN_D\varepsilon_{si}U_T}}\right)\right]\bigg|_S^D. \tag{3.67}$$

After inserting this expression in (3.65), the current in accumulation can be calculated from the values of the charge densities (mobile and fixed) evaluated from (3.43) at source and drain.

The Current in Depletion

Similarly, relying on relation (3.46), the integrand of (3.65) is:

$$Q_{sc}dV_{ch} \overset{dep}{\approx} \frac{Q_{sc}^2}{4qN_D\varepsilon_{si}}dQ_{sc} + \frac{Q_{sc}}{2C_{ox}}dQ_{sc} + 2U_T\frac{\left(\dfrac{Q_{sc}}{qN_DT_{sc}}\right)^2}{1 - \left(\dfrac{Q_{sc}}{qN_DT_{sc}}\right)^2}dQ_{sc}, \tag{3.68}$$

and its integral is obtained analytically:

$$\int_S^D Q_{sc}dV_{ch}\bigg|_{dep} = \frac{1}{12qN_D\varepsilon_{si}}Q_{sc}^3\bigg|_S^D + \frac{1}{4C_{ox}}Q_{sc}^2\bigg|_S^D - 2U_TQ_{sc}\bigg|_S^D$$

$$+ U_T(qN_DT_{sc})\left[\ln\left(1 + \frac{Q_{sc}}{qN_DT_{sc}}\right) - \ln\left(1 - \frac{Q_{sc}}{qN_DT_{sc}}\right)\right]\bigg|_S^D. \tag{3.69}$$

Inserting this expression in (3.65) and evaluating the charge density in the semiconductor at source and drain from (3.46), the current is calculated for depletion.

3.1.9 General Treatment of the Current in Junctionless Double-Gate MOSFETs

So far, current in a uniformly accumulated or depleted silicon channel has been derived analytically. However, depending on the potentials applied to the device, a section of the channel can be in accumulation (close to the source) while the other can be in depletion (close to the drain). Whereas such a situation does not happen in inversion-mode FETs, this "hybrid" channel state requires a dedicated approach. For instance, since the transition occurs at the flat-band voltage, three different situations are identified.

Uniformly Accumulated Channel

The situation where the whole channel is in accumulation is ensured as soon as the inequality $V_{GS} - V_{DS} - \Delta\phi_{ms} \geq U_T \ln(N_D/n_i)$ is satisfied. Indeed, since $V_{DS} > 0$ and the source is used as reference, this condition imposes that $V_{GS} - \Delta\phi_{ms} \geq U_T \ln(N_D/n_i)$ as well. Then, the current reads:

$$I_{DS} = \frac{W}{L_G}\mu q N_D T_{sc} V_{DS} - \frac{W}{L_G}\mu \left(\int_S^D Q_{sc} dV_{ch} \Bigg|_{acc} \right). \tag{3.70}$$

Uniformly Depleted Channel

Conversely to the accumulated channel case, uniform depletion is satisfied as soon as $V_{GS} - \Delta\phi_{ms} \leq U_T \ln(N_D/n_i)$, which also implies that $V_{GS} - V_{DS} - \Delta\phi_{ms} \leq U_T \ln(N_D/n_i)$. The current can be written formally as:

$$I_{DS} = \frac{W}{L_G}\mu q N_D T_{sc} V_{DS} - \frac{W}{L_G}\mu \left(\int_S^D Q_{sc} dV_{ch} \Bigg|_{dep} \right). \tag{3.71}$$

But the case of a uniformly accumulated or depleted channel is not always satisfied, and a third situation may happen mixing both states is possible in junctionless FETs.

The Hybrid Channel

A peculiar situation inherent to junctionless FETs is when depletion and accumulation coexist inside the channel, what is called *hybrid* channel state. This happens when part of the channel near the source is in depletion whereas it turns into accumulation near the drain, implying that $V_{GS} - \Delta\phi_{ms} \geq U_T \ln(N_D/n_i)$ and $V_{GS} - V_{DS} - \Delta\phi_{ms} \leq U_T \ln(N_D/n_i)$. Additionally, it presumes that flat-band occurs inside the channel. Having identified the hybrid operation, the integral in (3.65) is split into two parts, below and above the flat-band. The current is then expressed as the sum of two contributions, each of which is evaluated either from (3.67) or (3.69):

$$I_{DS} = \frac{W}{L_G}\mu q N_D T_{sc} V_{DS} - \frac{W}{L_G}\mu \left(\int_S^{FB} Q_{sc} dV_{ch} \Bigg|_{acc} + \int_{FB}^D Q_{sc} dV_{ch} \Bigg|_{dep} \right). \tag{3.72}$$

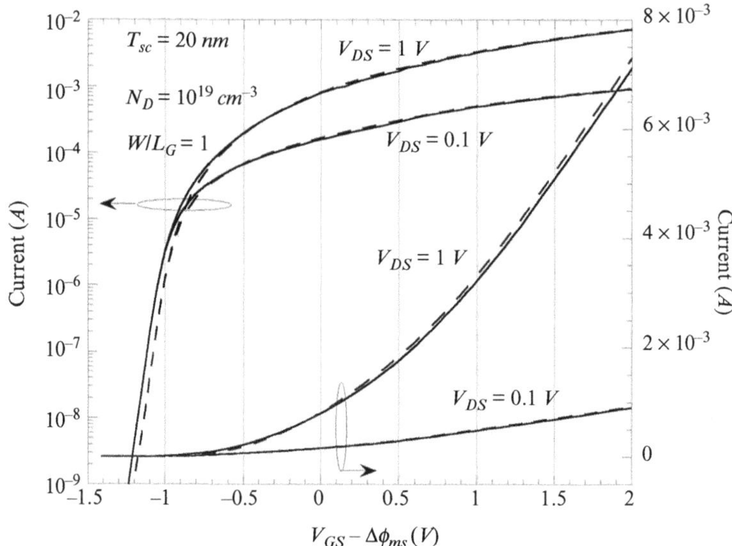

Figure 3.9 Drain current versus the gate voltage in linear and saturated regimes in a $20\ nm$ silicon thickness junctionless double-gate MOSFET doped at $1 \times 10^{19}\ cm^{-3}$. Solid lines: model and dashed lines: TCAD simulations. Reprinted from [86] with permission.

3.1.10   Simulation Results

In this section, current–voltage characteristics predicted by the analytical model are discussed and compared with numerical simulations. Figure 3.9 shows the channel current versus the gate voltage at low ($V_{DS} = 0.1\ V$) and high ($V_{DS} = 1\ V$) drain voltages for a highly doped junctionless FET ($10^{19}\ cm^{-3}$) with $20\ nm$ silicon thickness. When the drain potential is $0.1\ V$, the flat-band voltages at source and drain terminals are 0.53 and $0.63\ V$, respectively, and thus the channel (which is almost uniform) operates in depletion for $V_{GS} - \Delta\phi_{ms}$ lower than $0.53\ V$, and in accumulation otherwise (higher than $0.63\ V$). No discontinuity is evidenced when crossing depletion/accumulation modes. Increasing the drain potential up to $1\ V$ encompasses the hybrid case. For instance, when the gate voltage is between 0.53 and $1.53\ V$, the channel is partly depleted and partly accumulated.

The similarity to the numerical simulations reveals that the model is accurate. Similarly, Figure 3.10 confirms the model can predict current in $10\ nm$ silicon thickness channels despite the very high doping concentration of $5 \times 10^{19}\ cm^{-3}$ which exceeds the effective density of states for electrons, and could invalidate the use of Boltzmann statistics. Note that such a large value is met in working devices as reported by Colinge et al. [1].

It is worth noting that a shift in the gate voltage is present when the device operates in depletion mode, whereas such a discrepancy does not exist for the charges (see Figure 3.5; numerical integration of (3.64) with (3.39) and (3.40) also shows the same mismatch with respect to TCAD simulations).

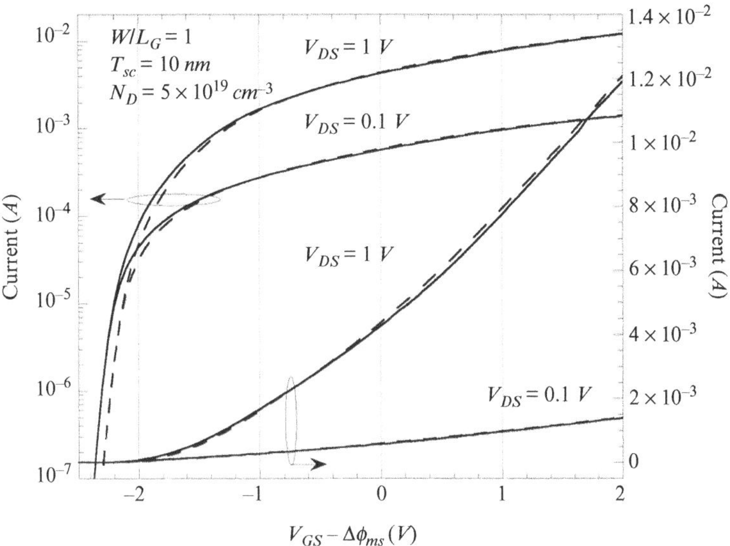

Figure 3.10 Drain current versus gate voltage 10 *nm* silicon thickness junctionless double-gate MOSFET doped $5 \times 10^{19}\ cm^{-3}$. Solid lines: model and dashed lines: numerical (TCAD) simulations. Reprinted from [86] with permission.

Current versus drain voltage has also been investigated in Figures 3.11 and 3.12 for 20 and 10 *nm* silicon thicknesses, respectively. Compared to I_{DS} versus V_{GS} characteristics, some discrepancy between the model and TCAD simulations exists, which is not expected since the charge densities were predicted accurately. Similarly, this mismatch still exists when the current is evaluated by numerical integration.

Small signal characteristics are also simulated. Figure 3.13 shows the gate transconductance (g_m) for a 10 *nm* silicon thickness with different doping densities i.e., 10^{19}, 3×10^{19}, and $5 \times 10^{19}\ cm^{-3}$. The drain potential is set to 1 *V*. The three curves represent the gate transconductance characteristics derived either from TCAD simulations (dots), from the current obtained from (3.64) (dotted line), or from numerical evaluation of the mobile charge density with (3.39) and (3.40), and then integration of the current from (3.65). No discontinuity is observed in g_m, meaning that not only the expressions for the mobile charge densities converge to the same values at flat-band, but also their first-order derivatives.

In addition, the difference in their gate transconductance reveals unambiguously that junctionless devices cannot be simulated with double-gate inversion-mode MOSFET models. The differences in the charge–voltage dependences are clearly evidenced through the gate transconductance trend, which is peculiar to junctionless devices.

Finally, concerning the output conductance g_{ds}, the analytical model still matches the TCAD simulations (see Figure 3.14).

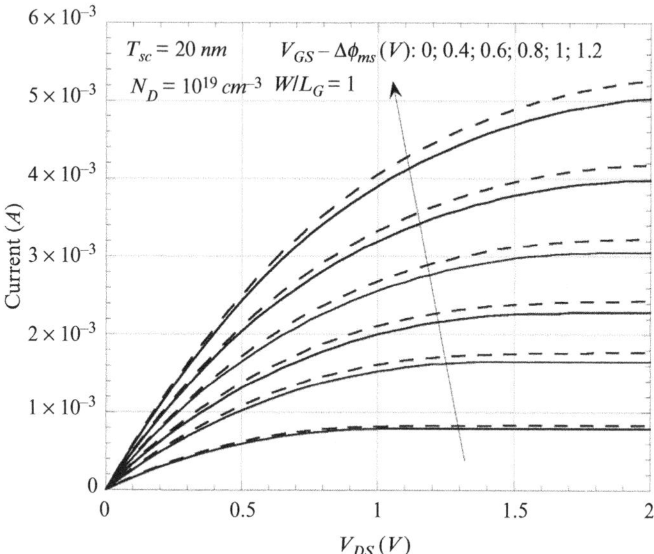

Figure 3.11 Drain current versus the drain voltage for various gate potentials in a $20\,nm$ silicon thickness junctionless double-gate MOSFET doped at $10^{19}\,cm^{-3}$. Solid lines: model and dashed lines: TCAD simulations. Reprinted from [86] with permission.

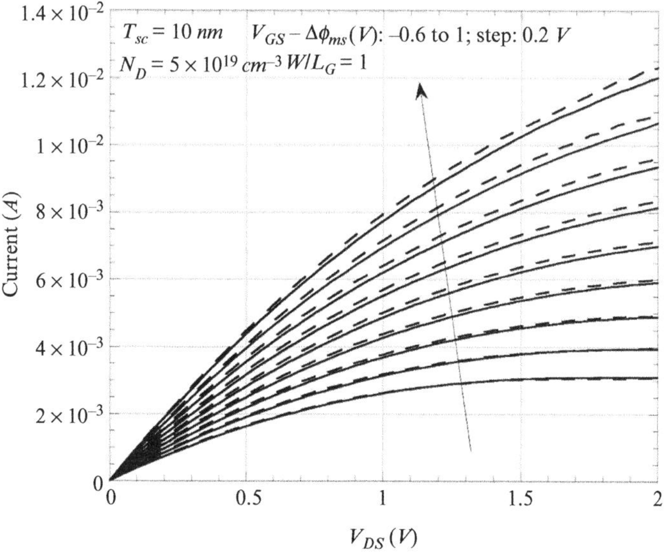

Figure 3.12 Drain current versus the drain voltage for various gate potentials in a $10\,nm$ silicon thickness junctionless double-gate MOSFET doped at $5 \times 10^{19}\,cm^{-3}$. Solid lines: model and dashed lines: TCAD simulations. Reprinted from [86] with permission.

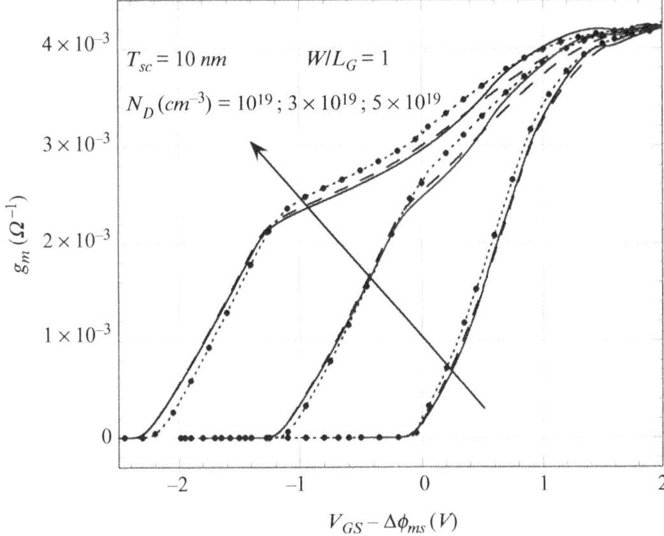

Figure 3.13 Gate transconductance versus the gate voltage at $V_{DS} = 1\ V$ in a $10\ nm$ silicon thickness junctionless double-gate MOSFET for different doping concentrations (10^{19}, 3×10^{19}, $5 \times 10^{19}\ cm^{-3}$). The dots are TCAD simulations and the solid lines are analytical model relying on (3.72) combined with (3.67) and (3.69). The dashed lines are based on numerical integration of (3.64). Note that $\Delta\phi_{ms}$ is the difference between the metal work function and an intrinsic reference semiconductor. Reprinted from [86] with permission.

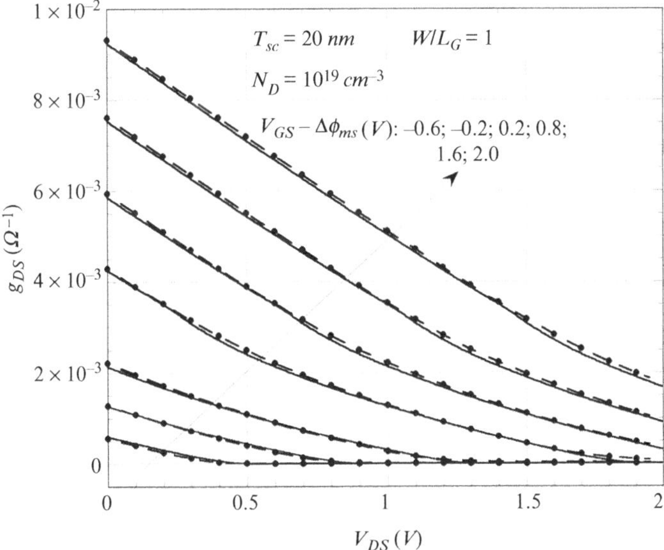

Figure 3.14 Drain transconductance versus the drain voltage for different gate potentials in a $20\ nm$ silicon thickness junctionless double-gate MOSFET doped at $10^{19}\ cm^{-3}$. Dots: TCAD simulations and solid lines: analytical model relying on (3.72) combined with (3.67) and (3.69). Reprinted from [86] with permission.

3.2 A Common Core Model for Junctionless Nanowires and Symmetric DG FETs

A formal link between double-gate and cylindrical junctionless MOSFET architectures is developed in this section. In particular, it will be shown that the analytical model for double-gate junctionless FETs can simulate charges and current in nanowire junctionless FETs.

3.2.1 Analysis of Electrostatics in Junctionless Nanowire FETs

The device is a junctionless nanowire FET with radius (R) and uniform channel donor concentration N_D and assumes a nondegenerate channel (Figure 3.15).

Approximate Poisson–Boltzmann Equation in Junctionless Nanowire FETs
Assuming a long channel, the Poisson–Boltzmann equation in cylindrical coordinates is:

$$\frac{\partial^2 \Psi(r)}{\partial r^2} + \frac{1}{r}\frac{\partial \Psi(r)}{\partial r} = \frac{q}{\varepsilon_{si}}\left[n_i \exp\left(\frac{\Psi(r) - V_{ch}}{U_T}\right) - N_D \right]. \tag{3.73}$$

As for double-gate junctionless FETs, the source Fermi potential is the reference. As in (3.28), relation (3.73) has no analytical solution and the derivation used to model the double-gate architecture is not transposable in nanowires since an additional term arises from the cylindrical shape. Likewise, for the double-gate geometry, derivatives and integrals are approximated by finite differences involving the center and surface coordinates. Multiplying (3.73) by $\partial\Psi(r)/\partial r$ gives after manipulation:

$$\frac{1}{2}\frac{\partial}{\partial r}\left(\frac{\partial \Psi(r)}{\partial r}\right)^2 + \frac{1}{r}\left(\frac{\partial \Psi(r)}{\partial r}\right)^2 = \frac{q}{\varepsilon_{si}}\left[n_i \exp\left(\frac{\Psi(r) - V_{ch}}{U_T}\right) - N_D \right]\frac{\partial \Psi(r)}{\partial r}. \tag{3.74}$$

Next, integrating from $r = 0$ to R and noting that the electric field vanishes at the center:

$$\frac{1}{2}E_s^2 + \int_0^R \frac{1}{r}\left(\frac{\partial \Psi(r)}{\partial r}\right)^2 dr = \frac{q}{\varepsilon_{si}}\int_{\Psi_0}^{\Psi_s}\left[n_i \exp\left(\frac{\Psi(r) - V_{ch}}{U_T}\right) - N_D \right] d\Psi. \tag{3.75}$$

Omitting the second term on the LHS of (3.75) leads to relation (3.29), meaning that this term is the main deviation to the double-gate architecture description. In a second step, the integral in (3.75) is evaluated using a trapezoidal integration scheme involving the nodes at the center and at the surface:

$$\int_0^R \frac{1}{r}\left(\frac{\partial \Psi(r)}{\partial r}\right)^2 dr \approx \frac{1}{2}\left(\frac{\partial \Psi(r)}{\partial r}\right)^2\bigg|_{r=R} + \frac{R}{2r}\left(\frac{\partial \Psi(r)}{\partial r}\right)^2\bigg|_{r=0}. \tag{3.76}$$

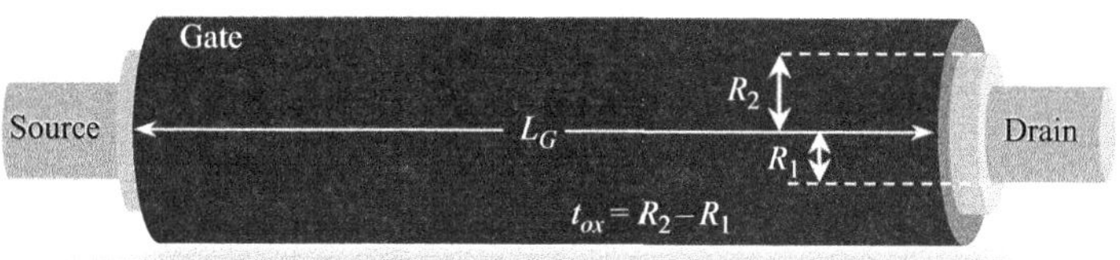

Figure 3.15 Schematic view of an n-type junctionless nanowire FET investigated in this chapter.

The first term on the RHS represents the electric field at the Si/SiO$_2$ interface. The second term can be rewritten as:

$$\frac{R}{2r}\left(\frac{\partial\Psi(r)}{\partial r}\right)^2\bigg|_{r=0} = \frac{Rr}{2}\left(\frac{1}{r}\frac{\partial\Psi(r)}{\partial r}\right)^2\bigg|_{r=0}. \tag{3.77}$$

In addition, relation (3.73) must be satisfied at the center ($r=0$), implying that $\dfrac{1}{r}\dfrac{\partial\Psi(r)}{\partial r}$ must take a finite value at $r=0$. Therefore, the LHS of (3.77) vanishes as well and a link with the surface electric field is evidenced (3.76):

$$\int_0^R \frac{1}{r}\left(\frac{\partial\Psi(r)}{\partial r}\right)^2 dr \approx \frac{E_s^2}{2}. \tag{3.78}$$

Replacing (3.78) in (3.75) links the surface electric field to the center and surface potentials Ψ_0 and Ψ_s for the nanowire:

$$2\times\left(\frac{1}{2}E_s^2\right)^{NW} \approx \frac{q}{\varepsilon_{si}}\int_{\Psi_0}^{\Psi_s}\left[n_i\exp\left(\frac{\Psi(r)-V_{ch}}{U_T}\right)-N_D\right]d\Psi. \tag{3.79}$$

In junctionless double-gate MOSFET the second term on the LHS of (3.73) is discarded:

$$\left(\frac{1}{2}E_s^2\right)^{DG} \approx \frac{q}{\varepsilon_{si}}\int_{\Psi_0}^{\Psi_s}\left[n_i\exp\left(\frac{\Psi(r)-V_{ch}}{U_T}\right)-N_D\right]d\Psi. \tag{3.80}$$

Comparing (3.80) and (3.79) highlights similar dependence between surface electric field and surface potential for planar and cylindrical shapes since these expressions differ by a factor of only two. Relation (3.79) can be written as:

$$\left(\frac{1}{2}E_s^2\right)^{NW} \approx \frac{q}{\varepsilon_{si}}\int_{\Psi_0}^{\Psi_s}\left[\frac{n_i}{2}\exp\left(\frac{\Psi(r)-V_{ch}}{U_T}\right)-\frac{N_D}{2}\right]d\Psi. \tag{3.81}$$

Relation (3.81) suggests that charge densities in a nanowire can be simulated with the model of a junctionless double-gate MOSFET provided the channel doping and intrinsic carrier densities are divided by two. When the mobile charge density is negligible, relation (3.73) simplifies to:

$$\frac{\partial^2 \Psi(r)}{\partial r^2} + \frac{1}{r}\frac{\partial \Psi(r)}{\partial r} = \frac{-qN_D}{\varepsilon_{si}}, \tag{3.82}$$

and $\Psi(r)$ is given by

$$\Psi(r) = \frac{-qN_D}{4\varepsilon_{si}}r^2 + \Psi(0). \tag{3.83}$$

On the other hand, for a double-gate structure, relation (3.28) gives:

$$\frac{\partial^2 \Psi(r)}{\partial r^2} = \frac{-qN_D}{\varepsilon_{si}}, \tag{3.84}$$

which solution is

$$\Psi(r) = \frac{-qN_D}{2\varepsilon_{si}}r^2 + \Psi(0). \tag{3.85}$$

Therefore, (3.83) can be obtained from (3.85) provided the doping N_D is divided by two (the intrinsic carrier density does not come into play because we are only considering electrostatics).

Accounting for the Gate Capacitance

In a nanowire with a radius R and a dielectric thickness t_{ox}, the Gauss theorem applied to the cylindrical shape gives (E_S is oriented from the surface toward the center of the nanowire):

$$C_{ox}^{NW}(V_{GS} - \Delta\phi_{ms} - \Psi_s) = \varepsilon_{si}E_s, \tag{3.86}$$

where the gate capacitance per unit surface of a cylindrical capacitor is

$$C_{ox}^{NW} = \frac{\varepsilon_{ox}}{R\ln\left(1 + \dfrac{t_{ox}}{R}\right)}. \tag{3.87}$$

In a double-gate like configuration, the gate capacitance should be expressed per unit of projected surface i.e., assuming half of the perimeter contributes one gate capacitance in the double-gate topology. Then, the total charge in the nanowire expressed per unit surface Q_{sc} is

$$Q_{sc} = -2C_{ox}^{NW}(V_{GS} - \Delta\phi_{ms} - \Psi_s). \tag{3.88}$$

Noting that $\Delta\phi_{ms}$ is the difference between the metal work function and an intrinsic reference semiconductor given by $\Delta\phi_{ms} = W_{ms} - U_T\ln(N_D/n_i)$, where W_{ms} is the metal-semiconductor work function difference. The equivalent parameters defined above would then give the same value for $\Delta\phi_{ms}$, supporting the consistency of the equivalent double-gate parameters.

In the same manner, the fixed-charge density Q_{fix} per unit of surface (in a double-gate sense) is given by

$$Q_{fix} = (qN_D\pi R^2 L)\frac{1}{\pi RL} = qN_DR. \tag{3.89}$$

Comparison of relations (3.34) and (3.89) shows that the radius of the nanowire must be used in place of the semiconductor thickness T_{sc} in a double-gate geometry. Then, the link between the total (Q_{sc}) and mobile (Q_m) charge densities (per unit surface in a "DG" formulation) gives:

$$Q_m = Q_{sc} - Q_{fix} = Q_{sc} - qN_D R. \tag{3.90}$$

Introducing the equivalent doping density N_D^{Eq-DG}, we obtain:

$$Q_m = Q_{sc} - qN_D^{Eq-DG} 2R. \tag{3.91}$$

According to (3.91), it is convenient to define an equivalent thickness that reverts to the wire diameter $T_{sc}^{Eq-DG} = 2R$, leading to a consistent expression in terms of equivalent double-gate parameters:

$$Q_m = Q_{sc} - qN_D^{Eq-DG} T_{sc}^{Eq-DG}. \tag{3.92}$$

Mapping between nanowire and double-gate topologies seems therefore appropriate when considering the device as a whole i.e., the semiconductor with the gate electrode. A complete description of junctionless nanowires must also include the current as addressed in the next section.

3.2.2 Derivation of the Current in a Junctionless Nanowire

Since the mobile charge density given by (3.92) represents the mobile carrier concentration per unit of "projected" surface in the double-gate sense, using half of the nanowire perimeter as the equivalent double-gate width, $W^{Eq-DG} = \pi R$ (which is meaningful), the current reads:

$$I_{DS} = -W^{Eq-DG} \mu Q_m \frac{\partial V_{ch}}{\partial x}. \tag{3.93}$$

Integrating (3.93) from source to drain, and making use of the definitions of the equivalent parameters in Table 3.1, the current can be readily calculated:

$$I_{DS} = \frac{W^{Eq-DG}}{L_G} \mu \left[qN_D^{Eq-DG} T_{sc}^{Eq-DG} V_{DS} - \int_S^D Q_{sc} dV_{ch} \right]. \tag{3.94}$$

Relation (3.94) is formally the same as that obtained in (3.65) for the double-gate topology. The equivalence between planar and cylindrical shapes is now in order for junctionless FETs.

3.2.3 Simulations

Long-channel silicon nanowires ($L_G = 1\,\mu m$) of 5 and $10\,nm$ radius and different doping densities (n-type channels) are considered. The oxide thickness is $1.5\,nm$ and the drift and diffusion model with a constant carrier mobility of $1,000\,cm^2 V^{-1} s^{-1}$ is used to simulate transport. TCAD simulations and the analytical model share

Table 3.1. Correspondence between junctionless nanowire physical parameters and equivalent junctionless double-gate FET model parameters

Physical parameters	Nanowire	Equivalent double-gate MOSFET
Radius and thickness	R (radius)	$T_{sc} = 2 \times R$
Oxide thickness	t_{ox}	$\dfrac{T_{sc}}{2} \times \ln\left(1 + 2\dfrac{t_{ox}}{T_{sc}}\right)$
Width	–	$W = \pi \times R$
Doping concentration	N_D	$N_D/2$
Intrinsic carrier concentration	n_i	$n_i/2$

the same technological and model parameters. Quantum mechanical effects are ignored, which is acceptable for the relatively large silicon radius considered here.

Figure 3.16a represents the mobile charge density as a function of the gate voltage for TCAD and model simulations based on the equivalent parameters listed in Table 3.1. The $5\,nm$ nanowire radius shows good agreement between the numerical and the analytical simulations for doping densities as large as $10^{19}\,cm^{-3}$. For higher doping values i.e., 2×10^{19} and $5 \times 10^{19}\,cm^{-3}$ a mismatch is evidenced in the subthreshold regime (continuous lines). The shift in the gate voltage is about $50\,mV$ for $2 \times 10^{19}\,cm^{-3}$ and $90\,mV$ for $5 \times 10^{19}\,cm^{-3}$. However, this is not a V_T shift as it has little impact above the threshold.

At this point, a correction factor "k" is introduced in such a way that ε_{si} will now be replaced by $k\varepsilon_{si}$, with $k = 1.2$, as though the silicon dielectric constant were overestimated by 20 percent.

Simulations with this parameter included are represented by the dotted lines in Figure 3.16. Interestingly, for both devices (Figure 3.16a and b), this correction – which is still empirical – is able to recover the mismatch in subthreshold without affecting the above threshold behavior. For lower doped channels, the model is also accurate (Figure 3.16a and b).

Note that for lower values of the doping, the equivalent double-gate model is accurate to calculate the mobile charge density in junctionless nanowire FETs without a correction factor i.e., $k = 1$. Transfer characteristics are shown in Figure 3.16c and d for a $5\,nm$ radius junctionless nanowire with two doping densities, $10^{19}\,cm^{-3}$ and $2 \times 10^{19}\,cm^{-3}$ in linear $(V_{DS} = 0.1\,V)$ and saturation modes $(V_{DS} = 1\,V)$. The junctionless nanowire doped at $10^{19}\,cm^{-3}$ is well modeled with the equivalent double-gate approach from deep depletion to accumulation and from linear to saturation. For the nanowire doped at $2 \times 10^{19}\,cm^{-3}$, the model agrees with TCAD simulations, especially above the threshold, and only in deep depletion some discrepancies are visible. Still, better accuracy is obtained using the modified silicon dielectric constant (dotted lines in Figure 3.16d).

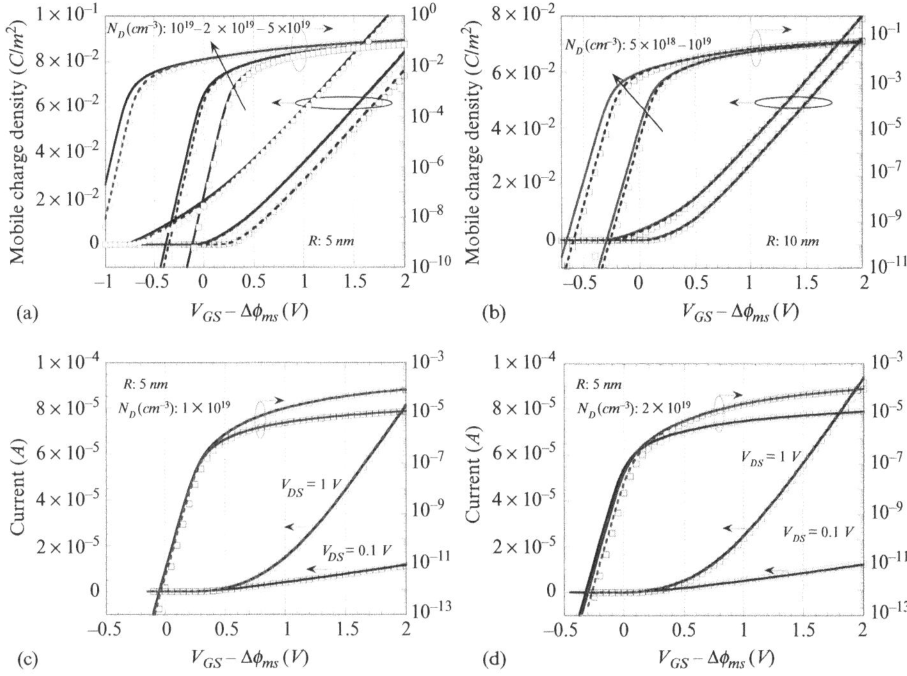

Figure 3.16 Mobile charge density versus gate voltage in (a) $5\,nm$ and (b) $10\,nm$ radius nanowire with various doping densities ($V_{DS} = 0$). Drain current versus gate voltage in a $5\,nm$ radius nanowire doped at (c) $10^{19}\,cm^{-3}$ and (d) $2 \times 10^{19}\,cm^{-3}$. Symbols: TCAD, full lines: model with no correction, and dotted line: model with $k = 1.2$. Note that $\Delta\phi_{ms}$ is the difference between the metal work function and an intrinsic reference semiconductor given by $\Delta\phi_{ms} = W_{ms} - U_T \ln(N_D/n_i)$, where W_{ms} is the metal-semiconductor work function difference. Reprinted from [124] with permission.

3.3 Explicit Model for Long-Channel Gate-All-Around Junctionless MOSFETs

In [125], Lime et al. proposed to solve the Poisson–Boltzmann equation in cylindrical coordinates to develop a core model for drain current in long-channel junctionless gate-all-around MOSFETs. Using $u = r^2$ as a new variable allows writing:

$$\frac{\partial \Psi(r)}{\partial r} = \frac{\partial \Psi(r)}{\partial u} 2r. \tag{3.95}$$

The Poisson–Boltzmann equation (relation 3.73) in cylindrical coordinates can be expressed as

$$\frac{\partial^2 \Psi(r)}{\partial r^2} + \frac{1}{r}\frac{\partial \Psi(r)}{\partial r} = \frac{1}{r}\frac{\partial}{\partial r}\left(r\frac{\partial \Psi(r)}{\partial r}\right) = 4\frac{\partial}{\partial u}\left(u\frac{\partial \Psi}{\partial u}\right)$$

$$= \frac{q}{\varepsilon_{si}}\left[n_i \exp\left(\frac{\Psi(r) - V_{ch}}{U_T}\right) - N_D\right]. \tag{3.96}$$

Multiplying both sides by $\partial\Psi(r)/\partial u$ and integrating from $r = 0 \to \Psi(r) = \Psi_0$ to $r = R \to \Psi(r) = \Psi_s$ yields

$$\int_0^{R^2} \frac{1}{u}\frac{\partial}{\partial u}\left(u\frac{\partial\Psi}{\partial u}\right)^2 du = \int_{\Psi_0}^{\Psi_s} \frac{q}{2\varepsilon_{si}}\left[n_i \exp\left(\frac{\Psi(r) - V_{ch}}{U_T}\right) - N_D\right] d\Psi. \tag{3.97}$$

As discussed in [125], integration by part gives

$$u\left(\frac{\partial\Psi}{\partial u}\right)^2\Bigg|_{r=0}^{R^2} + \int_0^{R^2}\left(\frac{\partial\Psi}{\partial u}\right)^2 du$$

$$= \frac{qn_iU_T}{2\varepsilon_{si}}\left[\exp\left(\frac{\Psi_s - V_{ch}}{U_T}\right) - \exp\left(\frac{\Psi_0 - V_{ch}}{U_T}\right) - \frac{N_D(\Psi_s - \Psi_0)}{n_iU_T}\right]. \tag{3.98}$$

Assuming a parabolic approximation for the potential profile in the semiconductor links the surface and center potentials:

$$\Psi(r) = \frac{\Psi_s - \Psi_0}{R^2}r^2 + \Psi_0. \tag{3.99}$$

From the Gauss theorem evaluated at $r = R$ ($\partial\Psi/\partial r = Q_{sc}/\varepsilon_{si}$), relation (3.99) leads to

$$\Psi(r) = U_T\frac{Q_{sc}}{Q_{cp}}\left(\frac{r}{R}\right)^2 + \Psi_0, \tag{3.100}$$

where Q_{cp} is given by $2\varepsilon_{si}U_T/R$. Replacing (3.100) in (3.98), the surface potential-based relationship for the mobile charge density is

$$Q_m = Q_{cp}Q_{fix}\frac{1 - \exp\left(\dfrac{-Q_{sc}}{Q_{cp}}\right)}{Q_{sc}}\exp\left(\frac{\Psi_s - V_{ch}}{U_T}\right), \tag{3.101}$$

where $Q_{sc} = Q_m - Q_{fix}$ is the algebraic sum of fixed (Q_{fix}) and mobile (Q_m) charges densities (per unit area) in the channel. Last, the continuity of the displacement vector at the semiconductor-gate insulator boundary ($\Psi_s = V_{GS} - \Delta\phi_{ms} - U_T\ln(N_D/n_i) - Q_{sc}/C_{ox}$) gives the mobile charge density versus the gate potential:

$$Q_m\exp\left(\frac{Q_m}{C_{ox}U_T}\right) = \frac{Q_{cp}Q_{fix}}{Q_{sc}}\frac{1 - \exp\left(\dfrac{-Q_{sc}}{Q_{cp}}\right)}{\exp\left[-\dfrac{V_{GS} - \Delta\phi_{ms} - V_{ch} - U_T\ln(N_D/n_i) - Q_{fix}/C_{ox}}{U_T}\right]}, \tag{3.102}$$

or equivalently:

$$V_{GS} - \Delta\phi_{ms} - U_T\ln\left(\frac{N_D}{n_i}\right) - V_{ch}$$

$$= -\frac{Q_{fix}}{C_{ox}} + U_T\ln\left(\frac{Q_m}{Q_{fix}}\right) + \frac{Q_m}{C_{ox}} - U_T\ln\left\{\frac{Q_{cp}\left[1 - \exp\left(-\dfrac{Q_m - Q_{fix}}{Q_{cp}}\right)\right]}{Q_m - Q_{fix}}\right\}. \tag{3.103}$$

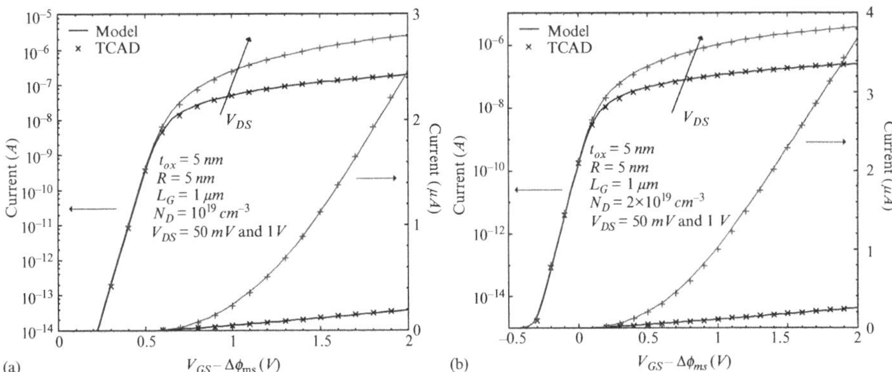

Figure 3.17 Drain current versus gate voltage in a $5\,nm$ radius nanowire doped at (a) $10^{19}\,cm^{-3}$ and (b) $2 \times 10^{19}\,cm^{-3}$. Note that $\Delta\phi_{ms}$ is the difference between the metal work function and an intrinsic reference semiconductor given by $\Delta\phi_{ms} = W_{ms} - U_T \ln(N_D/n_i)$, where W_{ms} is the metal-semiconductor work function difference. Reprinted from [125] with permission.

The drain current I_{DS} in the nanowire is given by

$$I_{DS} = 2\pi\mu\frac{R}{L_G}\int_0^{V_{DS}} Q_m(V_{ch})dV_{ch}, \tag{3.104}$$

where dV_{ch} is obtained by differentiation of (3.103) with respect to Q_m. Even though the mobile charge-density can be analytically obtained through (3.102), a closed-form solution of the current with respect to the mobile charge density is not possible. Conversely, numerical integration of (3.102) is plotted in Figure 3.17. The parabolic approximation used by the authors happens to be quite accurate for $5\,nm$ radius nanowires doped at 10^{19} and $2 \times 10^{19}\,cm^{-3}$. Additionally, F. Lime et al. proposed simplified relationships valid in depletion and accumulation modes of operation.

3.3.1 Approximated Solution in Depletion

In order to obtain a simpler expression of the mobile charge density in depletion mode (where $Q_{sc} \approx -Q_{fix}$ and $\exp(Q_{fix}/Q_{cp}) \gg 1$), relation (3.102) can be approximated by the following equation:

$$Q_m \exp\left(\frac{Q_m}{C_{ox}U_T}\right) = Q_{cp}\exp\left(\frac{Q_{fix}}{Q_{cp}}\right)$$
$$\times \exp\left[\frac{V_{GS} - \Delta\phi_{ms} - U_T\ln(N_D/n_i) - V_{ch} + Q_{fix}/C_{ox}}{U_T}\right]. \tag{3.105}$$

Alternatively, this relation can be rewritten as a charge-based relationship:

$$V_{GS} - \Delta\phi_{ms} - U_T\ln(N_D/n_i) - V_{ch} + \frac{Q_{fix}}{C_{ox}} = U_T\ln\left(\frac{Q_m}{Q_{cp}}\right) + \frac{Q_m}{C_{ox}}. \tag{3.106}$$

Comparing this approximated solution with relation (3.46), it is worth noting that a quadratic term $Q_m{}^2$ is missing in (3.106). This is roots in the parabolic approximation which has been used in [125].

3.3.2 Approximated Solution in Accumulation Mode

In an accumulation regime where $\exp(-Q_{sc}/Q_{cp}) \ll 1$. (3.102) can be approximated by

$$Q_m\left(Q_m - Q_{fix}\right)\exp\left(\frac{Q_m}{C_{ox}U_T}\right)$$
$$= Q_{cp}Q_{fix}\exp\left[\frac{V_{GS} - \Delta\phi_{ms} - U_T\ln(N_D/n_i) + Q_{fix}/C_{ox}}{U_T}\right], \qquad (3.107)$$

Again, this can be turned into a charge-based formulation:

$$V_{GS} - \Delta\phi_{ms} - U_T\ln\left(\frac{N_D}{n_i}\right) - V_{ch} + \frac{Q_{fix}}{C_{ox}} = U_T\ln\left(\frac{Q_m}{Q_{cp}}\left(\frac{Q_m}{Q_{fix}} - 1\right)\right) + \frac{Q_m}{C_{ox}}. \quad (3.108)$$

3.3.3 Approximated Solution in Weak Accumulation Mode

In their model, the authors introduce an additional regime that has weak accumulation. Neglecting the quadratic term in (3.108) the mobile charge density is given by

$$V_{GS} - \Delta\phi_{ms} - U_T\ln\left(\frac{N_D}{n_i}\right) - V_{ch} + \frac{Q_{fix}}{C_{ox}} = U_T\ln\left(\frac{Q_m}{Q_{cp}}\right) + \frac{Q_m}{C_{ox}}. \qquad (3.109)$$

As the solution is obtained by iterating between two equations that are valid in both regimes of operations, the transition between accumulation and depletion is preserved and does not need any interpolation functions. The agreement between the model and TCAD simulations is fine in depletion and accumulation regimes, but slightly degrades at the transition between both as well as for high values of the drain voltage V_{DS} and channel-doping concentrations.

3.4 Summary

This chapter introduced the basis of the EPFL charge-based model to calculate the charge density and the current in junctionless symmetric double-gate field-effect transistors. The model is valid in all regions of operation, from deep depletion to accumulation and from linear to saturated regimes, as confirmed by this chapter's detailed comparison with TCAD numerical simulations. In particular, the occurrence of two distinct slopes in the charge–voltage dependence was predicted with respect to inversion-mode double-gate MOSFETs. The intimate correspondence between junctionless nanowires and double-gate FET topologies was discussed in light of equivalent parameters. With the exception of a correction of the silicon dielectric constant for the nanowire, no empirical parameters were used. Similarly, a model based on the parabolic approximation of the potential in nanowires was introduced. The model can simulate charges and current in junctionless nanowire FETs by defining three distinct regions of operation.

4   Model-Driven Design-Space of Junctionless FETs

This chapter discusses the intimate link between the technological parameters of junctionless FET devices and some of their advantages. For a given channel thickness, there is an upper limit for the doping density above which the channel remains filled with majority carriers whatever the gate voltage.

Analytical expressions involving the silicon thickness and the doping density are derived and used to predict *off*-current performance, *on/off*-current ratio, and rail-to-rail supply voltage. Such analytical expressions can be effectively used as guidelines for technology optimization of junctionless FETs.

4.1   *Off*-Current and Inversion Layer in Junctionless FETs

Since the principle of operation of junctionless double-gate MOSFETs is based on majority carriers flowing in the volume of a semiconductor channel, turning off the device needs to deplete the channel of these carriers [1, 28, 143]. However, this is challenging when the doping concentration is large with respect to the channel thickness (full depletion issues and performance of junctionless FETs are discussed in [52]).

Considering an *n*-type doped channel, when the semiconductor layer is highly doped and/or quite thick, a large negative gate potential may be needed to fully deplete the channel. However, as shown in Appendix B [144], this is not only a matter of gate voltage: below some value, a hole-inversion layer builds up at the channel interface, further screening the gate electric field and preventing depletion of electrons from the channel.

This hypothesis is sustained by transient current measurements performed on double-gate junctionless FETs revealing that nonswitching devices could still be turned off in transient mode (see Appendix B).

Following the discussion in [144], what ultimately limits channel depletion and *off*-state current density is the inherent inversion-hole layer at the Si–SiO$_2$ interface (for *n*-channel junctionless FETs). Preliminary studies have attempted to evaluate the performance of junctionless double-gate MOSFETs integrated in bulk substrate [145].

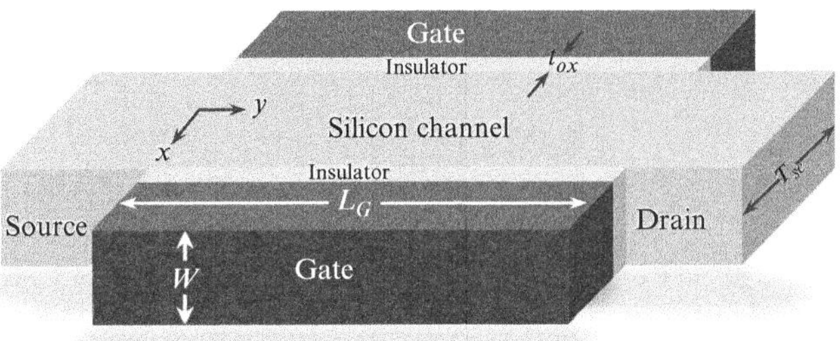

Figure 4.1 Schematic view of the *n*-type junctionless symmetric double-gate MOSFET.

4.2 Electrostatics in Junctionless Double-Gate MOSFET including Minority Carriers

The device of interest here is a *n*-type doped long-channel ($L_G = 100\,nm$) junctionless double-gate MOSFET as illustrated in Figure 4.1 (the symbols have their usual meaning, i.e., L_G and t_{ox} are the gate length and gate-oxide thickness and T_{sc} and N_D are the semiconductor thickness and the donor density in the channel). Accounting for minority carriers in the channel, the Poisson–Boltzmann equation in a nondegenerate semiconductor including holes is:

$$\frac{\partial^2 \Psi}{\partial x^2} = \frac{q n_i}{\varepsilon_{si}} \left[\exp\left(\frac{\Psi - V_{ch}}{U_T}\right) - \exp\left(-\frac{\Psi}{U_T}\right) - \frac{N_D}{n_i} \right], \tag{4.1}$$

where x is the coordinate across the two gates ($-T_{sc}/2 < x < T_{sc}/2$), U_T is the thermal voltage, n_i and ε_{si} are the silicon intrinsic carrier density and permittivity, and V_{ch} is the shift in the quasi-Fermi potential for electrons (potentials are referenced to the source Fermi potential). In addition, if the transverse current in the x-direction is negligible (no current across the gates), the Fermi potential V_{ch} has no x dependence and relation (4.1) can be symmetrized as follows:

$$\frac{\partial^2}{\partial x^2}\left(\Psi - \frac{V_{ch}}{2}\right) = \frac{2 q n_i}{\varepsilon_{si}} \exp\left(-\frac{V_{ch}}{2U_T}\right) \left[\sinh\left(\frac{\Psi - V_{ch}/2}{U_T}\right) - \frac{N_D}{2 n_i} \exp\left(\frac{V_{ch}}{2U_T}\right) \right]. \tag{4.2}$$

At this point, a symmetric formulation is obtained when introducing the generalized potential and intrinsic density:

$$\Psi^* = \Psi - \frac{V_{ch}}{2}, \tag{4.3}$$

$$n_i^* = n_i \exp\left(-\frac{V_{ch}}{2U_T}\right). \tag{4.4}$$

Using these definitions, relation (4.2) becomes:

$$\frac{\partial^2 \Psi^*}{\partial x^2} = \frac{2 q n_i^*}{\varepsilon_{si}} \left[\sinh\left(\frac{\Psi^*}{U_T}\right) - \frac{N_D}{2 n_i^*} \right]. \tag{4.5}$$

This differential equation has no analytical solution, but integrating once [86] relation (4.5) gives:

$$E_s^2 = \frac{4qn_i^* U_T}{\varepsilon_{si}} \left[\cosh\left(\frac{\Psi_s^*}{U_T}\right) - \cosh\left(\frac{\Psi_0^*}{U_T}\right) - \frac{N_D}{2n_i^* U_T}\left(\Psi_s^* - \Psi_0^*\right) \right], \qquad (4.6)$$

where $\Psi_s^* = \Psi\left(\frac{T_{sc}}{2}\right) - \frac{V_{ch}}{2} = \Psi_s - \frac{V_{ch}}{2}$, $\Psi_0^* = \Psi(0) - \frac{V_{ch}}{2} = \Psi_0 - \frac{V_{ch}}{2}$, with Ψ_s and Ψ_0 being the surface and center potentials, respectively.

The continuity of the displacement vector at the silicon/insulator interface imposes:

$$E_s = \frac{\varepsilon_{ox}}{\varepsilon_{si} t_{ox}} \left[\Psi_s + \Delta\phi_{ms} - V_{GS}\right] = \frac{\varepsilon_{ox}}{\varepsilon_{si} t_{ox}} \left[\Psi_s^* + \Delta\phi_{ms} - V_{GS}^*\right], \qquad (4.7)$$

with $V_{GS}^* = V_{GS} - V_{ch}/2$, where V_{GS} is the gate-to-source voltage and $\Delta\phi_{ms}$ is the difference between the metal work function and an intrinsic reference semiconductor given by $\Delta\phi_{ms} = W_{ms} - U_T \ln(N_D/n_i)$, where W_{ms} is the metal–semiconductor work function difference. Combining (4.6) and (4.7), the generalized surface and body center potentials become interrelated:

$$\left(\Psi_s^* + \Delta\phi_{ms} - V_{GS}^*\right)^2 = \frac{4qn_i^* U_T \varepsilon_{si} t_{ox}^2}{\varepsilon_{ox}^2} \left[\cosh\left(\frac{\Psi_s^*}{U_T}\right) - \cosh\left(\frac{\Psi_0^*}{U_T}\right) - \frac{N_D}{2n_i^* U_T}\left(\Psi_s^* - \Psi_0^*\right) \right].$$

$$(4.8)$$

Relation (4.8) reveals that once Ψ_0^* is given, Ψ_s^* can be calculated, which from (4.6) and (4.7) gives a unique value of the charge density in the channel (via the surface electric field E_s) and gate-to-source potential. As discussed in Chapter 3, the generalized surface and center potentials can be interdependent by considering the first four terms of a MacLaurin series of (4.5) at the center. The approximate solution of the potential distribution across the channel which results is:

$$\Psi(x) \approx \Psi_0 + \frac{qn_i^*}{\varepsilon_{si}} \left[2\sinh\left(\frac{\Psi_0 - V_{ch}/2}{U_T}\right) - \frac{N_D}{n_i^*} \right] \frac{x^2}{2}. \qquad (4.9)$$

Next, the surface potential is obtained by evaluating (4.9) at the channel interfaces:

$$\Psi_s^* = \Psi_0^* + K \left[2\sinh\left(\frac{\Psi_0^*}{U_T}\right) - \frac{N_D}{n_i^*} \right], \qquad (4.10)$$

where K is a parameter linked to the semiconductor thickness:

$$K = \frac{qn_i^* T_{sc}^2}{8\varepsilon_{si}}. \qquad (4.11)$$

Therefore, relation (4.10) is a generalization of (3.31) when including the contribution of minority carriers in the channel. Analytical expressions are compared to TCAD simulations in Figures 4.2 and 4.3. In Figure 4.2, the surface potential and center potentials are plotted as a function of the gate voltage for different values of the silicon thickness assuming $N_D = 10^{19}\ cm^{-3}$, $L_G = 100\ nm$, $V_{DS} = 0\ V$, and $t_{ox} = 1.5\ nm$.

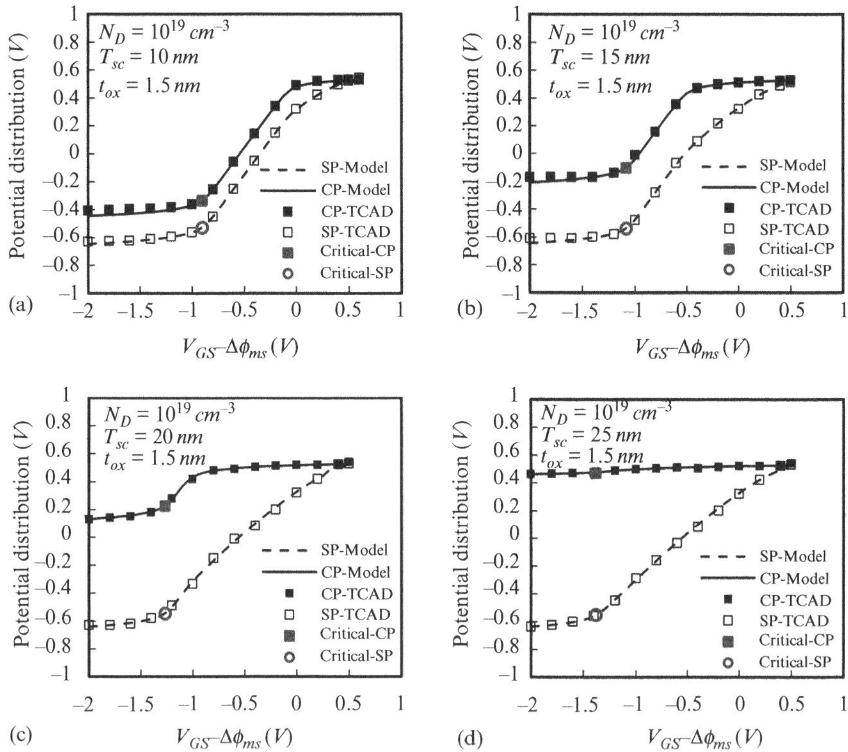

Figure 4.2 Dependence of the body center potential (CP) and surface potential (SP) versus the gate voltage ($V_{GS} - \Delta\phi_{ms}$) for different silicon thickness values (T_{sc}) ($V_{ch} = 0\,V$, $L_G = 100\,nm$, and $N_D = 10^{19}\,cm^{-3}$). Critical values of surface potential (Critical-SP) and center potential (Critical-CP) were obtained from (4.25) and (4.27). Reprinted from [146] with permission.

At flat-band, the channel is neutral and the center and surface potentials are equal, which from (4.8) gives $\Psi_{FB} + \Delta\phi_{ms} - V_{GFB} = 0\,V$. Inserting these conditions in (4.10) and noting that holes are negligible at flat-band in n-type silicon gives:

$$\Psi_{s,FB} = \Psi_{0,FB} = V_{GS,FB} - \Delta\phi_{ms} \approx \frac{V_{ch}}{2} + U_T \mathrm{asinh}\left(\frac{N_D}{2n_i^*}\right) \approx V_{ch} + U_T \ln\left(\frac{N_D}{n_i}\right). \quad (4.12)$$

Figure 4.2 shows the surface and center potential dependencies on the gate voltage for different values of the silicon thickness in silicon doped at $10^{19}\,cm^{-3}$ when the shift in quasi-Fermi potential for electrons with respect to the source is zero ($V_{ch} = 0\,V$). For $T_{sc} = 10\,nm$, when the gate voltage is decreased, the surface and center potentials are decreasing as the gate voltage (see Figure 4.2a), which is common for a bulk MOSFET operating in weak inversion. However, beyond some critical value of the gate voltage, both Ψ_s and Ψ_0 are reaching kinds of asymptotes. This happens when an inversion layer (holes) is created at the channel interface, further screening the electric field and preventing any change in the center potential.

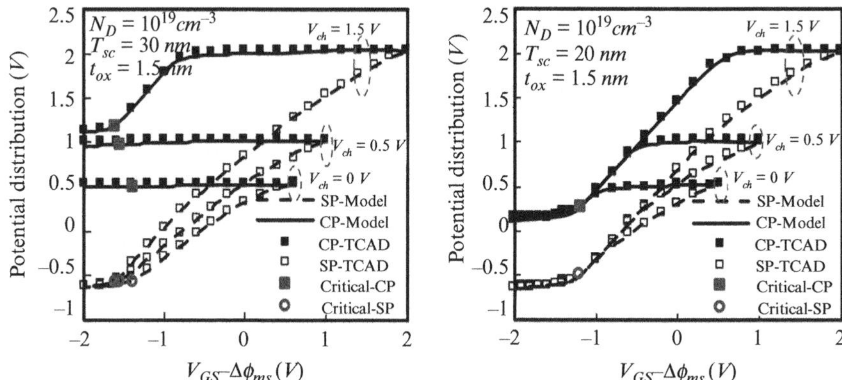

Figure 4.3 Dependence of the body center potential (CP) and surface potential (SP) versus the gate voltage ($V_{GS} - \Delta\phi_{ms}$) for $T_{sc}=$ (a) 30 nm and (b) 20 nm for different values of V_{ch}; $V_{ch} = 0, 0.5$, and $1.5\ V$, $L_G = 100\ nm$, and $N_D = 10^{19}\ cm^{-3}$. Critical values of surface potential (Critical-SP) and center potential (Critical-CP) were obtained from (4.25) and (4.27). Note that $\Delta\phi_{ms}$ is the difference between the metal work function and an intrinsic reference semiconductor given by $\Delta\phi_{ms} = W_{ms} - U_T \ln(N_D/n_i)$, where W_{ms} is metal–semiconductor work function difference. Reprinted from [146] with permission.

In addition, the separation between Ψ_s and Ψ_0 remains almost constant and close to KN_D/n_i, which is the potential drop between the center and the surface when the silicon is totally depleted. Therefore, since the electron density depends on the center potential, a lesser variation of the former will also increase the subthreshold swing of $60\ mV/decade$ in long-channel devices. According to Figure 4.2, below the threshold the center potential Ψ_0 changes a lot with the silicon thickness in contrast to the surface potential. The separation between Ψ_s and Ψ_0 increases significantly in thick-channel devices since an inversion layer (holes) is created before full depletion can be met. Ultimately, for $T_{sc} = 25\ nm$, the center potential remains almost constant whatever the gate voltage (see Figure 4.2d), meaning that the center of the channel is almost neutral for densities greater than $N_D = 1.5 \times 10^{19}\ cm^{-3}$.

4.2.1   Role of the Channel Potential

So far, center Ψ_0 and surface Ψ_s potentials have been estimated with $V_{ch} = 0\ V$, a special case where the generalized variables Ψ^* and n_i^* revert to their original definitions. The Fermi potential V_{ch} can effectively create some depletion in relatively thick and highly doped channels. For instance, decoupling the center and surface potentials using V_{ch} has been shown using 30 nm (Figure 4.3a) and 20 nm (Figure 4.3b) silicon channel thickness.

According to (4.12), at flat-band the surface and center potentials are shifted by V_{ch} (note that the shift in quasi-Fermi potential for holes is set to zero), and the channel potential has very little impact on the critical surface potential. Here, a change in V_{ch} only results in modifying the flat-band voltage. However, the majority carrier density will still depend on V_{ch}.

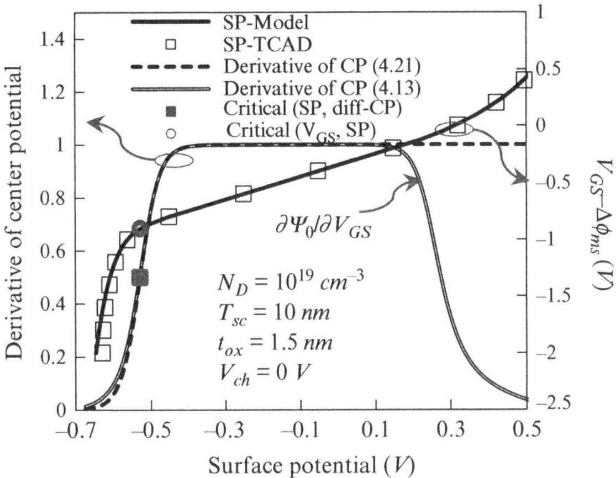

Figure 4.4 Derivative of the center potential $\partial\Psi_0^*/\partial V_{GS}$ with respect to the surface potential for $V_{ch} = 0\,V$, $N_D = 10^{19}\,cm^{-3}$, $t_{ox} = 1.5\,nm$, $L_G = 100\,nm$ in a $10\,nm$-silicon thick junctionless double-gate MOSFET (left axis). The surface potential according to gate voltage is assigned to the right axis. The lines and symbols are related to the analytical model and TCAD simulations, respectively. The critical surface potential (SP), critical gate voltage (V_{GS}), and critical value of the derivative (diff-CP) are shown as gray symbols. Noting that $\Delta\phi_{ms}$ is difference between the metal work function and an intrinsic reference semiconductor given by $\Delta\phi_{ms} = W_{ms} - U_T \ln(N_D/n_i)$, where W_{ms} is the metal–semiconductor work function difference. Reprinted from [146] with permission.

For $T_{sc} = 30\,nm$, the center potential remains close to the flat-band condition when V_{ch} is set to 0.5 V (see Figure 4.3a). However, when increasing V_{ch} up to 1.5 V the center potential departs from flat-band, which shows that the Fermi voltage effectively enhances the depletion. Introducing holes in the Poisson equation invalidates the apparent equivalence between the gate and the drain potentials (see relation (3.40)) by unpinning the center potential (and thus the mobile charge density). Therefore, in real devices, it is possible to inhibit inversion and achieve full depletion at the drain, even in thick and highly doped channels.

Note that for $20\,nm$ silicon thickness (see Figure 4.3b), since the channel is already depleted when $V_{ch} = 0\,V$, the surface and center potentials will not change when increasing V_{ch}. Here, the lowest intrinsic *off*-current that can be obtained for a given T_{sc} and doping density in a long-channel junctionless FET depends on the remaining majority carriers, which is a worst case for long-channel devices. Additionally, in short-channel devices, the current will also be affected by the drain voltage through the so-called Drain-Induced Barrier Lowering effect (DIBL), which will degrade the *off*-current even more (see Chapter 7).

4.2.2 Estimation of the Critical Potentials

The potential below which the mobile charge density is nearly unaffected by the gate voltage is shown in Figures 4.2 to 4.4, and is defined by where the slopes of the center and surface potentials exhibit the largest variation. This critical condition involves the derivative of the center potential with respect to gate voltage using relations (4.8) and (4.10):

$$\left(\frac{\partial \Psi_0^*}{\partial V_{GS}^*}\right)^{-1} = \frac{2K}{U_T}\cosh\left(\frac{\Psi_0^*}{U_T}\right) + 1$$

$$+ \sqrt{\eta}\,\frac{\sinh\left(\dfrac{\Psi_0^*}{U_T}\right) - \sinh\left(\dfrac{\Psi_s^*}{U_T}\right) - \dfrac{2K}{U_T}\cosh\left(\dfrac{\Psi_0^*}{U_T}\right)\sinh\left(\dfrac{\Psi_s^*}{U_T}\right) + \dfrac{KN_D}{n_i^* U_T}\cosh\left(\dfrac{\Psi_0^*}{U_T}\right)}{2U_T\sqrt{\cosh\left(\dfrac{\Psi_s^*}{U_T}\right) - \cosh\left(\dfrac{\Psi_0^*}{U_T}\right) - \dfrac{N_D}{2n_i^*}\left(\dfrac{\Psi_s^* - \Psi_0^*}{U_T}\right)}}\,,$$

$$\tag{4.13}$$

where η is given by

$$\eta = \frac{4\varepsilon_{si}qn_i^* U_T t_{ox}^2}{\varepsilon_{ox}^2}. \tag{4.14}$$

The $\partial \Psi_0^*/\partial V_{GS}^*$ can be obtained from relation (4.13), but a simpler analytical expression of the critical center potential with respect to device parameters is possible using an approximate form of (4.13): close to the critical center potential, the surface potential is negative (see Figures 4.2 and 4.3). Then, the hyperbolic-sine and cosine functions can be approximated by retaining the most relevant arguments around the critical point:

$$\Psi_{sc}^* < 0 \Rightarrow -2\sinh\left(\frac{\Psi_{sc}^*}{U_T}\right) \approx 2\cosh\left(\frac{\Psi_{sc}^*}{U_T}\right) \approx \exp\left(-\frac{\Psi_{sc}^*}{U_T}\right). \tag{4.15}$$

In addition the following inequalities are satisfied:

$$\left|\frac{\Psi_{sc}^*}{U_T}\right| \gg \left|\frac{\Psi_{0c}^*}{U_T}\right| \Rightarrow \left|\sinh\left(\frac{\Psi_{sc}^*}{U_T}\right)\right| \gg \left|\sinh\left(\frac{\Psi_{0c}^*}{U_T}\right)\right|, \tag{4.16}$$

$$\left|\frac{2K}{U_T}\cosh\left(\frac{\Psi_{0c}^*}{U_T}\right)\sinh\left(\frac{\Psi_{sc}^*}{U_T}\right)\right| \ll \left|\frac{KN_D}{n_i^* U_T}\cosh\left(\frac{\Psi_{0c}^*}{U_T}\right)\right| \ll \left|\sinh\left(\frac{\Psi_{sc}^*}{U_T}\right)\right|, \tag{4.17}$$

$$\left|\cosh\left(\frac{\Psi_{sc}^*}{U_T}\right) - \cosh\left(\frac{\Psi_{0c}^*}{U_T}\right)\right| \ll \left|-\frac{N_D}{2n_i^*}\left(\frac{\Psi_{sc}^* - \Psi_{0c}^*}{U_T}\right)\right|, \tag{4.18}$$

$$\frac{2K}{U_T}\cosh\left(\frac{\Psi_{0c}^*}{U_T}\right) \ll 1, \tag{4.19}$$

and,

$$\left(\frac{\Psi_{sc}^* - \Psi_{0c}^*}{U_T}\right) \approx -\frac{KN_D}{U_T n_i^*}. \tag{4.20}$$

Introducing these approximations in (4.13) leads to:

$$\frac{\partial \Psi_0^*}{\partial V_{GS}^*} \approx \cfrac{1}{1 + \cfrac{\sqrt{\eta}}{4U_T}\cfrac{1}{\sqrt{-\cfrac{N_D}{2n_i^*}\left(\cfrac{\Psi_{sc}^* - \Psi_{0c}^*}{U_T}\right)}}\exp\left(-\cfrac{\Psi_s^*}{U_T}\right)}$$

$$\approx \cfrac{1}{1 + \cfrac{\sqrt{\eta}}{4U_T}\cfrac{1}{\sqrt{-\cfrac{N_D}{2n_i^*}\left(-\cfrac{KN_D}{U_T n_i^*}\right)}}\exp\left(-\cfrac{\Psi_s^*}{U_T}\right)} \approx \cfrac{1}{1 + \xi\exp\left(-\cfrac{\Psi_s^*}{U_T}\right)}, \tag{4.21}$$

where ξ is given by

$$\xi = \frac{\sqrt{2\eta}n_i^*}{4\sqrt{U_T K N_D}} = \frac{2n_i^* t_{ox}\varepsilon_{si}}{N_D T_{sc}\varepsilon_{ox}} = 2\frac{n_i^*}{N_D}\frac{C_{si}}{C_{ox}}, \tag{4.22}$$

with C_{ox} being the gate-oxide capacitance and $C_{si} = \varepsilon_{si}/T_{sc}$. The approximate expression (4.21) only involves the surface potential, as illustrated in Figure 4.4.

The critical point can be extracted from the maximum of the second derivative of $\partial \Psi_0^*/\partial V_{GS}^*$ with respect to the surface potential (Ψ_s^*). Alternatively, imposing the second-order-derivative of (4.21) with respect to the surface potential to equal zero,

$$\frac{\partial^2}{\partial \Psi_s^{*2}}\left(\frac{\partial \Psi_0^*}{\partial V_{GS}^*}\right) = 0, \tag{4.23}$$

the critical surface (Ψ_{sc}), center (Ψ_{0c}), and gate potentials (V_{GScrit}) are calculated. Finally, the approximate solution of the second-order derivative of $\partial \Psi_0^*/\partial V_{GS}^*$ with respect to the surface potential is

$$\frac{\partial^2}{\partial \Psi_s^{*2}}\left(\frac{\partial \Psi_0^*}{\partial V_{GS}^*}\right) = -\frac{1 - \xi\exp\left(-\cfrac{\Psi_s^*}{U_T}\right)}{\left[1 + \xi\exp\left(-\cfrac{\Psi_s^*}{U_T}\right)\right]^3}\exp\left(-\frac{\Psi_s^*}{U_T}\right)\frac{\xi}{U_T^2}. \tag{4.24}$$

Imposing the condition given by (4.23) on (4.24), the critical surface potential is readily obtained:

$$\Psi_{sc} = \Psi_{sc}^* + \frac{V_{ch}}{2} = \frac{V_{ch}}{2} + U_T\ln(\xi) = U_T\ln\left(2\frac{n_i^*}{N_D}\frac{C_{si}}{C_{ox}}\right). \tag{4.25}$$

From (4.25), the critical surface potential (Ψ_{sc}) is independent of V_{ch} (see Figure 4.3) and is a function of the technological parameters only. According to (4.10), the corresponding critical center potential satisfies the implicit relationship:

$$\Psi_{0c}^* - U_T\ln(\xi) - K\frac{N_D}{n_i^*} = -2K\sinh\left(\frac{\Psi_{0c}^*}{U_T}\right) \cong -K\exp\left(\frac{\Psi_{0c}^*}{U_T}\right). \tag{4.26}$$

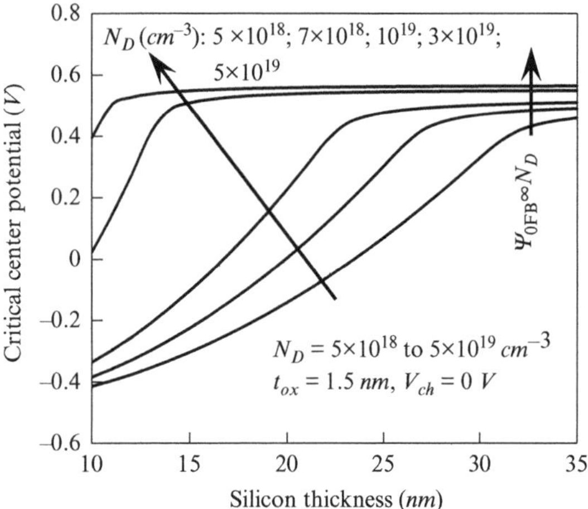

Figure 4.5 Critical body center potential (Ψ_{0c}) according to the silicon thickness (T_{sc}) for different values of doping concentration. Reprinted from [146] with permission.

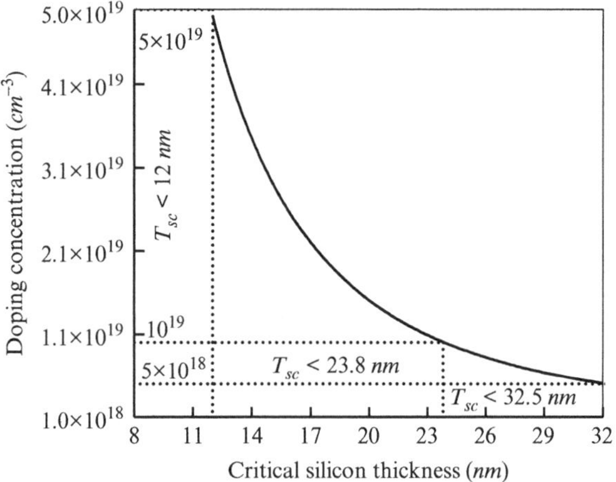

Figure 4.6 Critical space of junctionless double-gate MOSFET according to (4.27). Reprinted from [146] with permission.

Introducing the principal branch of the *Lambert W* function, Ψ_{0c} is given by

$$\Psi_{0c} = \frac{V_{ch}}{2} + U_T \ln(\xi) + K\frac{N_D}{n_i^*} - U_T \text{LambertW}\left[\frac{\xi K}{U_T}\exp\left(\frac{KN_D}{U_T n_i^*}\right)\right]. \qquad (4.27)$$

Based on these relationships, critical surface and center potentials are calculated as a function of the technological parameters, channel potential (V_{ch}), and temperature (see Figures 4.2 and 4.3). For highly doped and/or thick devices, when $V_{ch} = 0$ the critical body center potential is close to the flat-band potential ($\Psi_{0,FB}$), meaning that the center of the channel remains almost neutral and the channel cannot be turned off (see Figure 4.5). Figure 4.6 shows the limit needed to maintain neutral the center of the channel where $\Psi_0 = \Psi_{0,FB}$. For instance, for a doping density of $5 \times 10^{19}\ cm^{-3}$, the center of the channel cannot be depleted if the silicon thickness is greater than $12\ nm$, even by pushing the gate to large negative voltages.

4.2.3 Minimum Mobile Charge Density

The potential distribution $\Psi(x)$ given by (4.9) is used to calculate the total mobile charge density in the channel:

$$
\begin{aligned}
Q_m &= \int_{-\frac{T_{sc}}{2}}^{\frac{T_{sc}}{2}} n_i \exp\left(\frac{\Psi(x) - V_{ch}}{U_T}\right) dx \\[2ex]
&\approx \int_{-\frac{T_{sc}}{2}}^{\frac{T_{sc}}{2}} n_i \exp\left(\frac{\Psi_0 + \dfrac{q n_i^*}{\varepsilon_{si}}\left[2\sinh\left(\dfrac{\Psi_0 - V_{ch}/2}{U_T}\right) - \dfrac{N_D}{n_i^*}\right]\dfrac{x^2}{2} - V_{ch}}{U_T}\right) dx,
\end{aligned}
\tag{4.28}
$$

which gives

$$
Q_m = \sqrt{\pi}\frac{q n_i^*}{\lambda}\exp\left(\frac{\Psi_{0c} - \frac{V_{ch}}{2}}{U_T}\right)\mathrm{erf}\left(\frac{\lambda T_{sc}}{2}\right),
\tag{4.29}
$$

where Ψ_{0c} is obtained from (4.27) and λ depends on the center and Fermi potentials:

$$
\lambda = \sqrt{-\frac{q n_i^*}{\varepsilon_{si} U_T}\left[\sinh\left(\frac{\Psi_{0c} - \frac{V_{ch}}{2}}{U_T}\right) - \frac{N_D}{2 n_i^*}\right]}.
\tag{4.30}
$$

The carrier density obtained from relation (4.29) is illustrated in Figure 4.7 for different values of N_D and V_{ch}.

Relation (4.29) is accurate when the channel is depleted of mobile carriers. When the channel is almost neutral in the center, this relation can still be used as a first guess for the mobile charge density. The local mobile charge density at the critical point ($V_{ch} = 0\ V$) versus the doping concentration and for different values of silicon thickness reveals that the thickness is the most critical parameter in terms of residual mobile charge density (see Figure 4.8).

4.2.4 Estimation of *On/Off*-Current Ratio in Long-Channel Junctionless FETs

The design-space of junctionless double-gate MOSFETs is discussed through an explicit formulation of the I_{On}/I_{Off} figure of merit. In depletion mode, the current can still be expressed as a combination of relations (3.46), (3.68), and (3.71), even

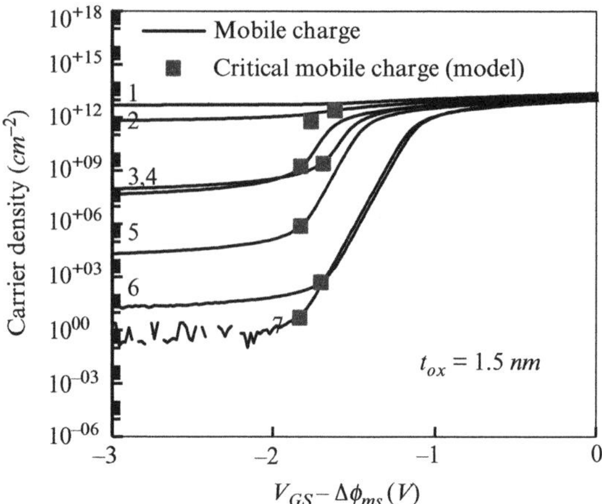

Figure 4.7 Mobile charge density for different values of doping concentration and thickness in junctionless double-gate MOSFET. The solid lines correspond to TCAD simulations and the symbols to the proposed analytical model for the critical current. $2 \times 10^{19}\ cm^{-3}$ and $T_{sc} = 16\ nm$ (4, 6); $1.5 \times 10^{19}\ cm^{-3}$ and $T_{sc} = 24\ nm$ (1, 2, 3, 5, 7); $V_{ch} = 0.1\ V$ (1, 4), $V_{ch} = 0.5\ V$ (2, 6), $V_{ch} = 0.8\ V$ (3), $V_{ch} = 1\ V$ (5), $V_{ch} = 1.3\ V$ (7), and $\Delta\phi_{ms} = W_{ms} - U_T \ln(N_D/n_i)$. Reprinted from [146] with permission.

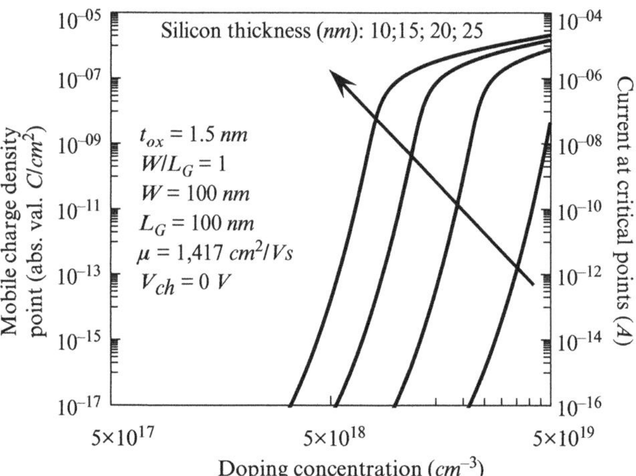

Figure 4.8 Mobile charge density and intrinsic *off*-current with respect to the doping concentration (N_D) for different values of silicon thickness (T_{sc}). The mobile charge density and *off*-current are assigned to the left axis and the right axis, respectively ($W/L_G = 1$). Reprinted from [146] with permission.

though the way they have been derived does not consider holes:

$$I_{DS}\frac{L_G}{W\mu} = -\frac{Q_{sc}^3}{12Q_{fix}C_{si}} - \frac{Q_{sc}^2}{4C_{ox}} + 2U_T Q_{sc} + Q_{fix}\left[V_{GS} - \Delta\phi_{ms} - U_T \ln\left(\frac{N_D}{n_i}\right)\right]$$

$$- 2U_T Q_{fix} \ln\left(1 + \frac{Q_{sc}}{Q_{fix}}\right) + \frac{Q_{sc}^2}{8C_{si}} + Q_{fix}\left.\frac{Q_{sc}}{2C_{ox}}\right|_{Source}^{Drain}, \tag{4.31}$$

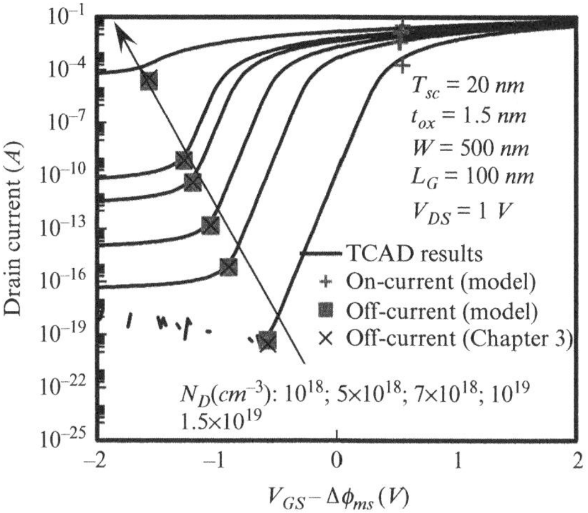

Figure 4.9 I–V characteristics of a *20 nm* silicon thickness junctionless double-gate MOSFET for different doping concentrations ($W/L_G = 5$). The solid lines correspond to the TCAD simulations and the symbols to the proposed analytical model for the critical current. (The crosses correspond to the threshold voltage defined in [86].) Note that $\Delta\phi_{ms}$ is the difference between the metal work function and an intrinsic reference semiconductor given by $\Delta\phi_{ms} = W_{ms} - U_T \ln(N_D/n_i)$, where W_{ms} is the metal–semiconductor work function difference. Reprinted from [146] with permission.

where $Q_{sc} = Q_{fix} + Q_m$, Q_m and $Q_{fix} = qN_D T_{sc}$ are the mobile and the total fixed-charge densities per unit surface, μ is the constant carrier mobility, and $C_{si} = \varepsilon_{si}/T_{sc}$.

On-Current

The *on*-state current is defined in order to deliver the highest current when the device operates in depletion mode. This is obtained once these conditions are satisfied:

- Channel at flat-band at the source i.e., $V_{GS} = V_{GS,FB}$, $Q_{sc} = 0$,
- Channel pinched at the drain i.e., $Q_{sc} = Q_{fix}$ at the drain.

Introducing these conditions in (4.31), we obtain:

$$I_{on}\frac{L_G}{W\mu} = Q_{fix}^2\left(\frac{1}{24C_{si}} + \frac{1}{4C_{ox}}\right) + 2U_T Q_{fix}\left[1 - \ln(2)\right]. \tag{4.32}$$

With these assumptions, the *on*-current is a simple function of the device parameters only.

Off-Current

Similarly, the *off*-state current corresponds to the current density when the source is biased at the critical point while the drain is kept depleted. These conditions translate to:

- $Q_m|_{Drain} = 0$,
- $Q_m|_{Source} = Q_{crit}$ and $V_{GS} = V_{GScrit}$, where V_{GScrit} is the critical gate voltage given from (4.8) imposing $V_{ch} = 0$.

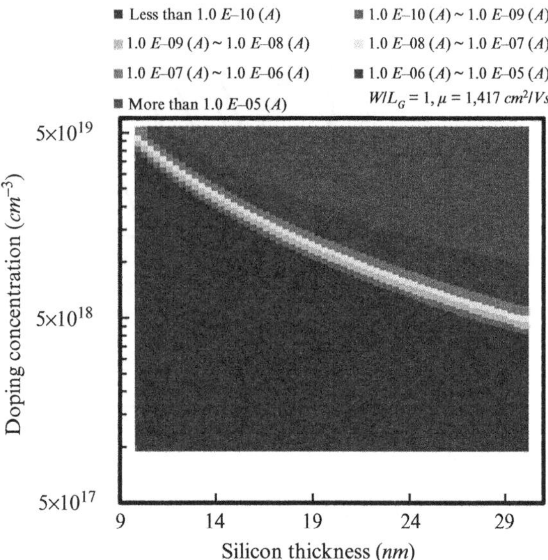

Figure 4.10 Pattern of normalized critical current (A) versus doping concentration (N_D) and silicon thickness (T_{sc}) in a junctionless double-gate MOSFET ($W = L_G = 100\,nm$). Reprinted from [146] with permission.

Again, inserting these quantities in (4.31) gives an estimation of the *off*-current. Figure 4.9 displays *on*- and *off*-currents on the transfer characteristics plotted in logarithmic scales. Figure 4.10 shows a pattern for the critical current parametrized in terms of the silicon thickness and doping density. The transition between low i.e., 10^{-10} A and large i.e., 10^{-8} A *off*-current densities takes place in a narrow region and is extremely sensitive to the channel thickness. Combining (4.32) with (4.31) provides a good estimation of the I_{on}/I_{off} ratio for long channels (see Figure 4.11). Such a representation is used to define an intrinsic characteristic for the junctionless architecture (note that the region below the dotted line indicates where calculations may be less accurate; see Section 4.2.3). An explicit expression for I_{on}/I_{off} is possible when assuming that the *off*-current is controlled by diffusion (still maintaining depletion at the drain, i.e., $Q_m|_{Drain} = 0$):

$$I_{off} = \frac{W}{L_G} D_n \left(Q_m|_{Source} - Q_m|_{Drain} \right) \approx \frac{W}{L_G} D_n Q_m|_{Source}, \tag{4.33}$$

where $D_n = \mu U_T$ is the diffusion coefficient. From (4.32) and (4.33), the *on/off*-current ratio is given by

$$\frac{I_{on}}{I_{off}} \cong \frac{Q_{fix}^2}{4 U_T Q_m} \left(\frac{1}{6 C_{si}} + \frac{1}{C_{ox}} \right) + 2 \frac{Q_{fix}}{Q_m} \left[1 - \ln(2) \right], \tag{4.34}$$

where Q_m is the mobile charge density at the source side obtained from (4.29). Tuning of the gate capacitance by modifying the SiO_2 insulator thickness is shown in

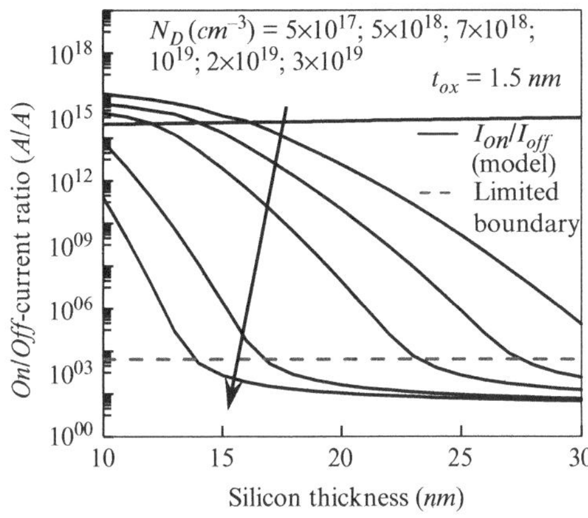

Figure 4.11 *On/Off*-current ratio of junctionless double-gate MOSFET with respect to silicon thickness for different doping concentrations. Reprinted from [146] with permission.

Figure 4.12. As can be seen I_{On}/I_{Off} is almost insensitive to the gate capacitance as far as long-channel devices are concerned.

4.2.5 Rail-to-Rail Supply Voltage Benchmark

From a circuit point of view, downscaling of CMOS technology imposes a new constraint on the maximum voltage that can be used to bias a circuit. It is possible to define as a relevant parameter the difference between the $V_{GS}(on)$ and V_{GScrit} i.e., the gate voltage at the critical point where the current almost reaches I_{Off}. As done for the current, we can deliberately identify V_{GScrit} as $V_{GS}(off)$ and $V_{GS}(on)$ as the gate voltage at flat-band from (4.12). The off-gate voltage $V_{GS}(off)$ is obtained from (4.8) using the values for the surface and center potentials taken at the critical conditions i.e., Ψ_{sc} and Ψ_{0c} (relations (4.25) and (4.27)):

$$V_{GS}(off) = V_{GScrit} = \Psi_{sc} + \Delta\phi_{ms}$$

$$- \sqrt{\eta\left[\cosh\left(\frac{\Psi_{sc}}{U_T}\right) - \cosh\left(\frac{\Psi_{0c}}{U_T}\right) - \frac{N_D}{2n_i U_T}\,(\Psi_{sc} - \Psi_{0c})\right]}. \quad (4.35)$$

Next, combining (4.35) with (4.12), the maximum gate-voltage swing $V_{GS}(on) - V_{GS}(off)$ is obtained:

$$V_{GS}(on) - V_{GS}(off) = U_T\mathrm{asinh}\left(\frac{N_D}{2n_i}\right) - \Psi_{sc}$$

$$+ \sqrt{\eta\left[\cosh\left(\frac{\Psi_{sc}}{U_T}\right) - \cosh\left(\frac{\Psi_{0c}}{U_T}\right) - \frac{N_D}{2n_i U_T}\,(\Psi_{sc} - \Psi_{0c})\right]}.$$

$$(4.36)$$

The maximum gate-voltage excursion $V_{GS}(on) - V_{GS}(off)$ is illustrated in Figure 4.13 using (4.36). For a given doping density, $V_{GS}(on) - V_{GS}(off)$ varies linearly with the

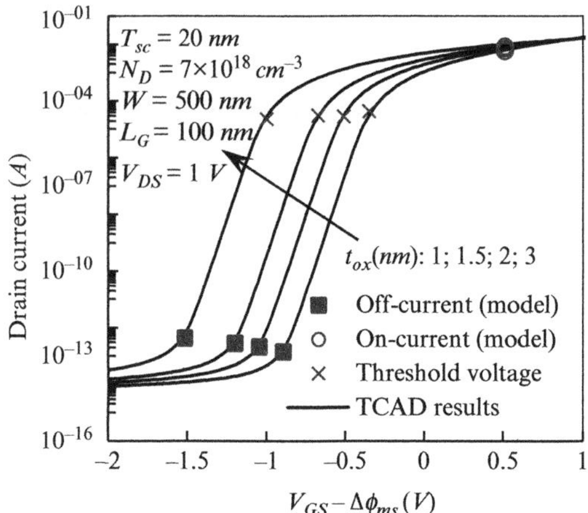

Figure 4.12 $I_{DS} - V_{GS}$ characteristics for $20\ nm$ silicon thickness junctionless double-gate MOSFET for different oxide thickness values. The solid lines correspond to the TCAD simulations and the symbols to the proposed analytical models for the critical *off*-current. (The crosses correspond to the threshold voltage defined in [86].) Note that $\Delta\phi_{ms}$ is the difference between the metal work function and an intrinsic reference semiconductor. Reprinted from [146] with permission.

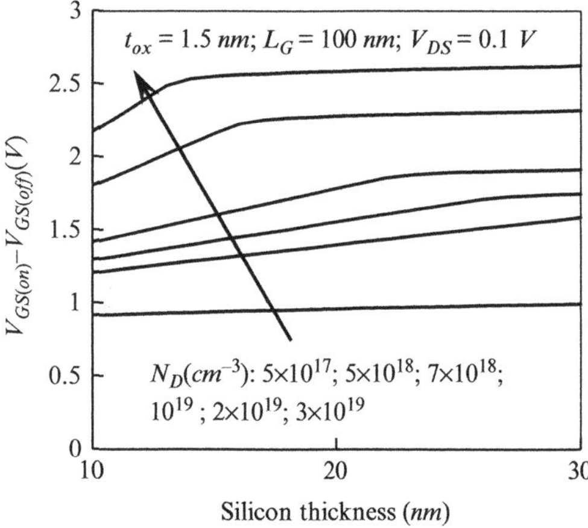

Figure 4.13 Rail-to-rail supply voltage $(V_{GS}(on) - V_{GS}(off))$ according to the silicon thickness for different doping concentrations in junctionless double-gate MOSFET. Reprinted from [146] with permission.

silicon thickness, before reaching a plateau. This limit happens with the occurrence of the hole-inversion layer, which screens the potential and freezes the depletion region. In this case, decreasing the gate potential further will not make the depletion region deeper, and this is independent of the channel thickness.

According to the definition of V_{GScrit}, which is assigned to $V_{GS}(off)$, its value will become constant above a channel thickness. $V_{GS}(on)$ corresponds to the flat-band voltage, a quantity not influenced by silicon thickness. This analysis explains why $V_{GS}(on) - V_{GS}(off)$ still reaches an upper limit with respect to T_{sc} even with doping. However, in practice $V_{GS}(on) - V_{GS}(off)$ can be reduced if higher *off*-currents are acceptable for the application of interest. This voltage excursion can be decreased by about $60\,mV$ (at room temperature) whenever the *off*-current increases by one decade. Using Figures 4.11 and 4.13, once the voltage range $V_{GS}(on) - V_{GS}(off)$, I_{on}/I_{off}- ratio, and I_{on} figures of merit are selected, the optimum long-channel device parameters can be determined.

4.3 Summary

Compared to junction-based MOSFETs, junctionless architectures have fewer technological constraints with regards to the process flow complexity of regular MOSFETs, in particular the source and drain implementation. However, designing these devices still requires special care since technological parameters may significantly impact device performance (e.g., the *off*-current density). Introducing critical points for the gate voltage and the surface and center potentials has been used to obtain the intrinsic *off*-current, the *on/off*-current ratio, and the maximum/minimum operating voltages using analytical expressions. The analytical approach in this chapter can be used to ensure that junctionless double-gate MOSFETs can effectively be set to a predefined current using the right technological parameters such as the silicon thickness, gate capacitance, and doping density.

5 Generalization of the Charge-based Model: Accounting for Inversion Layers

Considering only majority carriers when modeling junctionless FETs is justified as long as the depletion of the channel can be achieved without generating a substantial amount of minority carriers. Instead, when the channel thickness and/or doping density are too high, minority carriers will accumulate at the channel interface and impose a lower limit for the off-current [144]. This topic was thoroughly discussed in Chapter 4. Following the same trend, this chapter generalizes the model for junctionless field-effect transistors by including minority carriers in the core equations in order to simulate the full range of I–V characteristics.

5.1 Electrostatics including Minority Carriers

As shown in Figure 5.1, a model is developed for a double-gate structure. It consists of an n-type doped long-channel junctionless symmetric double-gate MOSFET where symbols have their usual meaning (L_G for the gate length, T_{sc} and N_D for the semiconductor thickness and donor concentration, and t_{ox} and W for the gate-oxide thickness and channel width). The Poisson equation and Boltzmann statistics, including holes, gives [146]

$$\frac{\partial^2 \Psi}{\partial x^2} = \frac{q n_i}{\varepsilon_{si}} \left[\exp\left(\frac{\Psi - V_{ch}}{U_T}\right) - \exp\left(-\frac{\Psi}{U_T}\right) - \frac{N_D}{n_i} \right], \tag{5.1}$$

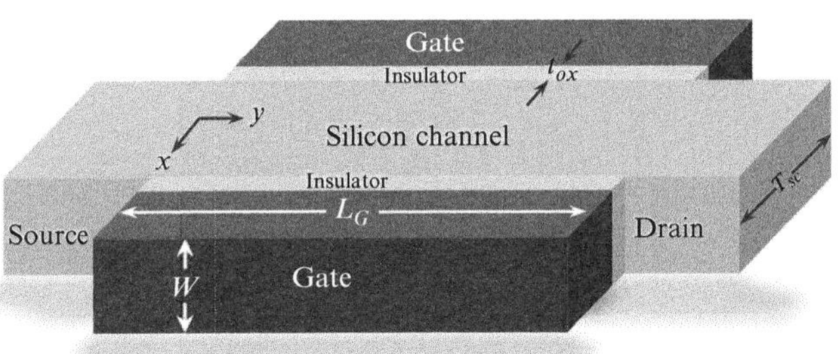

Figure 5.1 Schematic view of the n-type junctionless symmetric double-gate MOSFET.

where V_{ch} is the electron quasi-Fermi potential. In addition, assuming that the channel potential V_{ch} is constant between the gates (along x), it can be introduced in the second derivative of (5.1). After rearrangement, relation (5.1) becomes:

$$\frac{\partial^2}{\partial x^2}\left(\Psi - \frac{V_{ch}}{2}\right) = \frac{2qn_i}{\varepsilon_{si}}\,\exp\left(-\frac{V_{ch}}{2U_T}\right)\left[\sinh\left(\frac{\Psi - V_{ch}/2}{U_T}\right) - \frac{N_D}{2n_i}\exp\left(\frac{V_{ch}}{2U_T}\right)\right]. \quad (5.2)$$

This relation can be simplified when introducing the generalized surface potential $\Psi^* = \Psi - V_{ch}/2$ and generalized intrinsic carrier density $n_i^* = n_i\exp(-V_{ch}/2U_T)$, which involve the Fermi potential:

$$\frac{\partial^2 \Psi^*}{\partial x^2} = \frac{2qn_i^*}{\varepsilon_{si}}\left[\sinh\left(\frac{\Psi^*}{U_T}\right) - \frac{N_D}{2n_i^*}\right]. \quad (5.3)$$

Multiplying both members of (5.3) with $d\Psi/dx$ and integrating with respect to Ψ gives the surface electric field (and thus the charge in the semiconductor Q_{sc}) in terms of the surface and central potentials:

$$E_s^2 = (Q_{sc}/2\varepsilon_{si})^2 = \frac{4qn_i^* U_T}{\varepsilon_{si}}\left[\cosh\left(\frac{\Psi_s^*}{U_T}\right) - \cosh\left(\frac{\Psi_0^*}{U_T}\right) - \frac{N_D}{2n_i^* U_T}(\Psi_s^* - \Psi_0^*)\right], \quad (5.4)$$

where $\Psi_s^* = \Psi_s - V_{ch}/2$, $\Psi_0^* = \Psi_0 - V_{ch}/2$, and Ψ_s and Ψ_0 are the surface and center potentials. The total charge density in the semiconductor must be compensated with the charge on the gate electrodes:

$$2C_{ox}\left(\Psi_s^* + \Delta\phi_{ms} - V_{GS}^*\right) = Q_{sc}, \quad (5.5)$$

where $V_{GS}^* = V_{GS} - V_{ch}/2$ and $C_{ox} = \varepsilon_{ox}/t_{ox}$. Next, replacing Q_{sc} from (5.4) in (5.5) links the gate-to-source voltage V_{GS} to the center and surface potentials:

$$(\Psi_s^* + \Delta\phi_{ms} - V_{GS}^*)^2 = \eta\left[\cosh\left(\frac{\Psi_s^*}{U_T}\right) - \cosh\left(\frac{\Psi_0^*}{U_T}\right) - \frac{N_D}{2n_i^* U_T}(\Psi_s^* - \Psi_0^*)\right], \quad (5.6)$$

with $\eta = 4qn_i^* U_T\varepsilon_{si}/C_{ox}^2$ and $\Delta\phi_{ms} = W_{ms} - U_T\ln(N_D/n_i)$. Relation (5.6) is very general and introduces no assumption. It links the gate, the surface, and the center potentials.

As in Chapter 3, the second-order MacLaurin series (around $x = 0$) is used to approximate the Laplacian in (5.3). Following similar developments, a key relationship is obtained:

$$\Psi_s^* = U_T\mathrm{asinh}(C) + U_T\lambda, \quad (5.7)$$

where $C = \sinh\left(\dfrac{\Psi_0^*}{U_T}\right)$ and λ is given by

$$\lambda = \frac{(\Psi_s^* - \Psi_0^*)}{U_T} = K\left(2C - \frac{N_D}{n_i^*}\right), \quad (5.8)$$

with $K = \dfrac{qn_i^* T_{sc}^2}{8\varepsilon_{si} U_T}$.

This means that as soon as Ψ_0^* is known, Ψ_s^* and λ are known as well. Therefore, the surface electric field E_s and the generalized gate voltage V_{GS}^* obtained from relations (5.4) and (5.6) can be determined without ambiguity. In summary, there is a one-to-one correspondence between the gate voltage and the surface electric field i.e., the charge density in the semiconductor.

The "cosh" terms in (5.4) and (5.6) are now replaced with the parameter C using the properties of hyperbolic sine and cosine functions:

$$C = \sinh\left(\frac{\Psi_0^*}{U_T}\right) \Rightarrow \cosh\left(\frac{\Psi_0^*}{U_T}\right) = \sqrt{1 + C^2}, \tag{5.9}$$

similarly

$$\cosh\left(\frac{\Psi_s^*}{U_T}\right) = \cosh\left(\frac{\Psi_0^* + U_T\lambda}{U_T}\right) = \cosh\left(\frac{\Psi_0^*}{U_T}\right)\cosh(\lambda) + \sinh\left(\frac{\Psi_0^*}{U_T}\right)\sinh(\lambda)$$
$$= \sqrt{1 + C^2}\cosh(\lambda) + C\sinh(\lambda). \tag{5.10}$$

Making use of relations (5.7), (5.8), (5.9), and (5.10) into (5.6), a "center potential" based relationship is obtained:

$$V_{GS}^* - \Delta\phi_{ms} = U_T \mathrm{asinh}(C) + U_T\lambda$$
$$+ \sqrt{\eta}\,\mathrm{sign}(\lambda)\sqrt{\sqrt{1 + C^2}\cosh(\lambda) + C\sinh(\lambda) - \sqrt{1 + C^2} - \frac{N_D\lambda}{2n_i^*}}. \tag{5.11}$$

The total charge density in the semiconductor can be either positive or negative based on whether the junctionless FET operates in depletion or in accumulation (for n-doped channel). This is why we use the "sign" function. *Note that relations (5.7), (5.8), and (5.11) constitute a closed system that can be used to simulate charge densities in all the regimes of operation.*

Introducing (5.7) in (5.5), the quantity $V_{GS}^* - \Delta\phi_{ms}$ is linked to the center potential only:

$$V_{GS}^* - \Delta\phi_{ms} = -\frac{Q_{sc}}{2C_{ox}} + U_T \mathrm{asinh}(C) + U_T\lambda. \tag{5.12}$$

Eliminating $V_{GS}^* - \Delta\phi_{ms}$ between (5.11) and (5.12) gives the total charge density in the semiconductor (Q_{sc}) (this could also be obtained from (5.4) by applying the same analysis).

$$\frac{Q_{sc}^2}{4\eta C_{ox}^2} = \sqrt{1 + C^2}\cosh(\lambda) + C\sinh(\lambda) - \sqrt{1 + C^2} - \frac{N_D\lambda}{2n_i^*}. \tag{5.13}$$

5.1.1   Coexistence of Depletion and Inversion: General Treatment

When the channel is depleted, *the surface potential Ψ_s is lower than the center potential Ψ_0. This condition is fulfilled in the following analysis.* Depletion means that the center potential ψ_0 is lower than the flat-band potential ψ_{FB}. However, depending on the channel V_{ch}, the generalized center potential Ψ_0^* can take large or small values. These two situations are discussed in this section.

When $|\Psi_0^*| \ll U_T$: The *sinh* function can be approximated by the first term of the Taylor series $C = \sinh\left(\dfrac{\Psi_0^*}{U_T}\right) \approx \dfrac{\Psi_0^*}{U_T}$, implying $\sqrt{1 + C^2} \approx 1$.

It follows that the generalized surface potential Ψ_s^* becomes only a function of the device parameters and is independent of the gate voltage:

$$\Psi_s^* = \Psi_0^* + K\left[2\sinh\left(\frac{\Psi_0^*}{U_T}\right) - \frac{N_D}{n_i}\right] \approx -\frac{qN_D T_{sc}^2}{8\varepsilon_{si}} = -\frac{Q_{fix}}{8C_{sc}}, \tag{5.14}$$

where $C_{sc} = \dfrac{\varepsilon_{si}}{T_{sc}}$ and $Q_{fix} = qN_D T_{sc}$ is the total fixed-charge density. From relation (5.14), the surface and center potentials experience the same "shift" with the channel potential. In addition, the condition $|\Psi_0^*| \ll U_T$ reverts to the full depletion approximation as Ψ_0^* is almost $0\ V$ (there is no free carrier in relation (5.14)).

When $|\Psi_0^*| \gg U_T$: The case $|\Psi_0^*| \gg U_T$ happens when the channel is in inversion or in depletion. Then $\sqrt{1 + C^2} \approx |C|$ since $|\Psi_0^*| \gg U_T$, and relation (5.13) can be rewritten as:

$$\frac{Q_{sc}^2}{4\eta C_{ox}^2} = |C|\exp\left[\text{sign}(C)\lambda\right] - |C| - \frac{N_D\lambda}{2n_i^*} \approx |C|\exp\left[\text{sign}(C)\lambda\right] - \frac{N_D\lambda}{2n_i^*}. \tag{5.15}$$

Depending on the sign of C (which is the same as for Ψ_0^* (see relation (5.9))), additional simplifications can be proposed.

When $C > 0$, relation (5.9) gives $C \approx 0.5\exp\left(\dfrac{\Psi_0^*}{U_T}\right)$, and relation (5.15) becomes

$$\frac{Q_{sc}^2}{4\eta C_{ox}^2} = \frac{1}{2}\exp\left(\frac{\Psi_0^*}{U_T}\right)\exp\left(\frac{\Psi_s^* - \Psi_0^*}{U_T}\right) - \frac{N_D\lambda}{2n_i^*}. \tag{5.16}$$

When $C \leq 0$ relation (5.9) is approximated by $\approx -0.5\exp\left(-\dfrac{\Psi_0^*}{U_T}\right)$, (5.15) becomes

$$\frac{Q_{sc}^2}{4\eta C_{ox}^2} = \frac{1}{2}\exp\left(-\frac{\Psi_0^*}{U_T}\right)\exp\left(\frac{\Psi_0^* - \Psi_s^*}{U_T}\right) - \frac{N_D\lambda}{2n_i^*}. \tag{5.17}$$

As will be shown later, the condition $C < 0$ predicts the creation of an inversion layer of holes. As such, it defines the region of operation where depletion and inversion coexist. Therefore, when $|\Psi_0^*| \gg U_T$, depending on the sign of C a formulation of the charge density in the channel involving Ψ_0^* and Ψ_s^* is obtained:

$$\begin{aligned}
\frac{Q_{sc}^2}{4\eta C_{ox}^2} &= \frac{1}{2}\exp\left[\text{sign}(C)\left(\frac{\Psi_s^*}{U_T}\right)\right] - \frac{N_D\lambda}{2n_i^*} \\
&= \frac{1}{2}\exp\left[\text{sign}(C)\left(\frac{\Psi_s^*}{U_T}\right)\right] - \frac{N_D}{2n_i^*}K\left(2C - \frac{N_D}{n_i^*}\right).
\end{aligned} \tag{5.18}$$

Relation (5.18) is therefore valid as soon as $|\Psi_0^*| \gg U_T$. Once Ψ_0^* is known (and thus C), Ψ_s^* is calculated from relations (5.7) and (5.8) and gives the total charge density in the semiconductor Q_{sc}.

5.1.2 Charge–Voltage Relationships

For an n-type doped channel, inversion means creating a holes layer at the semiconductor/insulator interface. This situation occurs when C is negative, which also means $\Psi_0^* \leq 0$ (and thus for Ψ_s^*). Introducing this condition in relation (5.8) defines a higher limit for the generalized surface potential Ψ_s^*:

$$\Psi_s^* = \Psi_0^* + U_T \lambda = \Psi_0^* + U_T K\left(2C - \frac{N_D}{n_i^*}\right) \Rightarrow \Psi_s^* \leq \frac{-Q_{fix}}{8C_{si}}. \tag{5.19}$$

Note that this upper limit is also the difference in surface and center potentials using full-depletion approximation. Since when $C \leq 0$ the generalized surface and center potentials satisfy $\Psi_s^* \ll \Psi_0^* < 0$, the term "2C" can be neglected as compared to exponential in (5.18),

$$\frac{1}{2} \exp\left(\text{sign}(C)\frac{\Psi_s^*}{U_T}\right) \gg \frac{N_D}{2n_i^*} K(2C), \tag{5.20}$$

and relation (5.18) writes

$$\frac{Q_{sc}^2}{4\eta C_{ox}^2} = \exp\left[\text{sign}(C)\left(\frac{\Psi_s^*}{U_T}\right)\right] + \frac{K}{2}\left(\frac{N_D}{n_i^*}\right)^2 \Rightarrow Q_{sc}^2 = Q_{fix}^2 + 8\varepsilon_{si} U_T q n_i^* \exp\left(-\frac{\Psi_s^*}{U_T}\right). \tag{5.21}$$

Next, replacing the surface potential obtained from relation (5.5) into (5.21), a new charge-based relationship when depletion and inversion coexist is obtained:

$$V_{GS}^* - \Delta\phi_{ms} = -\frac{Q_{sc}}{2C_{ox}} - U_T \ln\left(\frac{Q_{sc}^2 - Q_{fix}^2}{8\varepsilon_{si} q n_i^* U_T}\right). \tag{5.22}$$

Finally, since $C \leq 0$, extracting $V_{GS}^* - \Delta\phi_{ms}$ from (5.5) and introducing it in (5.22) gives:

$$\frac{Q_{sc}^2 - Q_{fix}^2}{8\varepsilon_{si} q n_i^* U_T} = \exp\left(-\frac{\Psi_s^*}{U_T}\right). \tag{5.23}$$

Next, using the inequality stated in relation (5.19), the condition on the charge density in the semiconductor is:

$$\frac{Q_{sc}^2 - Q_{fix}^2}{8\varepsilon_{si} q n_i^* U_T} \geq \exp\left(-\frac{Q_{fix}}{8C_{si}}\right). \tag{5.24}$$

A condition can also be found for the gate potential as follows: since $C \leq 0$ and $Q_{sc} > Q_{fix}$, using the inequality (5.24) into (5.22) leads to an equivalent statement in terms of the generalized gate voltage:

$$V_{GS}^* \leq \Delta\phi_{ms} - Q_{fix}\left(\frac{1}{2C_{ox}} + \frac{1}{8C_{si}}\right) = V_T - U_T \ln\left(\frac{N_D}{n_i}\right) = V_{GS0}^*, \tag{5.25}$$

where V_T is introduced as the threshold voltage in [86].

Note that once the charge density in the channel Q_{sc} is known, the generalized surface potential Ψ_s^* is obtained from (5.21). Consequently, the concentration of holes at the channel interface, $p_s = n_i^* \exp\left(-\Psi_s^*/U_T\right)$, is only a function of the total charge density.

On the other hand, when $C > 0$, the generalized potentials satisfy

$$\Psi_s^* \ll 0 < \Psi_0^*. \tag{5.26}$$

and since $\Psi_s^* \ll \Psi_0^*$, relation (5.18) can be approximated by

$$\frac{Q_{sc}^2}{2\eta C_{ox}^2} = -\frac{N_D}{n_i^*}\lambda = -\frac{N_D}{n_i^*}K\left(2C - \frac{N_D}{n_i^*}\right). \tag{5.27}$$

From (5.27), C can be expressed as a function of the total charge density in the semiconductor:

$$C = \frac{N_D}{2n_i^*}\left(1 - \frac{Q_{sc}^2}{Q_{fix}^2}\right). \tag{5.28}$$

Replacing (5.28) in (5.12) provides a charge-based relationship given by (recalling that $C > 0$):

$$V_{GS}^* - \Delta\phi_{ms} = -\frac{Q_{sc}^2}{8qN_D\varepsilon_{si}} - \frac{Q_{sc}}{2C_{ox}} + U_T\text{asinh}\left[\frac{N_D}{2n_i^*}\left(1 - \frac{Q_{sc}^2}{Q_{fix}^2}\right)\right]. \tag{5.29}$$

Note that $\text{asinh}(z) = \ln\left(z + \sqrt{1 + z^2}\right)$ and neglecting the density of holes when $C > 0$, which means $\text{asinh}(z) = \ln(2z)$, relation (5.29) gives the same charge-based relation as in Chapter 3, as expected from the generalization approach.

It should be noted that the surface and center potentials are obtained by combining the continuity of the displacement vector and (5.22). In this case, when the generalized gate potential satisfies $V_{GS}^* > V_{GS0}^*$, relations (5.6) and (5.29) can be used.

Finally, when the total and fixed-charge densities are equal, i.e., $Q_{sc} = Q_{fix}$, the LHS in relation (5.29) reverts to the term $-Q_{fix}\left(1/2C_{ox} + 1/8C_{si}\right)$, which from (5.25) gives back $V_{GS}^* = V_{GS0}^*$.

5.1.3 Inversion Layer-Induced Capacitance in Junctionless Double-Gate MOSFET

According to the condition expressed in (5.25), the electron and hole densities in depletion mode are given by (5.22) and (5.29), respectively. It should be noted that in order to calculate the drain current, only (5.29) should be used. However, (5.22) is needed to model the inversion layer, holes in our case, which can build up at the channel interface below the critical gate potential given by (5.25). This hole layer is modulated by the gate potential and contributes to the intrinsic gate capacitance (C_{GG}), the latter proceeding from the bare derivative of the charge density (5.22) with respect to the generic potentials. Assuming $V_{DS} = 0V$, C_{GG} is given by

$$C_{GG} = WL_G\left(\frac{1}{2C_{ox}} + \frac{2U_TQ_{sc}}{Q_{sc}^2 - Q_{fix}^2}\right)^{-1}. \tag{5.30}$$

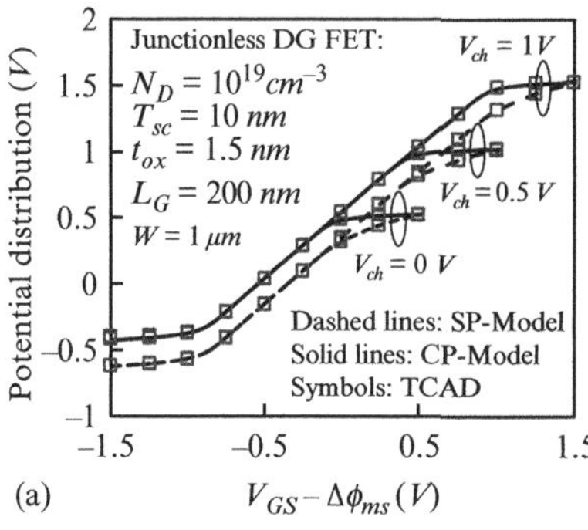

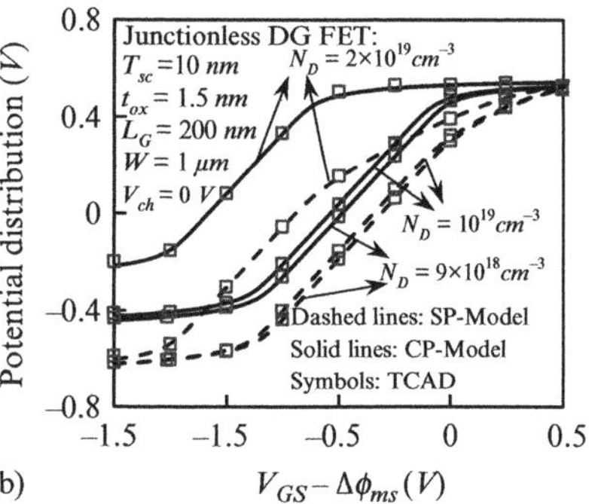

Figure 5.2 Dependence of the body center potential (CP) and surface potential (SP) versus the gate voltage ($V_{GS}^* - \Delta\phi_{ms}$) for different values of (a) channel potential and (b) doping concentration. Reprinted from [147] with permission.

It is worth noting that the modulation of holes can only be effective at low frequency since there is no source for them, unlike in bulk MOSFETs. Indeed, as for the MOS capacitor, only thermal generation and trap-assisted generation/recombination mechanisms are effective here.

5.2   Simulations and Model Assessments

The surface and center potentials are displayed versus the gate voltage in the junctionless double-gate FET in Figure 5.2a and b for different values of the electron quasi-Fermi potentials and doping concentrations. Numerical simulations (TCAD) validate the analytical solution for a wide range of potentials.

Figure 5.3 shows the total charge densities in a junctionless double-gate MOSFET versus the gate potential. Below a certain value of the gate potential (V_{G0}^*), a hole-inversion layer is created and the total charge density increases. Although the electron density is strongly influenced by the channel potential, it does not influence the total hole density as expected from relation (5.1).

Next, the hole density at the surface and at the center of the channel is depicted in Figure 5.4 for different values of silicon thickness. The thinner the silicon, the quicker the depletion and inversion in terms of the gate potential. However, thinner channels will already be fully depleted when inversion occurs, which is not necessarily the case for thick channels. The exponential dependence with V_{GS} corresponds to the weak inversion regime of an inversion-mode FET. When the gate voltage becomes more negative, a linear dependence is obtained as for inversion-mode FET biased in strong inversion. For clarity, break it up into more than one sentence because meaning unclear here.

The gate capacitances attributed to the intrinsic hole layer in a junctionless double-gate FET are plotted in Figure 5.5a for long-channel devices ($L_G = 1\ \mu m$). Figure 5.5b shows the gate capacitances arising from the hole layer in a $100\ nm$ gate

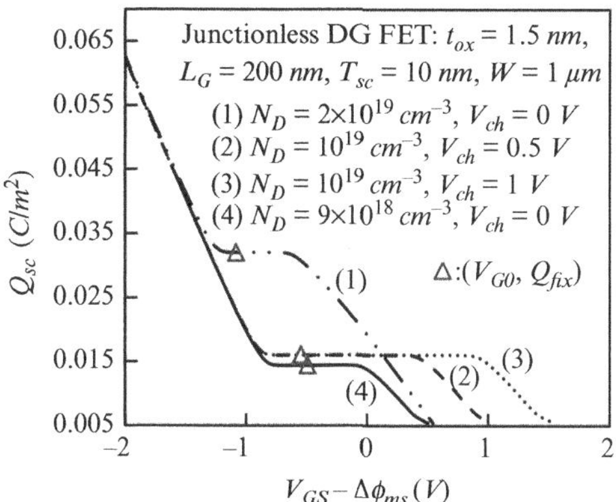

Figure 5.3 Total charge density (Q_{sc}) versus gate potential ($V^*_{GS} - \Delta\phi_{ms}$) for different physical parameters and channel potential in junctionless double-gate MOSFETs. Note that $\Delta\phi_{ms}$ is the difference between the metal work function and an intrinsic reference semiconductor given by $\Delta\phi_{ms} = W_{ms} - U_T \ln(N_D/n_i)$, where W_{ms} is the metal-semiconductor work function difference. Reprinted from [147] with permission.

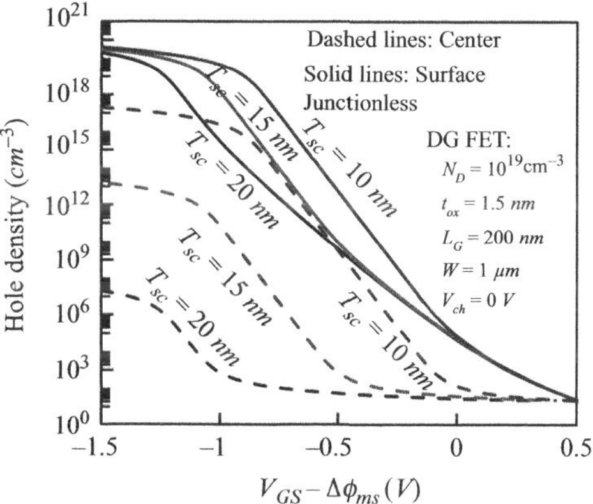

Figure 5.4 Hole concentration at the surface and center of the channel with respect to the gate potential ($V^*_{GS} - \Delta\phi_{ms}$) for different values of silicon thickness (T_{sc}). Note that $\Delta\phi_{ms}$ is the difference between the metal work function and an intrinsic reference semiconductor given by $\Delta\phi_{ms} = W_{ms} - U_T \ln(N_D/n_i)$, where W_{ms} is the metal-semiconductor work function difference. Reprinted from [147] with permission.

length junctionless double-gate FET and junctionless nanowire for different technological parameters. Note that the generalized model is still able to simulate the nanowire with the equivalent double-gate parameters.

5.2.1 Impact of Hole Layer on Drain Current

For large doping densities and silicon thicknesses, the channel cannot be depleted beyond a limit and therefore will not switch off. Figure 5.6 shows the drain current

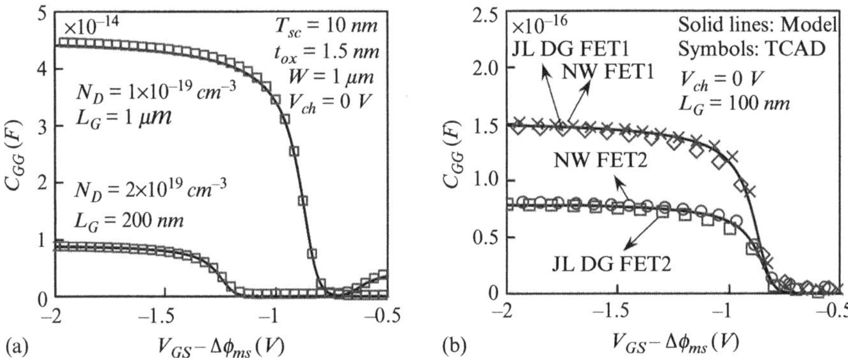

Figure 5.5 Dependence of the intrinsic gate capacitance versus the gate voltage ($V_{GS}^* - \Delta\phi_{ms}$) for different doping concentration values and gate lengths in junctionless double-gate MOSFETs (a). Dependence of the intrinsic gate capacitance versus the gate voltage ($V_{GS}^* - \Delta\phi_{ms}$) for different physical parameter values in junctionless double-gate and nanowire FETs (b). Junctionless double-gate FET1: $N_D = 5 \times 10^{18}\ cm^{-3}$, $T_{sc} = 20\ nm$, $t_{ox} = 1.39\ nm$, $L_G = 100\ nm$, $W = 31.4\ nm$, $V_{ch} = 0\ V$; Nanowire1: $N_D = 10^{19}\ cm^{-3}$, $R = 10\ nm$, $t_{ox} = 1.5\ nm$, $L_G = 100\ nm$, $V_{ch} = 0\ V$; junctionless double-gate FET2: $N_D = 10^{19}\ cm^{-3}$, $T_{sc} = 10\ nm$, $t_{ox} = 1.31\ nm$, $L_G = 100\ nm$, $W = 15.7\ nm$, $V_{ch} = 0\ V$; Nanowire2: $N_D = 10^{19}\ cm^{-3}$, $R = 10\ nm$, and $t_{ox} = 1.5\ nm$, $L_G = 100\ nm$, $V_{ch} = 0\ V$. Reprinted from [147] with permission.

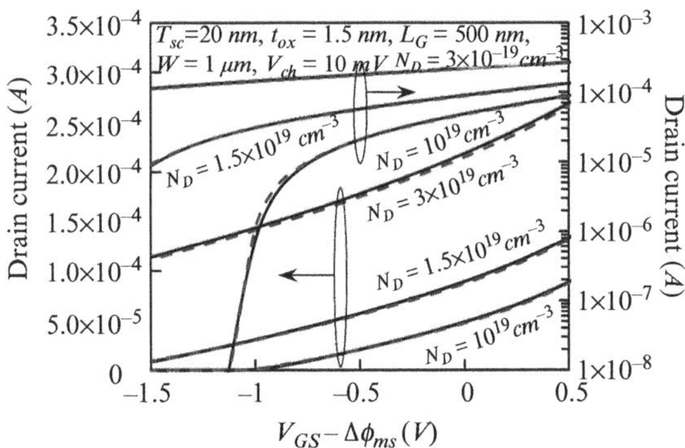

Figure 5.6 Drain current versus gate potential ($V_{GS}^* - \Delta\phi_{ms}$) in linear and logarithmic scales for different doping concentration values in an *n*-type junctionless double-gate MOSFET. The dashed and the solid lines correspond to the TCAD simulations and the model, respectively. Note that the channel is not switched off for $N_D = 3 \times 10^{19}\ cm^{-3}$, as predicted from the model when including the inversion layer. Reprinted from [147] with permission.

versus the gate potential for two different values of doping concentrations in junctionless double-gate FETs. The generalized charge-based approach (relation (5.29)) correctly predicts the behavior of the current even at large negative gate voltages. This is clearly seen in linear and logarithmic representations.

5.3 Summary

An analytical model for junctionless FETs including minority carriers was developed in this chapter. Relying on two charge-based equations arising from a common root, the potential distribution across the gates can be obtained, together with the minority and majority carrier charge densities. The full current–voltage and capacitance–voltage characteristics were simulated, including the intrinsic off-state current. The equivalence between the double-gate and the nanowire geometries was also preserved.

6 Predicted Performances of Junctionless FETs

Due to the principle of operation relying on majority carriers, it was predicted [1] that junctionless FETs would perform better than inversion-mode FETs. Adopting a simple approach based on the full-depletion approximation, this chapter aims to investigate the expected performance of junctionless FETs in terms of speed and power consumption, and in particular how the intrinsic delay and *on/off* voltage compare with inversion-mode FETs.

6.1 Device Scaling Principle

Initially, it was argued [1, 74] that when the junctionless FET is set as a normally-on device i.e., when *on*-state is equivalent to flat-band, the drain current I_{on} can be approximated by a resistor-like relationship, independent of the gate-oxide capacitance (C_{ox}):

$$I_{on} \approx q\mu n \frac{T_{sc}W}{L} V_{DS} \approx q\mu N_D \frac{T_{sc}W}{L} V_{DS}, \tag{6.1}$$

where T_{sc} is the semiconductor thickness, W and L are the device width and the length, V_{DS} is the drain-to-source voltage, and n is the free-electron concentration assumed equal to the doping concentration N_D at flat-band (see Figure 6.1).

Relation (6.1) is valid at low V_{DS} only. For large drain voltages, charges are nonuniformly distributed along the channel and the evaluation of the current requires additional developments (such as in Chapter 3). Nevertheless, expression (6.1) is still useful since it gives an upper limit for the drain current (assuming the devices operate in depletion, not in accumulation).

The intrinsic delay (CV/I) has been shown to be improved in junctionless FETs [1]. In a full-depletion approximation scheme, the *off*-state current is reached for a given gate voltage V_{off} where the width of the depletion layer equals the silicon thickness i.e., T_{sc}. Conversely, *on*-state is attained when the gate voltage V_{on} preserves neutrality of the channel i.e., the depletion-charge density Q_{sc} is negligible and the surface potential reverts to the flat-band potential Ψ_{FB}. These two conditions result in the following relationship:

$$(V_{GS} - \Delta\phi_{ms} - \Psi_s)|_{on} - (V_{GS} - \Delta\phi_{ms} - \Psi_s)|_{off}$$

$$= (V_{on} - V_{off}) + (\Psi_s|_{off} - \Psi_s|_{on}) \approx qN_D \frac{T_{sc}}{2C_{ox}}, \tag{6.2}$$

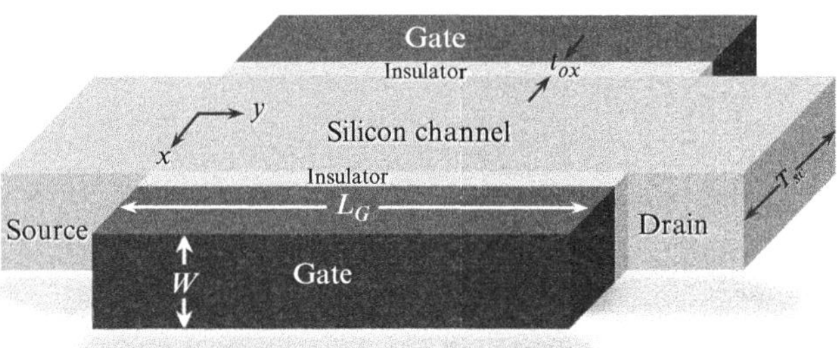

Figure 6.1 Schematic view of the *n*-type junctionless symmetric double-gate MOSFET.

where the difference in the surface potential between flat-band and depletion modes is obtained by applying the full-depletion approximation scheme:

$$\Psi_s|_{off} - \Psi_s|_{on} \approx -\frac{qN_D T_{sc}^2}{8\varepsilon_{si}}, \tag{6.3}$$

with $\Psi_s|_{off}$ and $\Delta\phi_{ms}$ being, respectively, the surface potential of the fully depleted silicon and the work function difference between the gate and the silicon. Therefore, from (6.2) the gate oxide, doping concentration, and silicon thickness are strongly interrelated:

$$t_{ox} \approx (V_{on} - V_{off})\frac{2\varepsilon_{ox}}{qN_D T_{sc}} - \frac{\varepsilon_{ox} T_{sc}}{4\varepsilon_{si}}. \tag{6.4}$$

Relation (6.4) ensures that the device can be switched *on* and *off* within a gate-voltage range $V_{on} - V_{off}$. Apparently, the current at flat-band can be increased by increasing the channel doping (6.1). However, this could require relatively high voltages that are not compatible with the technology. Replacing the term $N_D T_{sc}$ in (6.1), the current derived from (6.4) is:

$$I_{DS} \approx 2\mu\frac{W}{L_G}\left(\frac{1}{C_{ox}} + \frac{1}{4C_{si}}\right)^{-1} V_{DS}(V_{on} - V_{off}), \tag{6.5}$$

where C_{si} is the silicon capacitance i.e., ε_{si}/T_{sc}.

Figure 6.2 plots the oxide thickness calculated from (6.4) in order to achieve full depletion with $V_{on} - V_{off}$ set to $1\,V$ (defining $V_{on}=V_{FB}$). The higher N_D, the thinner t_{ox} needs to be in order to maintain a given voltage latitude. For instance, increasing N_D from 10^{19} to $2 \times 10^{19}\,cm^{-3}$ with $T_{sc} = 8\,nm$, the gate-oxide thickness t_{ox} will shift from 5 to $2\,nm$. Considering lower voltages makes full depletion even more difficult to achieve. When $V_{on} - V_{off} = 0.5\,V$, t_{ox} must be lower than $0.5\,nm$, which is not feasible unless high-*k* dielectrics are used.

These scaling constraints also affect the silicon channel. Starting from the fixed values for $V_{on} - V_{off}$ and t_{ox}, relation (6.4) implies that any increase in the doping concentration must be compensated for by a decrease in the silicon thickness. For example, if N_D changes from 10^{19} to $2 \times 10^{19}\,cm^{-3}$ still keeping $V_{on} - V_{off} = 1\,V$ and $t_{ox} = 5\,nm$, T_{sc} must decrease from about 8 to $4\,nm$ (see Figure 6.2). Furthermore,

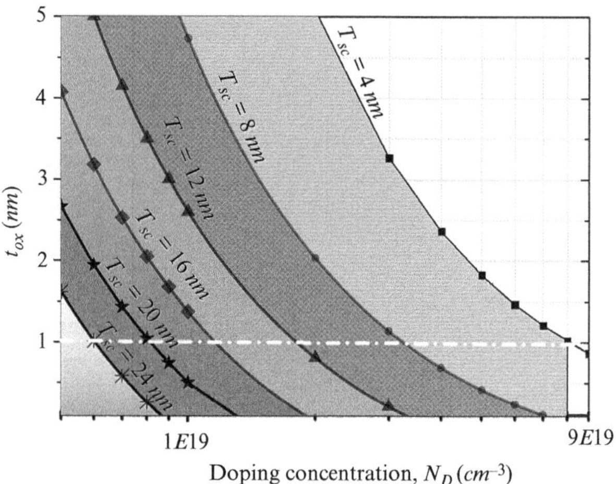

Figure 6.2 Required oxide thickness to achieve full depletion at V_{off} with the constraint $V_{on} - V_{off} = 1\,V$. Reprinted from [52] with permission.

the *on*-current might also not increase since the higher mobile charge concentration i.e., N_D is counterbalanced by the downscaling of the silicon-channel area. In addition, carrier mobility is significantly degraded in ultrathin heavily doped silicon due to the occurrence of ionized impurity scattering, surface roughness scattering, and quantization effects [54]. On the other hand, decreasing the silicon thickness exacerbates fluctuations in the threshold voltage [148].

6.2   Considerations on Intrinsic-Delay Scaling

The intrinsic delay ($\tau = CV/I$) is the most widely used figure of merit for the evaluation of new technologies and related benchmarking [1, 149]. This metric is an indicator of the overall device-switching speed since it accounts for the device capacitance, the voltage swing, and the drive current supplied from the device. The capacitance of junctionless FET can be estimated from (6.4), which can be rewritten as:

$$V_{on} - V_{off} = \frac{1}{2} Q_m \left(\frac{1}{C_{ox}} + \frac{1}{4C_{si}} \right), \tag{6.6}$$

where Q_m is the mobile charge density at flat-band i.e., $qN_D T_{sc}$. This equation indicates that the equivalent gate capacitance of junctionless biased in the *on*-state is $C_{eq} = \left(C_{ox}^{-1} + (4C_{si})^{-1} \right)^{-1}$ and not C_{ox} as for inversion-mode MOSFETs. Figure 6.3 shows the simulated gate capacitance versus the gate voltage for a drain voltage of $1\,V$ (the source being the reference). The average capacitance $\left(C_{ox}^{-1} + (4C_{si})^{-1} \right)^{-1}$ defined above indicates (Figure 6.3) that in the worst case (for low doping densities), it overestimates the flat-band capacitance by 50 percent, which is acceptable to estimate the intrinsic delay. Conversely, this average capacitance is quite accurate for

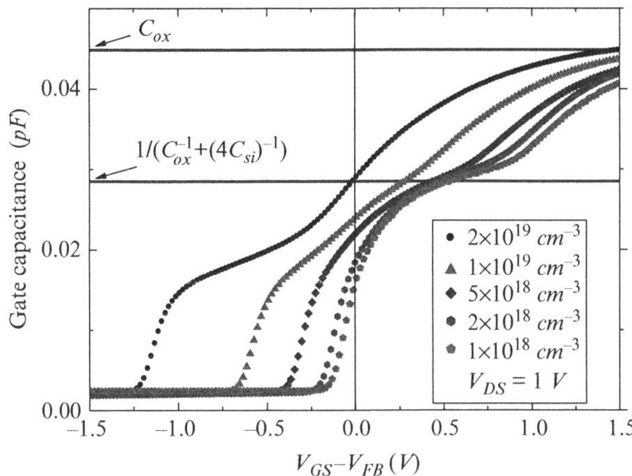

Figure 6.3 Simulated (TCAD) gate capacitance versus gate voltage for drain voltage set to $1\ V$ when N_D varies from 10^{18} to $2 \times 10^{19}\ cm^{-3}$. Reprinted from [52] with permission.

high doping densities. The depletion at the drain side induced by the drain voltage and its influence on the capacitance (neglected in (6.6)) is obviously more important for low doping than for high doping channels. The intrinsic delay (τ) of the junctionless FET is thus expressed as

$$\tau \approx \frac{C_{eq}V_{DD}}{I_{DS}} \approx \frac{\left(C_{ox}^{-1}+(4C_{si})^{-1}\right)^{-1}2WL_G(V_{on}-V_{off})}{2\mu\dfrac{W}{L_G}\left(C_{ox}^{-1}+(4C_{si})^{-1}\right)^{-1}V_{DS}(V_{on}-V_{off})} = \frac{L_G^2}{\mu V_{DD}}, \qquad (6.7)$$

where V_{DS} and V_{DD} are assumed equal to $V_{on} - V_{off}$. In comparison, MOSFETs obey the well-known intrinsic delay [1, 149] given by

$$\tau \approx \frac{CV_{DD}}{I_{DS}} \approx \frac{C_{ox}2WL_GV_{DD}}{\dfrac{1}{2}\mu C_{ox}\dfrac{2W}{L_G}(V_{DD}-V_{th})^2} \approx \frac{2L_G^2}{\mu V_{DD}}. \qquad (6.8)$$

Even though some simplification regarding the charge distribution along the channel relations has been introduced, (6.7) and (6.8) reveal that the speed of junctionless FET-based logic circuits is expected to follow almost the same scaling rule as for MOSFETs, improving this figure of merit by a factor of two at best since the *on*-current was overestimated by neglecting "pinch-off" at the drain. Note that the implicit dependence of the supply voltage with doping, oxide, and silicon thickness (6.4) implies that the intrinsic delay cannot be decreased by increasing the doping concentration.

The simulated drain currents for the same configurations (adopting a constant mobility of $\mu = 0.1\ m^2/Vs$) with $W = L_G = 1\ \mu m$, $t_{ox} = 1.5\ nm$, and $T_{sc} = 10\ nm$ are plotted in Figure 6.4. The current at flat-band (I_{on}) increases with the doping concentration, but at the cost of a higher $V_{on} - V_{off}$ when t_{ox} is fixed (e.g., $t_{ox} = 1.5\ nm$).

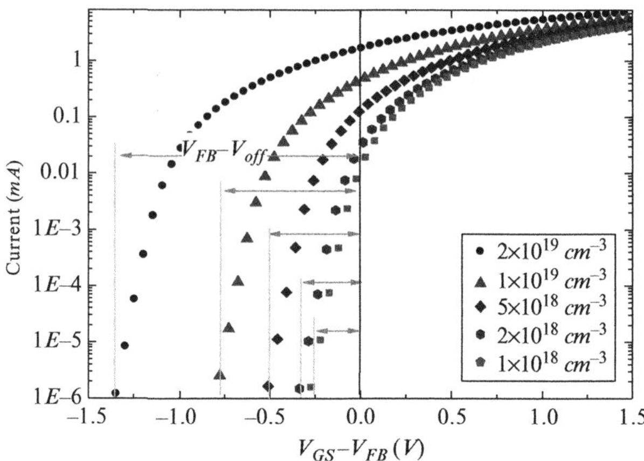

Figure 6.4 Simulated drain current versus gate voltage with a drain voltage of 1 V when N_D varies from 10^{18} to 2×10^{19} cm^{-3}. Reprinted from [52] with permission.

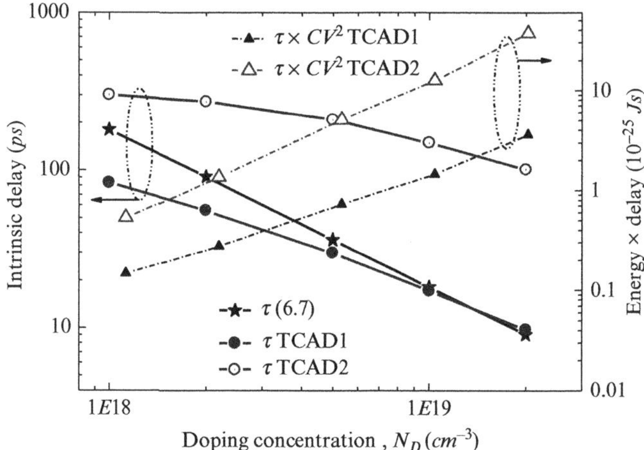

Figure 6.5 Calculated and simulated intrinsic delay and switching energy of the junctionless FETs versus silicon doping concentration. TCAD1 with constant mobility ($\mu = 0.1$ m^2/Vs) and TCAD2 with bias-dependent mobility. Reprinted from [52] with permission.

Relation (6.7) and TCAD simulations are shown in Figure 6.5 assuming constant mobility ($\mu = 0.1$ m^2/Vs). Also shown is the case of a field mobility dependence taking into account the ionized impurity scattering proposed by Masetti et al. [150].

Instead of using the CV product, a more accurate estimation of the intrinsic delay could be obtained from $(Q_{on} - Q_{off})/I_{on}$, where Q_{on} and Q_{off} represent, respectively, the total charge densities in the silicon in *on* and *off* states (I_{on} and $V_{on} - V_{off}$ are determined from Figure 6.4). These quantities are directly obtained by integrating the $C_{GG}(V_{GS})$ characteristic from V_{off} to V_{on} (Figure 6.3).

Figure 6.5 compares the intrinsic delay calculated from (6.7) with the one derived from TCAD simulations, still assuming constant mobility (TCAD1 in Figure 6.5) when the doping of the silicon is varied. The agreement between calculated and TCAD results confirm that the approximation of the intrinsic delay given by relation (6.7) is fairly accurate, especially for high doping concentrations. We notice that the intrinsic delay is divided by about 8 when N_D is increased from 10^{18} to $2 \times 10^{19} cm^{-3}$. However, the cost for this gain in speed is significant degradation (by about 25) of the switching energy. This tradeoff between speed and energy is even worse if the mobility reduction due to the Coulomb scattering in heavily doped silicon is considered. This is shown in Figure 6.5 (curves TCAD2) where we used the mobility model of Masetti in the TCAD simulations [150].

6.3   Summary

A study of the scaling performance of junctionless FETs considering the importance of the *on*/*off* voltage constraint here established that junctionless FETs require almost the same gate-oxide thickness scaling trends and intrinsic time delay as MOSFETs. These trends can be estimated through explicit analytical relationships, and can illustrate the tradeoffs between speed and switching power performances with regard to the junctionless FET technological parameters.

7   Short-Channel Effects in Symmetric Junctionless Double-Gate FETs

With the downscaling of the supply voltage of advanced CMOS technology, modeling the subthreshold behavior of junctionless FETs is essential to assess the switching performance and low-power capabilities of these devices. This chapter investigates how short-channel junctionless symmetric double-gate MOSFETs operate in the subthreshold regime and how they can be modeled.

Junctionless double-gate MOSFETs [1, 28] can compete with inversion-mode (IM) FETs with regard to overall performance. According to [143], they are also expected to be more immune to the DIBL, despite the fact that subthreshold behavior could be significantly impacted by random dopant fluctuations (RDF) for doping levels greater than 10^{19} cm^{-3} [56].

Among the most critical issues regarding SCEs, the DIBL is of major concern as it affects both static and dynamic operation at low gate voltages. Numerical simulations provide some insight into SCE (see [37–41]).

Different strategies to derive analytical solutions are given in [42–48]. For instance, a variable-separation technique resulting in a series expansion of the 2D potential distribution through the channel was developed in [103], whereas [151] solves the potential distribution by splitting the 2D Poisson's equation into 1D Poisson and 2D Laplace's equations. Recently, Chiang [98] proposed an analytical model to calculate the threshold voltage in short-channel junctionless double-gate MOSFETs assuming a parabolic potential across the gates, but no analytical expression for subthreshold current and subthreshold swing and SCEs were given. A rigorous approach using the conformal mapping technique is also possible despite the complex relationships involved [49, 152].

7.1   Electrostatics in Short-Channel Junctionless DG MOSFETS in the Subthreshold

A schematic of the n-type doped junctionless double-gate MOSFET is illustrated in Figure 7.1, with symbols having their usual meaning. In order to minimize the extrinsic access resistance, the doping concentration for the source/drain regions is intentionally increased to $N^+ = 10^{20}$ cm^{-3}, higher than for the channel N_D that must still switch off down to some *off*-current density (the doping profile is then $N^+/N/N^+$).

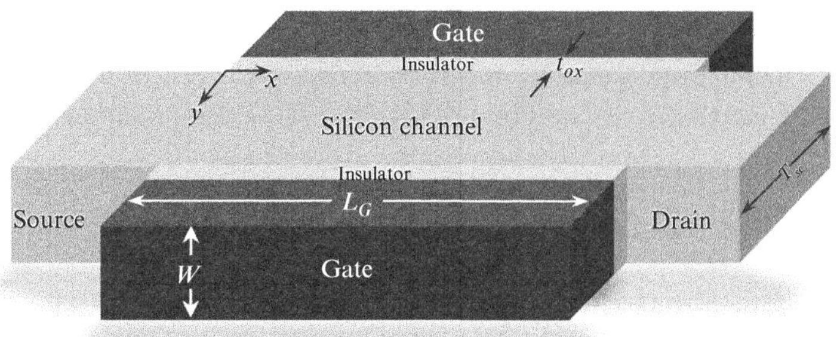

Figure 7.1 Schematic view of the *n*-type junctionless double-gate MOSFET investigated in this chapter.

7.1.1 Approximate Solution of the Potential Distribution

The electrostatic potential satisfies the 2D Poisson–Boltzmann relationship:

$$\frac{\partial^2 \Psi(x,y)}{\partial x^2} + \frac{\partial^2 \Psi(x,y)}{\partial y^2} = \frac{qn_i}{\varepsilon_{si}}\left[\exp\left(\frac{\Psi(x,y) - V_{ch}}{U_T}\right) - \frac{N_D}{n_i}\right], \tag{7.1}$$

where x and y are the orientations along and normal to the channel ($0 < x < L_G$ and $0 < y < T_{sc}$), V_{ch} and U_T are the shift in the electron quasi-Fermi potential and the thermal voltage, and n_i and ε_{si} are the intrinsic carrier density and permittivity for the semiconductor. While a closed form of (7.1) is not available, in a subthreshold operation the contribution of the mobile charges density is negligible in (7.1), leading to

$$\frac{\partial^2 \Psi(x,y)}{\partial x^2} + \frac{\partial^2 \Psi(x,y)}{\partial y^2} = -\frac{qN_D}{\varepsilon_{si}}. \tag{7.2}$$

An approximate solution of the 2D potential distribution, $\Psi(x,y)$, is obtained while assuming the potential parabolic along the y-direction [153]:

$$\Psi(x,y) = \Psi_s(x) + \varphi_1(x)y + \varphi_2(x)y^2, \tag{7.3}$$

where $\Psi_s(x)$ is the surface potential at $y = 0$ and $y = T_{sc}$ is the only symmetric operation considered.

Note that in conventional IM MOSFETs, the minimum potential barrier in the subthreshold is located somewhere between source and drain but at the semiconductor–insulator interface. Conversely, in junctionless devices this minima stands at the center of the semiconductor, between the gates.

Introducing symmetry considerations leads to:

$$\Psi(x,y) = \Psi_s(x) + \varphi_1(x)y\left(1 - \frac{y}{T_{sc}}\right). \tag{7.4}$$

In addition, the boundary conditions arising from the continuity of the displacement vector at the silicon–insulator interfaces must satisfy

$$\frac{\partial}{\partial y}\Psi(x,y)\bigg|_{y=0} = \varphi_1(x) = \frac{\varepsilon_{ox}}{t_{ox}\varepsilon_{si}}\left[\Psi_s(x) - V_{GS} + \Delta\phi_{ms}\right], \tag{7.5}$$

where V_{GS} is the gate-to-source voltage and $\Delta\phi_{ms}$ is the difference between the metal work function and a reference in the semiconductor so that $\Delta\phi_{ms} = W_{ms} - U_T \ln(N_D/n_i)$. Combining (7.4) and (7.5), the potential distribution based on the parabolic approximation can be expressed as:

$$\Psi(x, y) = \Psi_s(x)\left[1 + y\left(1 - \frac{y}{T_{sc}}\right)\frac{\varepsilon_{ox}}{t_{ox}\varepsilon_{si}}\right] + (\Delta\phi_{ms} - V_{GS})\, y\left(1 - \frac{y}{T_{sc}}\right)\frac{\varepsilon_{ox}}{t_{ox}\varepsilon_{si}}. \qquad (7.6)$$

Substitution of (7.6) in (7.2) leads to a differential equation in terms of the surface potential only:

$$\frac{\partial^2}{\partial x^2}\Psi_s(x) - \frac{2\varepsilon_{ox}}{T_{sc}t_{ox}\varepsilon_{si}}\Psi_s(x) = -\frac{q}{\varepsilon_{si}}N_D + \frac{2\varepsilon_{ox}}{T_{sc}t_{ox}\varepsilon_{si}}(\Delta\phi_{ms} - V_{GS}), \qquad (7.7)$$

whose solution is

$$\Psi_s(x) = \alpha \exp(\delta x) + \beta \exp(-\delta x) + \gamma, \qquad (7.8)$$

where the coefficients α and β will be derived later and δ and γ are given by

$$\delta = \sqrt{\frac{2\varepsilon_{ox}}{T_{sc}t_{ox}\varepsilon_{si}}}, \qquad (7.9)$$

$$\gamma = V_{GS} - \Delta\phi_{ms} + \frac{qN_D}{\delta^2 \varepsilon_{si}}. \qquad (7.10)$$

7.1.2   Assessment of the Center Potential with Regard to Numerical Simulations

The principle of operation of junctionless devices is different from regular junction-based double-gate devices as the current flows through the volume instead of flowing at the Si–SiO$_2$ interfaces. From (7.6) it can be seen that below threshold, the potential at the surface ($y = 0$ and $y = T_{sc}$) is lower than at the center ($y = T_{sc}/2$) where the electron concentration peaks. Therefore, below threshold, the center of the silicon channel is the leakiest pathway between the source and the drain, which is why DIBL must be evaluated by analyzing the center potential $\Psi_{BCP}(x) = \Psi(x, T_{sc}/2)$.

On the other hand, from (7.6) the body center potential is linked to the surface potential:

$$\Psi_{BCP}(x) = \Psi\left(x, \frac{T_{sc}}{2}\right) = a\Psi_s(x) + b, \qquad (7.11)$$

where "a" and "b" coefficients are given by

$$a = 1 + \frac{1}{8}\delta^2 T_{sc}^2, \qquad (7.12)$$

$$b = \frac{1}{8}\delta^2 T_{sc}^2 (\Delta\phi_{ms} - V_{GS}). \qquad (7.13)$$

Relation (7.11) links the surface to the center potential at any x-coordinate along the channel (it will be shown that while $\Psi_s(x)$ is less accurate for gate lengths below 30 nm, the center potential along y is still well predicted).

The parabolic approximation for the potential distribution along the y-direction as given in (7.11) cannot satisfy accurately limit conditions for surface and center potentials simultaneously. Depending on which condition is imposed, the center potential, which is the parameter of interest for the *off*-state current, will take different values.

Assigning Limit Conditions to the Surface Potential

The boundary conditions for the surface potential at the Si–SiO$_2$ interfaces evaluated at source and drain limits still inside the channel are:

$$\Psi_s(0, 0) = \Psi_s(0, T_{sc}) = V_{bi}, \tag{7.14}$$

$$\Psi_s(L_G, 0) = \Psi_s(L_G, T_{sc}) = V_{bi} + V_{DS}, \tag{7.15}$$

which gives

$$\alpha = -\beta - \gamma, \tag{7.16}$$

$$\beta = \frac{-\gamma\left[\exp(\delta L_G) - 1\right] - V_{DS}}{2\sinh(\delta L_G)}. \tag{7.17}$$

For gate lengths above $30\,nm$, this condition can also accurately obtain the body center potential (see Figure 7.2).

Here, V_{bi} is the built-in potential of the source/channel and drain/channel (N^+/N) junctions. Ideally, in a junctionless double-gate MOSFET, when the doping of the channel is similar to that of the drain and source. In that case, the built-in potential is simply zero. In this analysis, the doping concentration in the channel and in the S-D regions is $10^{19}\,cm^{-3}$. In these developments, it is assumed that the transition from the fully depleted body to the neutral contact is abrupt.

Assigning Limit Conditions to the Center Potential

It is also possible to impose boundary conditions on the center potential instead of the surface potential, which is supported by numerical simulations (see Figure 7.3):

$$\Psi_{BCP}(0) = V_{bi}, \tag{7.18}$$

$$\Psi_{BCP}(L_G) = V_{bi} + V_{DS}. \tag{7.19}$$

From (7.8) and (7.11), the body center potential $\Psi_{BCP}(x)$ is expressed as:

$$\Psi_{BCP}(x) = a\left[\alpha' \exp(\delta x) + \beta' \exp(-\delta x) + \gamma\right] + b. \tag{7.20}$$

Hence, α' and β' represent the new coefficients that determine the center potential from the boundary conditions in (7.18) and (7.19) [27, 98, 153]:

$$\alpha' = -\frac{b}{a} - \beta' - \gamma, \tag{7.21}$$

$$\beta' = \frac{(-b - a\gamma)\left[\exp(\delta L_G) - 1\right] - V_{DS}}{2a\sinh(\delta L_G)}. \tag{7.22}$$

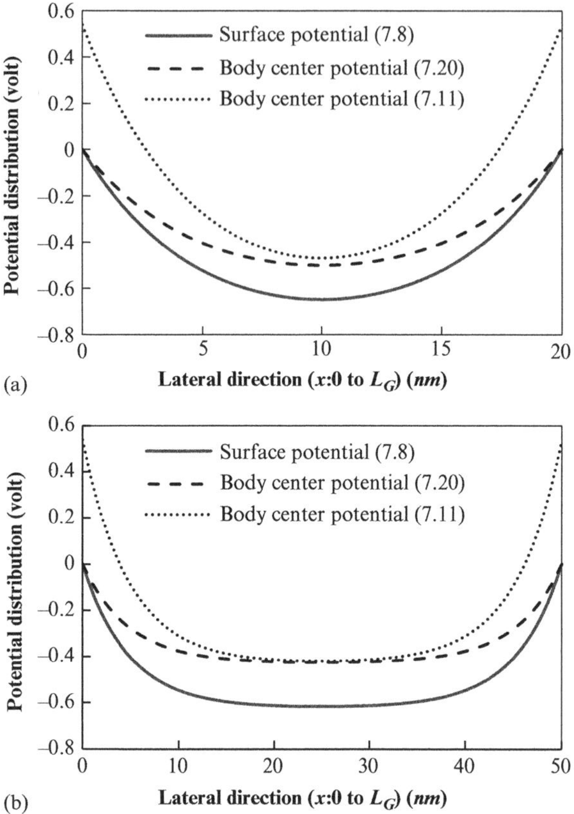

Figure 7.2 Body center potential and surface potential along the lateral direction for (a) 20 and (b) 50 *nm* gate lengths. All voltages have been set to 0 *V* and $\Delta\phi_{ms} = 0.54$ *V*. The doping concentration, gate-oxide thickness, and silicon thickness are 1×10^{19} *cm*$^{-3}$, 1.5 *nm*, and 10 *nm*, respectively. Surface and body center potentials are from (7.8) and (7.20). Note that for gate lengths above 30 *nm*, the body center potential obtained from (7.20) and (7.11) become equivalents. Reprinted from [111] with permission.

As shown in Figure 7.2, some mismatch is seen at the source and drain, particularly for the surface potentials, which is an intrinsic limitation of this approach. Nevertheless, in junctionless FETs, this drawback can be alleviated by assigning limit conditions to the center potential. Indeed, for devices less than 30 *nm* in length, imposing limit conditions on the center potential is more accurate while both kinds of conditions (on Ψ_s or Ψ_{BCP}) give almost the same value for the center potential in long-channel devices.

2D Potential Profiles

The parabolic approximation of the potential distribution is compared to numerical TCAD simulations in Figure 7.3 for 10 and 50 *nm* channel length devices doped at 1×10^{19} *cm*$^{-3}$. The silicon thickness is set to 10 *nm*, justifying that quantum effects are ignored [154, 155].

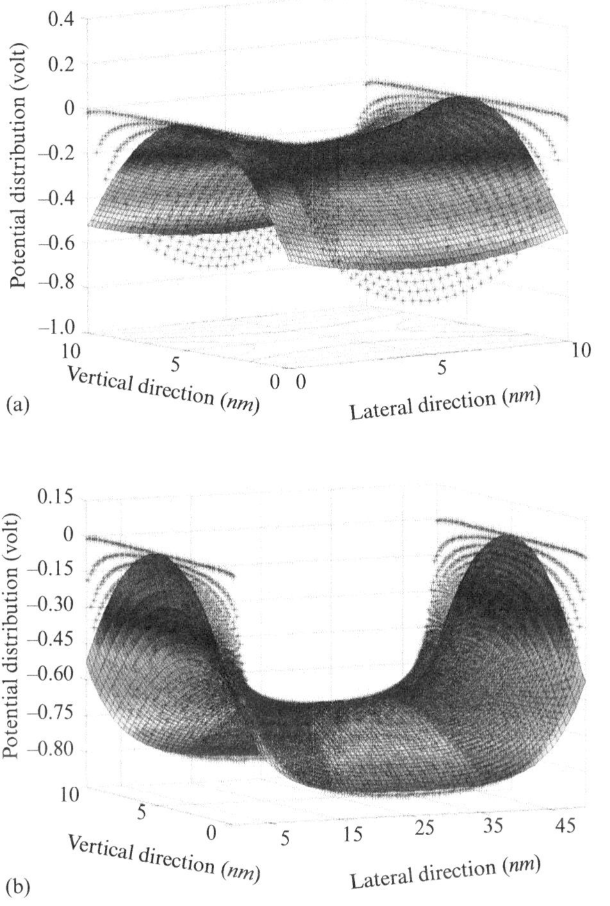

Figure 7.3 Variation of the electrostatic potential along lateral and vertical directions in a 10 nm silicon thickness junctionless double-gate MOSFET with donor concentration of 1×10^{19} cm^{-3} in the channel. The drain bias is set to 0 V and $V_{GS} = 0$ V and $\Delta\phi_{ms} = 0.54$ V. Two gate lengths are represented: (a) 10 and (b) 50 nm. The symbols (*) correspond to TCAD simulations and the surface meshing corresponds to the model. Note that $\Delta\phi_{ms}$ is the difference between the metal work function and an intrinsic reference semiconductor given by $\Delta\phi_{ms} = W_{ms} - U_T \ln(N_D/n_i)$, where W_{ms} is the metal semiconductor work function difference. Reprinted from [111] with permission.

According to these 2D representations, the channel potential distribution is well captured by the model along the channel and also across the gates. Even for the 10 nm channel length, the potential distribution at the center is well predicted, which is the most critical region in regard to the current.

As already stated, accuracy is degraded for the surface potential, but such a situation represents a kind of worst case since the device would be as long as thick. Figure 7.4 shows the distribution of the body center potential obtained from (7.20) along the channel for gate lengths of 10 and 22 nm with a 10 nm silicon thickness.

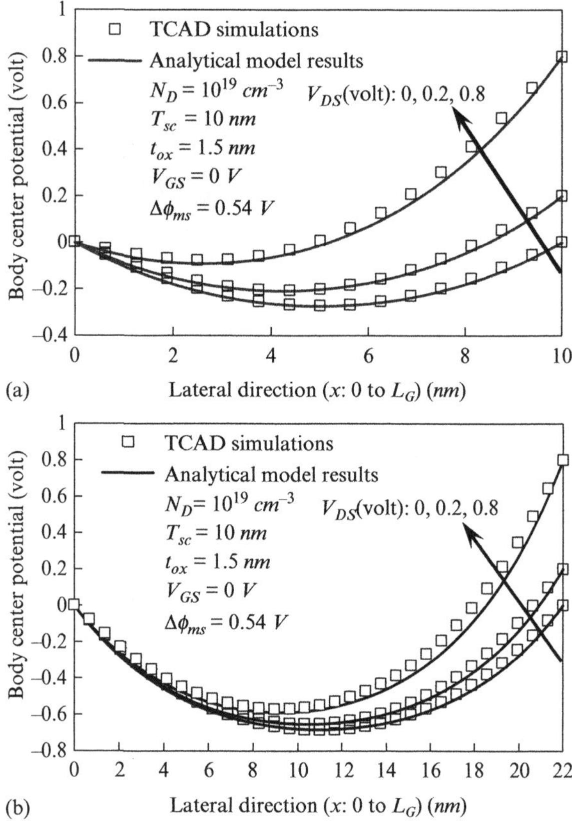

Figure 7.4 Distribution of the body center potential along the lateral direction in $10\ nm$ silicon thickness junctionless double-gate MOSFET devices for different drain voltages, with $V_{GS} = 0\ V$. Two gate lengths are represented: (a) 10 and (b) $22\ nm$. The doping concentration is $1 \times 10^{19}\ cm^{-3}$. Reprinted from [111] with permission.

The agreement of the model with TCAD simulations is reasonably accurate for high doping concentration ($1 \times 10^{19}\ cm^{-3}$) and for drain biases between 0 and $0.8\ V$. The slight discrepancy near the drain at higher V_{DS} will not affect the DIBL parameter given that the DIBL is extracted from the lowest potential region, where the model is accurate. Prediction of the body center potential for the shortest $10\ nm$ channel-length device is accurate from source to drain (see Figure 7.4).

7.2 Subthreshold Current, Subthreshold Swing, and DIBL

7.2.1 The Minimum Potential

The minimum value of the channel potential that controls the current in the subthreshold regime is used to derive the subthreshold current, subthreshold swing, and DIBL. The minimum body center potential ($\Psi_{BCP,min}$) at x_{min} must satisfy the

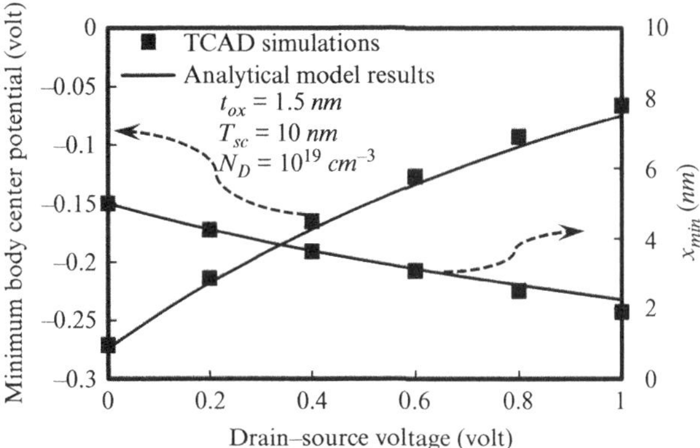

Figure 7.5 Minimum body center potential and its location according to drain–source voltage, imposing $V_{GS} = 0\ V$, $\Delta\phi_{ms} = 0.54\ V$, and $N_D = 10^{19}\ cm^{-3}$ in $10\ nm$ channel-length junctionless double-gate MOSFET. Reprinted from [111] with permission.

following condition of extrema:

$$\frac{\partial \Psi_{BCP}}{\partial x} = 0. \tag{7.23}$$

According to (7.20), this gives:

$$x_{min} = \frac{1}{2\delta} \ln\left(\frac{\beta'}{\alpha'}\right). \tag{7.24}$$

From (7.20), explicit relationships for the minimum center potential ($\Psi_{BCP,min}$) and minimum surface potential ($\Psi_{s,min}$) at x_{min} are obtained:

$$\Psi_{BCP,min} = 2a \sqrt{\left(-\frac{b}{a} - \beta' - \gamma\right)\beta' + a\gamma + b}, \tag{7.25}$$

$$\Psi_{s,min} = 2\sqrt{(-\beta' - \gamma)\beta'} + \gamma. \tag{7.26}$$

Figure 7.5 shows how the minimum body center potential varies with the drain voltage for a $10\ nm$ channel length. The center potential is a function of V_{DS}, moving from the center (at low V_{DS}) to the source side (when $V_{DS} = 1\ V$). This is evidenced by rewriting (7.25):

$$\Psi_{BCP,min}(V_{DS} + \Delta V_{DS}) = 2a\sqrt{\alpha'\beta'} + a\gamma + b, \tag{7.27}$$

where α' and β' are calculated from:

$$\alpha'|_{V_{DS}+\Delta V_{DS}} = \alpha'|_{V_{DS}} + \frac{\Delta V_{DS}}{2a\sinh(\delta L_G)} = \alpha' + \Delta, \tag{7.28}$$

$$\beta'|_{V_{DS}+\Delta V_{DS}} = \beta'|_{V_{DS}} - \frac{\Delta V_{DS}}{2a\sinh(\delta L_G)} = \beta' - \Delta, \tag{7.29}$$

$$\Delta = \frac{\Delta V_{DS}}{2a\sinh(\delta L_G)}. \tag{7.30}$$

7.2.2 Channel Current in Subthreshold

The 2D potential distribution $\Psi(x, y)$ obtained from (7.6) and (7.8) can be used to calculate the current in the subthreshold by adopting the general expression of the drift-diffusion model:

$$I_{DS} = -W\mu Q_m \frac{dV_{ch}}{dx}, \tag{7.31}$$

where total mobile charge density Q_m for a channel potential (V_{ch}) is obtained by

$$Q_m = -q \int_0^{T_{sc}} n_i \exp\left[\frac{\Psi(x, y) - V_{ch}}{U_T}\right] dy. \tag{7.32}$$

Therefore,

$$
\begin{aligned}
I_{DS} &= q\mu W \int_0^{T_{sc}} n_i \exp\left[\frac{\Psi(x, y) - V_{ch}}{U_T}\right] dy \frac{dV_{ch}}{dx} \\
&= q\mu W \int_0^{T_{sc}} n_i \exp\left[\frac{\Psi(x, y) - V_{ch}}{U_T}\right] \frac{dV_{ch}}{dx} dy \\
&= q\mu W \exp\left(-\frac{V_{ch}}{U_T}\right) \frac{dV_{ch}}{dx} \int_0^{T_{sc}} n_i \exp\left[\frac{\Psi(x, y)}{U_T}\right] dy \\
&\Rightarrow \frac{I_{DS}dx}{\int_0^{T_{sc}} n_i \exp\left[\dfrac{\Psi(x, y)}{U_T}\right] dy} = q\mu W \exp\left(-\frac{V_{ch}}{U_T}\right) dV_{ch}.
\end{aligned}
\tag{7.33}
$$

Integrating over x and V_{ch}, we can write:

$$
\begin{aligned}
I_{DS} \int_0^{L_G} \frac{dx}{\int_0^{T_{sc}} n_i \exp\left[\dfrac{\Psi(x, y)}{U_T}\right] dy} &= q\mu W \int_0^{V_{DS}} \exp\left(-\frac{V_{ch}}{U_T}\right) dV_{ch} \\
\Rightarrow I_{DS} = q\mu W U_T \left[1 - \exp\left(-\frac{V_{DS}}{U_T}\right)\right] &\left\{\int_0^{L_G} \frac{dx}{\int_0^{T_{sc}} n_i \exp\left[\dfrac{\Psi(x, y)}{U_T}\right] dy}\right\}^{-1}.
\end{aligned}
\tag{7.34}
$$

In general, relation (7.34) cannot be integrated analytically. Noting that the flow of electrons into the channel is limited by the minimum value of the potential along the transport direction at x_{min}, I_{DS} can be estimated by imposing $\Psi(x, y) = \Psi(x_{min}, y)$ in (7.34):

$$
\begin{aligned}
I_{DS} &= q\mu W U_T \left[1 - \exp\left(-\frac{V_{DS}}{U_T}\right)\right] \left\{\int_0^{L_G} \frac{dx}{\int_0^{T_{sc}} n_i \exp\left[\dfrac{\Psi(x_{min}, y)}{U_T}\right] dy}\right\}^{-1} \\
&= q\mu \frac{W}{L_G} U_T \left[1 - \exp\left(-\frac{V_{DS}}{U_T}\right)\right] \left\{\int_0^{T_{sc}} n_i \exp\left[\frac{\Psi(x_{min}, y)}{U_T}\right] dy\right\}.
\end{aligned}
\tag{7.35}
$$

From (7.6) and (7.8) evaluated at x_{min}, the minimum potential along the channel is:

$$\Psi(x_{min}, y) = \Psi_s(x_{min}) \left[1 + y\left(1 - \frac{y}{T_{sc}}\right) \frac{\varepsilon_{ox}}{t_{ox}\varepsilon_{si}} \right]$$
$$+ (\Delta\phi_{ms} - V_{GS})\, y\left(1 - \frac{y}{T_{sc}}\right)\frac{\varepsilon_{ox}}{t_{ox}\varepsilon_{si}}. \tag{7.36}$$

The integral in (7.35) can be solved using the error function. The subthreshold current is readily obtained:

$$I_{DS} = \frac{I_s}{\sqrt{C}} \exp\left[\frac{\Psi_s(x_{min})}{U_T} + \frac{CT_{sc}}{4} \right] \mathrm{erf}\left(\frac{\sqrt{CT_{sc}}}{2} \right), \tag{7.37}$$

where I_s is given by

$$I_s = q\mu n_i \left(\frac{W}{L_G}\right) U_T \left[1 - \exp\left(-\frac{V_{DS}}{U_T}\right) \right] \sqrt{\pi T_{sc}}, \tag{7.38}$$

and the gate potential is included in the parameter C:

$$C = \frac{\varepsilon_{ox}}{U_T t_{ox}\varepsilon_{si}} \left[\Psi_s(x_{min}) + \Delta\phi_{ms} - V_{GS} \right]. \tag{7.39}$$

Depending on the value of the argument $\dfrac{\sqrt{CT_{sc}}}{2}$, (7.37) can be simplified as follows:

$$I_{DS} = \frac{I_s}{2} \exp\left[\frac{\Psi_s(x_{min})}{U_T} + \frac{CT_{sc}}{4} \right] \sqrt{T_{sc}} \quad if \quad \frac{\sqrt{CT_{sc}}}{2} > 1, \tag{7.40}$$

$$I_{DS} = \frac{I_s}{\sqrt{C}} \exp\left[\frac{\Psi_s(x_{min})}{U_T} + \frac{CT_{sc}}{4} \right] \quad if \quad \frac{\sqrt{CT_{sc}}}{2} < 1. \tag{7.41}$$

In all the situations, the current has an exponential dependence on the surface potential calculated at the minimum potential coordinate. Based on these developments, relevant SCE performances can be obtained, namely the subthreshold swing of the drain current with respect to the gate voltage and the DIBL parameter.

Limit of the Validity of the Parabolic Approximation in Junctionless FETs
The parabolic distribution adopted in (7.3) has some intrinsic limitations when junctionless FETs are considered. For instance, along the channel i.e., in the x-direction, the potential at source/channel $(x = 0, y)$ and drain/channel $(x = L_G, y)$ frontiers must be higher than at (x_{min}, y). Since (7.23) is already satisfied, a sufficient condition for this extrema to be unique is ensured when:

$$\frac{\partial^2 \Psi}{\partial x^2} > 0 \quad while \quad 0 < x < L_G. \tag{7.42}$$

According to (7.7), this condition reverts to

$$\Psi_s(x) + \Delta\phi_{ms} - V_{GS} > \frac{qN_D}{\delta^2 \varepsilon_{si}}. \tag{7.43}$$

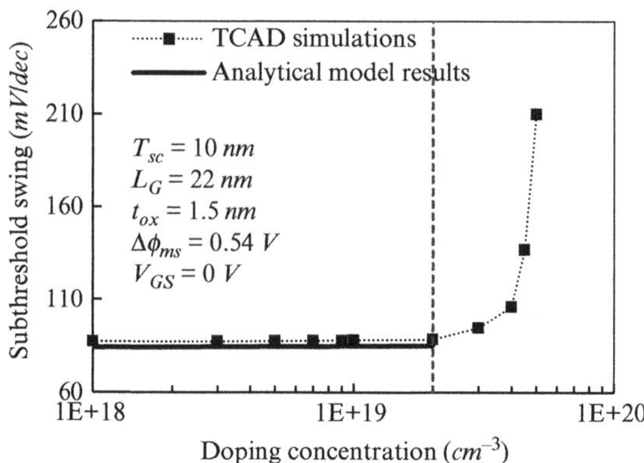

Figure 7.6 Subthreshold swing (mV/dec) versus the doping concentration through the channel for a $10\,nm$ silicon thickness junctionless double-gate MOSFET with $22\,nm$ gate length biased at $V_{DS} = 100\,mV$ and $V_{GS} = 0\,V$. The gray dashed line illustrates the limitation of the analytical model for a heavily doped channel where $\alpha\beta < 0$. Reprinted from [111] with permission.

Similarly, across the channel i.e., in the y-direction the body center potential is also an extrema, implying that

$$\frac{\partial^2 \Psi}{\partial y^2} < 0 \quad \text{while} \quad 0 < y < T_{sc}, \tag{7.44}$$

which from (7.6) gives:

$$\Psi_s(x) + \Delta\phi_{ms} - V_{GS} > 0. \tag{7.45}$$

Therefore, the most stringent condition is given by (7.43). According to the definition of γ (relation (7.10)), a sufficient condition is that the surface potential satisfies:

$$\Psi_s(x) > \gamma. \tag{7.46}$$

In particular, this must also be fulfilled by the minimum surface potential, leading to:

$$\Psi_{s,min} = 2\sqrt{\alpha\beta} + \gamma > \gamma. \tag{7.47}$$

Therefore, as soon as $\alpha\beta > 0$, the inequality (7.47) is satisfied. The case where $\alpha\beta < 0$ prevents using the parabolic approximation to simulate subthreshold characteristics in junctionless double-gate MOSFETs.

This condition happens when the drain voltage is large in comparison to the gate voltage, a situation where the minimum of the body center and surface potentials will have no extrema in the channel i.e., the x derivative of the potential never cancels and relation (7.23) can no longer be used (Figure 7.6).

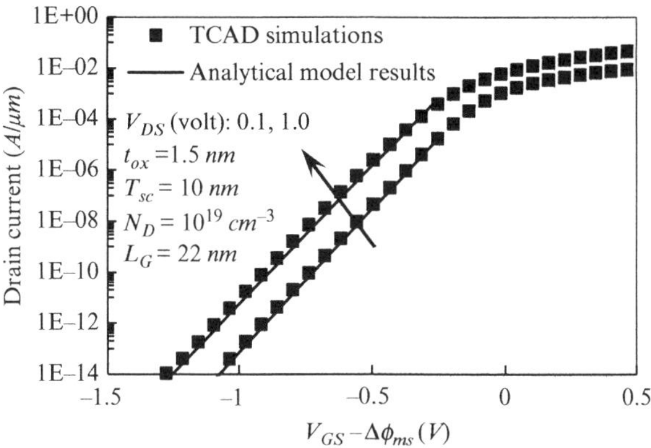

Figure 7.7 $I_{DS} - V_{GS}$ and its shift versus the shift in V_{DS} for a $10\ nm$ silicon thickness junctionless double-gate MOSFET doped at $1 \times 10^{19}\ cm^{-3}$ for a $22\ nm$ channel length. Note that $\Delta\phi_{ms}$ is the difference between the metal work function and an intrinsic reference semiconductor. Reprinted from [111] with permission.

7.2.3 Subthreshold Swing in Junctionless FETs

The analytical expression of the subthreshold current for different values of V_{GS} is used to extract the Subthreshold Swing (SS):

$$SS^{-1} = \frac{\partial \log(I_{DS})}{\partial V_{GS}} = \frac{1}{\ln(10)} \frac{\partial}{\partial V_{GS}} \left[\ln\left(\frac{I_s}{\sqrt{C}} \right) + \frac{\Psi_s(x_{min})}{U_T} + \frac{CT_{sc}}{4} \right]$$

$$= \frac{1}{U_T \ln(10)} \left[\eta \left(a - \frac{U_T}{U} \right) + 1 \right],$$

$$(7.48)$$

where

$$\eta = -\sqrt{\frac{\beta}{\alpha}} + \left[\sqrt{\frac{\beta}{\alpha}} - \sqrt{\frac{\alpha}{\beta}} \right] \frac{\exp(\delta L_G) - 1}{2 \sinh(\delta L_G)},$$

$$(7.49)$$

$$U = 4\sqrt{\alpha\beta} + \frac{2qN_D}{\delta^2 \varepsilon_{si}}.$$

$$(7.50)$$

Note that although the subthreshold current can be expressed through (7.48), this approach is not valid when $\alpha\beta < 0$. Figures 7.7 and 7.8 show the subthreshold current versus the gate voltage in a $10\ nm$ silicon thickness junctionless double-gate MOSFET doped at $N_D = 1 \times 10^{19}\ cm^{-3}$ for $L_G = 22$ and $30\ nm$, at low (0.1 V) and high (1 V) V_{DS}. Also shown are the TCAD numerical simulations.

The subthreshold swing as a function of the channel length is shown in Figure 7.9. The degradation of the subthreshold swing with channel shortening in highly doped devices ($N_D = 1 \times 10^{19}\ cm^{-3}$) is well captured by the analysis and the mismatch never exceeds 2.6 percent ($L_G = 20\ nm$, $T_{sc} = 10\ nm$, $t_{ox} = 1.5\ nm$, and $N_D = 1 \times 10^{19}\ cm^{-3}$). Note that the slope i.e., $\partial \log(I_{DS})/\partial V_{GS}$ is not strictly constant with V_{GS}, and an average value has been used for TCAD simulations.

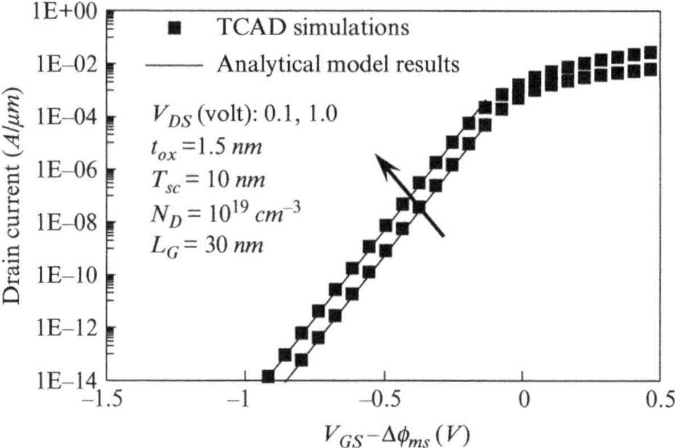

Figure 7.8 $I_{DS} - V_{GS}$ and its shift versus the shift in V_{DS} for a $10\ nm$ silicon thickness junctionless double-gate MOSFET doped at $10^{19}\ cm^{-3}$ for a $30\ nm$ channel length. Note that $\Delta\phi_{ms}$ is the difference between the metal work function and an intrinsic reference semiconductor. Reprinted from [111] with permission.

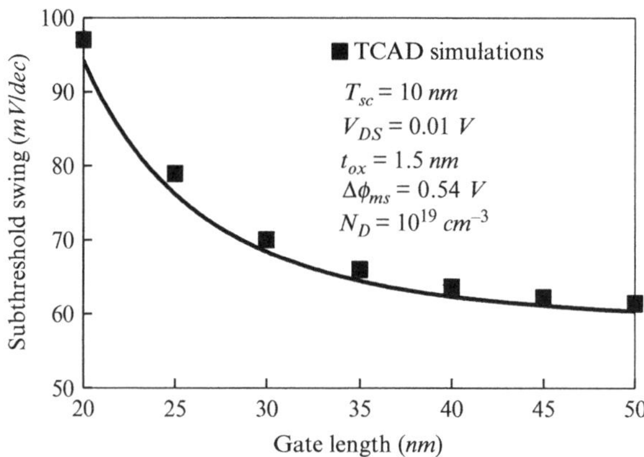

Figure 7.9 Subthreshold swing (mV/dec) versus the gate length for a $10\ nm$ silicon thickness junctionless double-gate MOSFET doped at $1 \times 10^{19}\ cm^{-3}$ and biased at $V_{DS} = 10\ mV$. Reprinted from [111] with permission.

The variation of Subthreshold swing with respect to doping concentration is illustrated in Figure 7.6. For heavily doped channel, the analytical approach becomes invalid above a certain doping level i.e., $N_D > 2 \times 10^{19}\ cm^{-3}$ for $10\ nm$ silicon thickness with $L_G = 22\ nm$.

7.2.4   DIBL in Junctionless FETs

Drain-induced barrier lowering is a key parameter used to estimate the influence of drain voltage on the drain current in the subthreshold (and saturation). From a circuit design point of view, DIBL represents the decrease in the threshold voltage after a 1 V increase in the drain potential, which is the definition commonly adopted in the literature [152]. However, the way it is calculated is not always consistent since

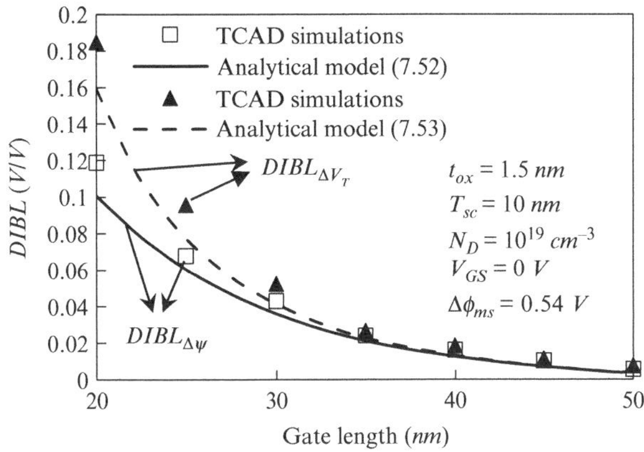

Figure 7.10 Definitions of DIBL according to (7.52) and (7.53) versus gate length for a $10\,nm$ silicon thickness junctionless double-gate MOSFET doped at $1 \times 10^{19}\,cm^{-3}$ ($V_{GS} = 0\,V$). Reprinted from [111] with permission.

it usually refers to the variation in minimum body potential with the drain voltage, including IM and depletion-mode FETs.

Regarding junction-based devices, DIBL is related to the shift in the surface potential since this is where most of the current flows. In contrast, in junctionless devices (operating below threshold), the Si–SiO$_2$ interface no longer represents the lowest energy across the gates and the DIBL parameter must be calculated from the center potential shift upon the drain voltage.

Combining (7.25) and (7.27), the variation in the minimum body center potential ($\Delta\Psi_{BCP,min}$) when the drain voltage changes by (ΔV_{DS}) is:

$$\Delta\Psi_{BCP,min} = \Psi_{BCP,min}(V_{DS} + \Delta V_{DS}) - \Psi_{BCP,min}(V_{DS})$$

$$= 2a\,\sqrt{\alpha'\beta'}\left[\sqrt{1 + \frac{\Delta}{\alpha'} - \frac{\Delta}{\beta'} - \frac{\Delta^2}{\alpha'\beta'}} - 1\right]. \tag{7.51}$$

The barrier lowering is also readily obtained:

$$DIBL_{\Delta\Psi} = \frac{\Delta\Psi_{BCP,min}}{\Delta V_{DS}} = \frac{2a\,\sqrt{\alpha'\beta'}}{\Delta V_{DS}}\left[\sqrt{1 + \frac{\Delta}{\alpha'} - \frac{\Delta}{\beta'} - \frac{\Delta^2}{\alpha'\beta'}} - 1\right]. \tag{7.52}$$

Relation (7.52) defines the DIBL in terms of the minimum center potential shift upon V_{DS}. We call it $DIBL_{\Delta\Psi}$. The parameters a, α', β', and δ were obtained from (7.12), (7.21), (7.23), and (7.30), respectively.

The DIBL$_{\Delta\Psi}$ is plotted in Figure 7.10 as a function of the gate length and compared with numerical simulations. Some slight discrepancy appears at very short dimensions but the error is still less than 10 percent for $L_G > 25\,nm$, which is acceptable given that this parameter is very sensitive to minor changes in the minimum body center potential.

According to relations (7.10), (7.14), and (7.52), increasing the doping concentration in the channel region will also increase $DIBL_{\Delta\Psi}$, meaning that higher doping concentrations might be deleterious for SCEs in these devices. On the other hand, from a designer point of view, DIBL is merely defined in terms of a threshold voltage shift as shown in (7.53) [152] i.e., $DIBL_{\Delta V_T}$:

$$DIBL_{\Delta V_T} = \frac{\Delta V_T}{\Delta V_{DS}}. \tag{7.53}$$

The ambiguity between (7.52) and (7.53) needs to be discussed in more detail. The variation in the current upon V_{DS} when the gate potential is maintained at a fixed bias is used to calculate the shift in V_T (ΔV_T) using the subthreshold swing. Assuming that i'_{ds} and i_{ds} are the subthreshold currents at $V_{DS} + \Delta V_{DS}$ and V_{DS} (obtained from (7.48) at constant V_{GS}), the threshold voltage shift can also be expressed as:

$$\Delta V_T = SS \left[\log\left(\frac{i'_{ds}}{i_{ds}} \right) \right]. \tag{7.54}$$

Combining (7.37) with (7.54), and considering that the current is fixed at $I_s(V_{DS} + \Delta V_{DS}) = I_s(V_{DS})$, ΔV_T can be obtained as follows (note that this is also how DIBL is measured graphically):

$$\Delta V_T = \frac{SS}{\ln(10)\,U_T} \left[a\Delta\Psi_{s,min} - \frac{U_T}{2} \ln\left(1 + \frac{\Delta\Psi_{s,min}}{\Psi_{s,min} + \Delta\phi_{ms} - V_{GS}} \right) \right], \tag{7.55}$$

where the value for "a" is obtained from (7.12) and $\Delta\Psi_{s,min}$ is the shift in the minimum surface potential upon V_{DS}. Assuming that

$$\frac{U_T}{2} \ln\left(1 + \frac{\Delta\Psi_{s,min}}{\Psi_{s,min} + \Delta\phi_{ms} - V_{GS}} \right) \ll a\Delta\Psi_{s,min}, \tag{7.56}$$

relation (7.55) can be simplified as:

$$\Delta V_T = \frac{SS}{\ln(10)\,U_T} a\Delta\Psi_{s,min}. \tag{7.57}$$

On the other hand, according to (7.11), $a\Delta\Psi_{s,min}$ reverts to $\Delta\Psi_{BCP,min}$, meaning that $DIBL_{\Delta V_T}$ is proportional to the shift in the minimum body center potential $\Delta\Psi_{BCP,min}$:

$$DIBL_{\Delta V_T} = \frac{\Delta V_T}{\Delta V_{DS}} = \frac{SS}{U_T \Delta V_{DS} \ln(10)} \Delta\Psi_{BCP,min}. \tag{7.58}$$

From (7.53) and (7.58), the link between the two definitions is seen:

$$DIBL_{\Delta V_T} = \frac{SS}{U_T \ln(10)} DIBL_{\Delta\Psi}. \tag{7.59}$$

Relation (7.59) is instructive from a conceptual point of view. As can be seen, for relatively long channels the value of the subthreshold swing (SS) is close to the theoretical limit of $U_T \ln(10)$, and both definitions become almost equivalent:

$$DIBL_{\Delta V_T} \cong DIBL_{\Delta\Psi}. \tag{7.60}$$

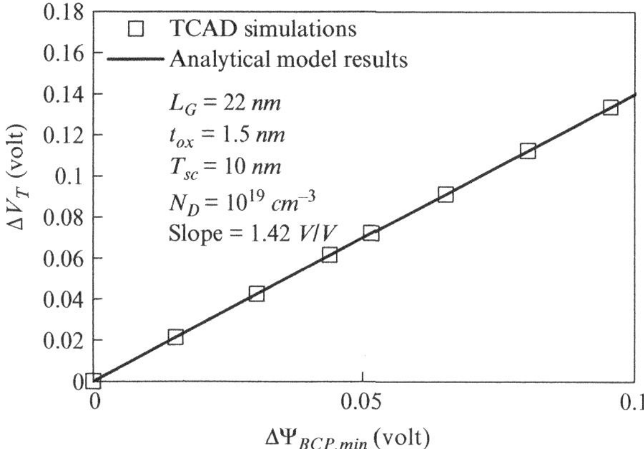

Figure 7.11 Threshold voltage shift versus the shift in the minimum center potential for a $22\ nm$ gate length in a $10\ nm$ silicon thickness junctionless double-gate MOSFET doped at $N_D = 1 \times 10^{19}\ cm^{-3}$ ($V_{GS} = 0\ V$, $\Delta\phi_{ms} = 0.54\ V$). Reprinted from [111] with permission.

While in relatively long-channel devices the two definitions of DIBL are equivalent, this is not true in short channels where the factor SS departs from the ideal value of $60\ mV/decade$ (at 20°C) (the mismatch can be as high as 30 percent when SS = $80\ mV/decade$).

Figure 7.11 depicts the link between the threshold-voltage shift (ΔV_T) and the shift in the minimum body center potential ($\Delta\Psi_{BCP,min}$) for a $22\ nm$ channel-length device. Note the linear dependence between the shift in V_T and the shift in the center potential, even though $\Psi_{BCP,min}$ is not proportional to V_{DS} (see Figure 7.5), as well as the sensitivity of ΔV_T upon $\Delta\Psi_{BCP,min}$.

7.3   Summary

An analytical model to calculate the 2D potential distribution and related SCEs in ultrathin body junctionless double-gate MOSFET with channel length down to $10\ nm$ was developed in this chapter based on the concept of minimum center potential. The model was used to calculate the sensitivity of the minimum potential shift along the channel with applied drain voltage when the device operates below threshold. The definitions of DIBL and subthreshold swing were discussed in detail and analytical expressions were derived for these two central concepts in short-channel junctionless FETs.

8   Modeling AC Operation in Symmetric Double-Gate and Nanowire JL FETs

In addition to analytical DC models of junctionless FETs, the design of analog and digital circuits requires accurate modeling of AC characteristics as well. Adopting the charge–voltage relationships presented in Chapter 3, a complete analytical model for transcapacitances valid in all regions of operation is derived for double-gate and nanowire junctionless FETs here.

The charge-based model presented in Chapter 3 covers depletion and accumulation modes. However, at flat-band the derivative of the charge densities with respect to the potential equals $2 \times C_{ox}$, exceeding the theoretical value (but continuity of the charge and derivatives is still preserved). This mismatch around the flat-band prevents obtaining accurate derivatives of the mobile charge density with respect to the applied voltages. In addition, as soon as V_{DS} is increased, the nonuniformity of the channel must also be taken into account to evaluate the equivalent charge densities on the different nodes. Here, a detailed treatment of the charges distribution and partitioning scheme inside the channel of junctionless FETs is given and used to derive a complete small signal-equivalent circuit.

8.1   Transcapacitance Matrix in Symmetric Double-Gate FETs

A general analysis of transadmittances is reviewed for symmetric double-gate field-effect transistors (see Figure 8.1), no matter the principle of operation (inversion,

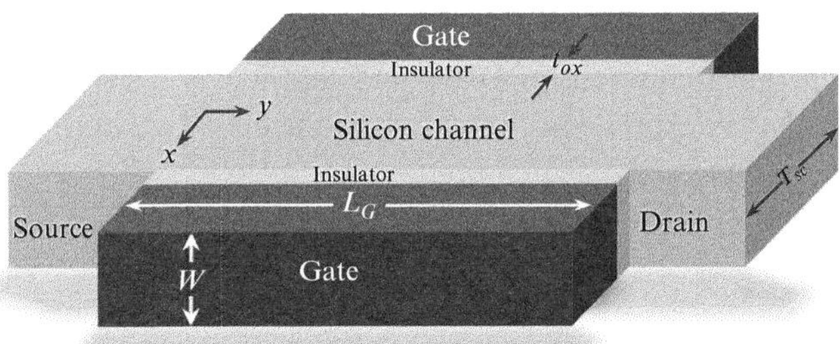

Figure 8.1 Schematic view of an *n*-type junctionless double-gate MOSFET investigated in this chapter.

accumulation, and depletion). The inherent symmetry and the lack of a bulk reference for these architectures results in some unique properties of the matrix transadmittance.

In a harmonic analysis, small-signal current and voltages (δi) and (δv) are linearly dependent through the so-called Y-matrix transadmittance:

$$\delta \mathbf{i} = \bar{\mathbf{Y}}.\delta \mathbf{v} = (\bar{\mathbf{A}} + j.\omega.\bar{\mathbf{C}}).\delta \mathbf{v}. \tag{8.1}$$

This Y-matrix can be split into two parts: a real number contribution $\bar{\mathbf{A}}$ dealing with conductances (matrix conductance) for the in-phase response of the signal and an imaginary number contribution $\bar{\mathbf{C}}$ dealing with capacitances for the out-phase contribution (capacitance matrix) (j denotes the imaginary unit and ω the angular frequency). For a typical MOSFET device with four terminals, gate (g), drain (d), source (s), and body (b), the $\bar{\mathbf{A}}$ and $\bar{\mathbf{C}}$ matrices have the following elements:

$$\begin{bmatrix} i_g \\ i_d \\ i_s \\ i_b \end{bmatrix} = \begin{bmatrix} A_{GG} & A_{GD} & A_{GS} & A_{GB} \\ A_{DG} & A_{DD} & A_{DS} & A_{DB} \\ A_{SG} & A_{SD} & A_{SS} & A_{SB} \\ A_{BG} & A_{BD} & A_{BS} & A_{BB} \end{bmatrix} \begin{bmatrix} v_g \\ v_d \\ v_s \\ v_b \end{bmatrix}$$

$$+ j\omega \begin{bmatrix} C_{GG} & C_{GD} & C_{GS} & C_{GB} \\ C_{DG} & C_{DD} & C_{DS} & C_{DB} \\ C_{SG} & C_{SD} & C_{SS} & C_{SB} \\ C_{BG} & C_{BD} & C_{BS} & C_{BB} \end{bmatrix} \begin{bmatrix} v_g \\ v_d \\ v_s \\ v_b \end{bmatrix}, \tag{8.2}$$

where the A_{ij} element represents the in-phase current response at node i upon a (small) voltage change at node j, which is a transconductance. Similarly, C_{ij} represents the out-phase current at node i in response to a change in voltage at node j. This quantity corresponds to a capacitance between nodes i and j (note that depending on the sign for currents, C_{ij} can be negative).

Concerning symmetric double-gate MOSFETs, since there is no substrate node, the matrix transadmittance is comprised of 9 elements instead of 16 for the bulk MOSFET. Indeed, using the source as reference for the potentials, only the drain-to-source and gate-to-source potentials are meaningful. Thus:

$$\begin{bmatrix} i_g \\ i_d \end{bmatrix} = \left(\bar{\mathbf{A}} + j\omega \begin{bmatrix} C_{GG} & C_{GD} \\ C_{DG} & C_{DD} \end{bmatrix} \right) \begin{bmatrix} v_{gs} \\ v_{ds} \end{bmatrix}. \tag{8.3}$$

Noting that charge neutrality implies $i_s = -(i_g + i_d)$, the imaginary components of the current at the gate (i_g^*), drain (i_d^*), and source (i_s^*) nodes are given by (note that with this rule, the current flowing out of the source of an n-channel junctionless FET will be negative):

$$i_g^* = j\omega \left[C_{GG} v_g + C_{GD} v_d - (C_{GG} + C_{GD}) v_s \right], \tag{8.4}$$

$$i_d^* = j\omega \left[C_{DG} v_g + C_{DD} v_d - (C_{DG} + C_{DD}) v_s \right], \tag{8.5}$$

$$i_s^* = j\omega \left[-(C_{GG} + C_{DG}) v_g - (C_{GD} + C_{DD}) v_d + (C_{GG} + C_{GD} + C_{DG} + C_{DD}) v_s \right]. \tag{8.6}$$

Therefore, the transcapacitance matrix becomes:

$$\bar{C} = \begin{bmatrix} C_{GG} & C_{GD} & -(C_{GG}+C_{GD}) \\ C_{DG} & C_{DD} & -(C_{DG}+C_{DD}) \\ -(C_{GG}+C_{DG}) & -(C_{GD}+C_{DD}) & (C_{GG}+C_{GD}+C_{DG}+C_{DD}) \end{bmatrix}, \tag{8.7}$$

where

$$-(C_{GG}+C_{GD}) = C_{GS}, \tag{8.8}$$

$$-(C_{DG}+C_{DD}) = C_{DS}, \tag{8.9}$$

$$-(C_{GG}+C_{DG}) = C_{SG}, \tag{8.10}$$

$$-(C_{GD}+C_{DD}) = C_{SD}, \tag{8.11}$$

$$C_{GG}+C_{GD}+C_{DG}+C_{DD} = C_{SS}. \tag{8.12}$$

As for bulk MOSFETs, the Ward–Dutton partitioning scheme is applicable for the double-gate junctionless FET [156, 157] where the bulk node is omitted. In this approach, transcapacitances are evaluated through averaged charge densities defined at the drain, source, and gate terminals. These are recalled hereafter:

$$\overline{Q}_{m,d} = W \int_0^{L_G} \left(\frac{y}{L_G}\right) Q_m(y) dy, \tag{8.13}$$

$$\overline{Q}_{m,s} = W \int_0^{L_G} \left(1 - \frac{y}{L_G}\right) Q_m(y) dy. \tag{8.14}$$

In addition, charge neutrality imposes:

$$Q_G + \overline{Q}_{m,s} + \overline{Q}_{m,d} + WL_G Q_{fix} = 0. \tag{8.15}$$

From [156], any variation of the applied voltage results in a change in these averaged charges from which transcapacitances can be defined according to derivative of the global mobile charge at node i with respect to the applied potential at node j:

$$C_{ij} = \partial \overline{Q}_{m,i} / \partial V_j, \tag{8.16}$$

8.2   General Case

Solutions of (8.13) and (8.14) presume that the distribution of the mobile charge density along the channel $Q_m(y)$ is known, which in our case is not true. In addition, y/L_G and dy can be expressed as a function of $Q_m(y)$ and dV_{ch}, respectively, which alleviates this drawback. These transformations are detailed in the following subsections.

8.2.1 Expressing dy versus dV_{ch}

Using the definition of the current assuming drift-diffusion model, the quantities dy and dV_{ch} can be linked to each other as follows (V_{ch} is the shift in electrons quasi-Fermi potential with the source taken as reference):

$$dy = -\frac{WQ_m\mu}{I_{DS}}dV_{ch}. \tag{8.17}$$

Introducing (8.17) in (8.13) and (8.14):

$$\overline{Q}_{m,d} = W\int_0^{L_G}\left(\frac{y}{L_G}\right)Q_m(y)dy = -\frac{W^2}{I_{DS}}\mu\int_0^{V_{DS}}\left(\frac{y}{L_G}\right)Q_m^2(y)dV_{ch}, \tag{8.18}$$

$$\overline{Q}_{m,s} = W\int_0^{L_G}\left(1-\frac{y}{L_G}\right)Q_m(y)dy = -\frac{W^2}{I_{DS}}\mu\int_0^{V_{DS}}\left(1-\frac{y}{L_G}\right)Q_m^2(y)dV_{ch}. \tag{8.19}$$

Looking back at the model detailed in Chapter 3, the link between $Q_m(y)$ and V_{ch} depends on whether the channel is in depletion or in accumulation.

The most general situation is a channel exhibiting depletion (near the drain) and accumulation (close to the source) (Section 3.1.9). As discussed the transition between these two regions crosses at a flat-band position (Y_{FB}) where $Q_m|_{FB} = Q_{m,FB} = -Q_{fix}$, so that the channel between the source and Y_{FB} is in accumulation while between Y_{FB} and drain it is in depletion. Splitting integrals (8.18) and (8.19) at the flat-band locus Y_{FB} gives:

$$\overline{Q}_{m,d} = W\int_0^{L_G}\left(\frac{y}{L_G}\right)Q_m(y)dy$$

$$= W\left[\int_0^{Y_{FB}}\left(\frac{y}{L_G}\right)Q_m(y)|_{acc}\,dy + \int_{Y_{FB}}^{L_G}\left(\frac{y}{L_G}\right)Q_m(y)|_{dep}\,dy\right]$$

$$= -\frac{W^2}{I_{DS}}\mu\left[\int_0^{V_{ch,FB}}\left(\frac{y}{L_G}\right)Q_m^2(y)|_{acc}\,dV_{ch} + \int_{V_{ch,FB}}^{V_{DS}}\left(\frac{y}{L_G}\right)Q_m^2(y)|_{dep}\,dV_{ch}\right], \tag{8.20}$$

$$\overline{Q}_{m,s} = W\int_0^{L_G}\left(1-\frac{y}{L_G}\right)Q_m(y)dy$$

$$= W\left[\int_0^{Y_{FB}}\left(1-\frac{y}{L_G}\right)Q_m(y)|_{acc}\,dy - \int_{Y_{FB}}^{L_G}\left(1-\frac{y}{L_G}\right)Q_m(y)|_{dep}\,dy\right]$$

$$= \frac{W^2}{I_{DS}}\mu\left[\int_0^{V_{ch,FB}}\left(1-\frac{y}{L_G}\right)Q_m^2(y)|_{acc}\,dV_{ch} + \int_{V_{ch,FB}}^{V_{DS}}\left(1-\frac{y}{L_G}\right)Q_m^2(y)|_{dep}\,dV_{ch}\right]. \tag{8.21}$$

8.2.2   Expressing $Q_m(y)\,dV_{ch}$

The charge-based model developed in Chapter 3 proposes explicit and continuous relationships for charges which are recalled hereafter for depletion (8.22) and accumulation (8.23) cases:

$$V_{GS} - V_{ch} - \Delta\phi_{ms} - U_T \ln\left(\frac{N_D}{n_i}\right) \overset{dep}{\approx} -\frac{Q_{sc}^2}{8qN_D\varepsilon_{si}} - \frac{Q_{sc}}{2C_{ox}} + U_T \ln\left[1 - \left(\frac{Q_{sc}}{Q_{fix}}\right)^2\right],$$

$$\tag{8.22}$$

$$V_{GS} - V_{ch} - \Delta\phi_{ms} - U_T \ln\left(\frac{N_D}{n_i}\right) \overset{acc}{\approx} -\frac{Q_{sc}}{2C_{ox}} + U_T \ln\left(1 + \frac{Q_{sc}^2}{8qN_D\varepsilon_{si}U_T}\right). \tag{8.23}$$

Recalling that Q_{sc} is the sum of the mobile (Q_m) and fixed ($Q_{fix} = qN_DT_{sc}$) charge densities, the differential forms of (8.22) and (8.23) are:

$$-Q_m dV_{ch} \overset{dep}{\approx} \left[-\frac{Q_m\left(Q_m + Q_{fix}\right)}{4qN_D\varepsilon_{si}} - \frac{Q_m}{2C_{ox}} + 2U_T\frac{Q_m + Q_{fix}}{Q_m + 2Q_{fix}}\right]dQ_m, \tag{8.24}$$

$$-Q_m dV_{ch} \overset{acc}{\approx} \left[-\frac{Q_m}{2C_{ox}} + 2U_T Q_m \frac{Q_m + Q_{fix}}{8U_T Q_{fix}C_{si} + \left(Q_m + Q_{fix}\right)^2}\right]dQ_m. \tag{8.25}$$

The last step is to establish a link between the channel coordinate y and the mobile charge density $Q_m(y)$.

8.2.3   Expressing y/L_G versus $Q_m(y)$

The way mobile charges propagate from the source to the drain is intimately linked to the current continuity equation. Under quasistatic operation, which is assumed in this analysis i.e., time constants are well below the carrier transit time, the current remains constant along the channel. Integrating the current from source ($V_{ch} = 0$) to a given coordinate y along the channel and treating it as a constant value gives:

$$I_{DS} = -\frac{W\mu}{y}\left(\int_0^{V_{ch}} Q_m dV_{ch}\right). \tag{8.26}$$

As detailed in Section 3.1.9, the link between $Q_m(y)$ and V_{ch} depends on whether the channel is in depletion or in accumulation. Still considering the general situation of a hybrid channel, the transition between these two regions crosses the flat-band position. Recalling that the channel between the source and Y_{FB} is in accumulation while between Y_{FB} and drain the channel is in depletion, the integrals in relations (8.26), (8.18), and (8.19) are split as follows:

$$I_{DS} = -\frac{W\mu}{y}\left(\int_0^{FB} Q_m|_{acc}\,dV_{ch} + \int_{FB}^y Q_m|_{dep}\,dV_{ch}\right). \tag{8.27}$$

Considering a coordinate y along the channel beyond the flat-band position ($Y_{FB} < y < L_G$), and replacing (8.24) and (8.25) in (8.27), after integration from $y = 0$ to y, the current satisfies:

$$I_{DS} = \frac{W\mu}{y} \left[F(Q_{m,FB}) - F(Q_{m,s}) + f(Q_m) - f(Q_{m,FB}) \right], \tag{8.28}$$

where $F(Q_m)$ and $f(Q_m)$ correspond to the integrals in (8.27) in accumulation and depletion regions, respectively, and are given by

$$F(Q_m) = -\frac{1}{4C_{ox}} Q_m^2 + 2U_T \left\{ Q_m - \frac{Q_{fix}}{2} \ln \left[8U_T Q_{fix} C_{si} + \left(Q_{fix} + Q_m \right)^2 \right] \right\}$$
$$- 2U_T \sqrt{8U_T Q_{fix} C_{si}} \arctan \left(\frac{Q_{fix} + Q_m}{\sqrt{8U_T Q_{fix} C_{si}}} \right), \tag{8.29}$$

$$f(Q_m) = -\frac{Q_m^3}{12qN_D\varepsilon_{si}} - Q_m^2 \left(\frac{1}{8C_{si}} + \frac{1}{C_{ox}} \right) + 2U_T Q_m - 2U_T Q_{fix} \ln \left(Q_m + 2Q_{fix} \right). \tag{8.30}$$

On the other hand, setting $y = L_G$ in relation (8.27) gives the current as a function of the charge densities at source $Q_{m,s}$ and drain $Q_{m,d}$ only (still the charge density at flat-band $Q_{m,FB}$ is merely $-Q_{fix}$):

$$I_{DS} = -\frac{W\mu}{L_G} \left(\int_0^{V_{DS}} Q_m dV_{ch} \right) = \frac{W\mu}{L_G} \left[F(Q_{m,FB}) - F(Q_{m,s}) + f(Q_{m,d}) - f(Q_{m,FB}) \right]. \tag{8.31}$$

Depending on the channel state, the quantity y/L_G will take different expressions: A first case is when $Y_{FB} < y < L_G$ i.e., when the channel is in depletion. The current simplifies as (8.28). Next, combining (8.28) with (8.31) leads to

$$\frac{y}{L_G} = \frac{F(Q_{m,FB}) - F(Q_{m,s}) + f(Q_m) - f(Q_{m,FB})}{F(Q_{m,FB}) - F(Q_{m,s}) + f(Q_{m,d}) - f(Q_{m,FB})}$$
$$= \frac{W\mu}{I_{DS}L_G} \left[F(Q_{m,FB}) - F(Q_{m,s}) + f(Q_{m,y}) - f(Q_{m,FB}) \right]. \tag{8.32}$$

The second case is when $0 < y < Y_{FB}$ i.e., the channel in accumulation. The current simplifies to:

$$I_{DS} = -\frac{W\mu}{y} \int_0^{V_{ch}} Q_m(y)|_{acc} dV_{ch} = \frac{W\mu}{y} \left[F(Q_m) - F(Q_{m,s}) \right]. \tag{8.33}$$

Similarly, combining (8.33) with (8.31) gives:

$$\frac{y}{L_G} = \frac{W\mu}{I_{DS}L_G} \left[F(Q_m) - F(Q_{m,s}) \right]. \tag{8.34}$$

Replacing in the general relation (8.20) the proper expressions for y/L_G and $Q_m dV_{ch}$ derived for the depleted ((8.32), (8.24)) and accumulated ((8.34), (8.25)) regimes of the channel, the equivalent charge density at the drain node is obtained:

$$\overline{Q}_{m,d} = W \int_0^{L_G} \left(\frac{y}{L_G}\right) Q_m(y)\, dy =$$

$$-\frac{W^3\mu^2}{I_{DS}^2 L_G} \int_0^{V_{ch,FB}} [F(Q_m) - F(Q_{m,s})]\, Q_m^2(y)\,|_{acc}\, dV_{ch} +$$

$$-\frac{W^3\mu^2}{I_{DS}^2 L_G} \int_{V_{ch,FB}}^{V_{DS}} [F(Q_{m,FB}) - F(Q_{m,s}) + f(Q_m) - f(Q_{m,FB})]\, Q_m^2(y)\,|_{dep}\, dV_{ch}.$$

$$(8.35)$$

Next, inserting (8.29), (8.30), (8.24), and (8.25) into (8.35) gives

$$\overline{Q}_{m,d} = \eta I_{F,acc} - F(Q_{m,s})\eta I_{acc} + [F(Q_{m,FB}) - F(Q_{m,s}) - f(Q_{m,FB})]\,\eta I_{dep} + \eta I_{f,dep}, \quad (8.36)$$

where

$$\eta = -\frac{W^3\mu^2}{I_{DS}^2 L_G}, \quad (8.37)$$

Relation (8.36) involves various integrals defined as $I_{F,acc}$, I_{acc}, $I_{f,dep}$, and I_{dep} (note that the equivalence $Q_{m,FB} = -Q_{fix}$ is used):

$$I_{F,acc} = \int_0^{V_{ch,FB}} F(Q_m) Q_m^2(y)\,|_{acc}\, dV_{ch}$$

$$= -\int_{Q_{m,s}}^{-Q_{fix}} \left[-\frac{1}{4C_{ox}} Q_m^2 + 2U_T\left\{ Q_m - \frac{Q_{fix}}{2} \ln\left[8U_T Q_{fix} C_{si} + \left(Q_{fix} + Q_m\right)^2 \right] \right\} \right] Q_m$$

$$\left[-\frac{Q_m}{2C_{ox}} + 2U_T Q_m \frac{Q_m + Q_{fix}}{8U_T Q_{fix} C_{si} + \left(Q_m + Q_{fix}\right)^2} \right] dQ_m,$$

$$(8.38)$$

$$I_{acc} = \int_0^{V_{ch,FB}} Q_m^2(y)\,|_{acc}\, dV_{ch}$$

$$= -\int_{Q_{m,s}}^{-Q_{fix}} Q_m \left[-\frac{Q_m}{2C_{ox}} + 2U_T Q_m \frac{Q_m + Q_{fix}}{8U_T Q_{fix} C_{si} + \left(Q_m + Q_{fix}\right)^2} \right] dQ_m$$

$$= \frac{Q_m^3}{6C_{ox}} - Q_m^2 U_T - \alpha \log\left(Q_m^2 + 2Q_m Q_{fix} + Q_{fix}^2 + 8C_{si}U_T Q_{fix}\right) + 2Q_m Q_{fix} U_T$$

$$-8\sqrt{2C_{si}Q_{fix}^3 U_T^3}\, \arctan\left(\frac{8\sqrt{2C_{si}Q_{fix}^5 U_T^3} + 8Q_m\sqrt{2C_{si}Q_{fix}^3 U_T^3}}{2Q_{fix}^3 U_T - 2Q_{fix}\alpha + 16C_{si}Q_{fix}^2 U_T^2} \right)\Bigg|_{Q_{m,s}}^{-Q_{fix}},$$

$$(8.39)$$

where α is given by $Q_{fix}^2 U_T - 8 C_{si} Q_{fix} U_T^2$, and

$$
\begin{aligned}
I_{f,dep} &= \int_{V_{ch,FB}}^{V_{DS}} f(Q_m) Q_m^2(y)\, |_{dep}\, dV_{ch} \\[2mm]
&= -\int_{-Q_{fix}}^{Q_{m,d}} \left[-\frac{Q_m^3}{12 q N_D \varepsilon_{si}} - Q_m^2 \left(\frac{1}{8 C_{si}} + \frac{1}{C_{ox}} \right) + 2 U_T Q_m - 2 U_T Q_{fix} \ln\left(Q_m + 2 Q_{fix} \right) \right] \\[2mm]
&\qquad Q_m \left[-\frac{Q_m \left(Q_m + Q_{fix} \right)}{4 q N_D \varepsilon_{si}} - \frac{Q_m}{2 C_{ox}} + 2 U_T \frac{Q_m + Q_{fix}}{Q_m + 2 Q_{fix}} \right] dQ_m.
\end{aligned}
$$

(8.40)

It should be noted that neglecting the "logarithmic" and "arctangent" terms in (8.30) and (8.29) does not introduce any substantial mismatch and leads to explicit solutions for transcapacitance matrix elements.

$$
\begin{aligned}
I_{dep} &= \int_{V_{ch,FB}}^{V_{DS}} Q_m^2(y)\, |_{dep}\, dV_{ch} \\[2mm]
&= -\int_{-Q_{fix}}^{Q_{m,d}} Q_m \left[-\frac{Q_m \left(Q_m + Q_{fix} \right)}{4 q N_D \varepsilon_{si}} - \frac{Q_m}{2 C_{ox}} + 2 U_T \frac{Q_m + Q_{fix}}{Q_m + 2 Q_{fix}} \right] dQ_m \\[2mm]
&= \frac{W^2 \mu}{I_{DS}} \left[\frac{Q_m^4}{16 q N_D \varepsilon_{si}} + \frac{Q_m^3 Q_{fix}}{12 q N_D \varepsilon_{si}} + \frac{Q_m^3}{6 C_{ox}} \right. \\[2mm]
&\qquad \left. - U_T Q_m^2 + 2 U_T Q_m Q_{fix} + 4 U_T Q_{fix}^2 \ln\left(\frac{1}{Q_m + 2 Q_{fix}} \right) \right] \Bigg|_{-Q_{fix}}^{Q_{m,d}}.
\end{aligned}
$$

(8.41)

Concerning the total charge density on the gate (Q_G), this is expressed as follows:

$$
\begin{aligned}
Q_G &= -W L_G Q_{fix} - W \int_0^{L_G} Q_m dy = -W L_G Q_{fix} - W \int_{Q_{m,s}}^{Q_{m,FB}} Q_m\, |_{acc}\, dy - W \int_{Q_{m,FB}}^{Q_{m,d}} Q_m\, |_{dep}\, dy \\[2mm]
&= -W L_G Q_{fix} + \frac{W^2 \mu}{I_{DS}} \int_0^{V_{ch,FB}} Q_m^2\, |_{acc}\, dV_{ch} + \frac{W^2 \mu}{I_{DS}} \int_{V_{ch,FB}}^{V_{DS}} Q_m^2\, |_{dep}\, dV_{ch} \\[2mm]
&= -W L_G Q_{fix} + \frac{W^2 \mu}{I_{DS}} \left(I_{acc} + I_{dep} \right).
\end{aligned}
$$

(8.42)

Following the same idea, the global charge density at the source terminal results from Q_G and $\overline{Q}_{m,d}$ [156]:

$$\overline{Q}_{m,s} = W \int_0^{L_G} \left(1 - \frac{y}{L_G}\right) Q_m(y) dy = W \int_0^{L_G} Q_m(y) dy - W \int_0^{L_G} \left(\frac{y}{L_G}\right) Q_m(y) dy$$

$$= -Q_G - Q_{fix} - \overline{Q}_{m,d}, \qquad (8.43)$$

where $\overline{Q}_{m,d}$ and Q_G are given by (8.36) and (8.42), respectively. Next, recalling that in the Ward–Dutton approach transcapacitances are defined according to (8.16), C_{GG}, and C_{GD} and are given by ($C_{ij} = \partial \overline{Q}_{m,i}/\partial V_j$)

$$C_{GG} = -\left(\frac{\partial \overline{Q}_{m,s}}{\partial V_{GS}} + \frac{\partial \overline{Q}_{m,d}}{\partial V_{GS}}\right) = -(C_{SG} + C_{DG}), \qquad (8.44)$$

$$C_{GD} = -\left(\frac{\partial \overline{Q}_{m,s}}{\partial V_{DS}} + \frac{\partial \overline{Q}_{m,d}}{\partial V_{DS}}\right) = -(C_{SD} + C_{DD}), \qquad (8.45)$$

$\overline{Q}_{m,s}$ and $\overline{Q}_{m,d}$ can be calculated analytically.

Neglecting the *arctg* and *log* terms in (8.29) and (8.30) introduces a negligible mismatch. However, it is convenient to evaluate these derivatives numerically. This is not a drawback since it is a common approach for electrical simulators. This point will be discussed in Section 8.4.

8.3   Special Case of a Channel Uniformly Depleted/Accumulated

When the whole channel is in either depletion or in accumulation, the general solution developed for the hybrid case is simplified.

8.3.1   Channel in Depletion Mode

When the whole channel operates in depletion, the gate-to-source voltage must satisfy $V_{GS} - \Delta\phi_{ms} < U_T \ln(N_D/n_i)$. Relations (8.20) and (8.21) simplify to

$$\overline{Q}_{m,d} = W \int_0^{L_G} \left(\frac{y}{L_G}\right) Q_m(y)|_{dep}\, dy = -\frac{W^2}{I_{DS}}\mu \int_0^{V_{DS}} \left(\frac{y}{L_G}\right) Q_m^2(y)|_{dep}\, dV_{ch}, \qquad (8.46)$$

$$\overline{Q}_{m,s} = W \int_0^{L_G} \left(1 - \frac{y}{L_G}\right) Q_m(y)|_{dep}\, dy = -\frac{W^2}{I_{DS}}\mu \int_0^{V_{DS}} \left(1 - \frac{y}{L_G}\right) Q_m^2(y)|_{dep}\, dV_{ch}. \qquad (8.47)$$

The channel in depletion means that the flat-band locus Y_{FB} is before the source. This is equivalent to assigning $Q_{m,FB}$ to $Q_{m,s}$ in (8.32):

$$y = L_G \frac{f(Q_{m,s}) - f(Q_{m,y})}{f(Q_{m,s}) - f(Q_{m,d})}. \qquad (8.48)$$

Similarly, the channel current I_{DS} is obtained from (8.28) while imposing $y = L_G$:

$$I_{DS} = \frac{W\mu}{L_G} \left[f(Q_{m,d}) - f(Q_{m,s}) \right].$$
(8.49)

Finally, combining (8.49) with (8.48) gives:

$$\frac{y}{L_G} = \frac{W\mu}{I_{DS}L_G} \left[f(Q_{m,y}) - f(Q_{m,s}) \right].$$
(8.50)

Rewriting equivalent charge at drain as defined from (8.46) but now inserting the simplified expressions in depletion for y/L_G gives:

$$\overline{Q}_{m,d} = W \int_0^{L_G} \left(\frac{y}{L_G} \right) Q_m dy =$$

$$\eta \int_{Q_{m,s}}^{Q_{m,d}} \left[f(Q_{m,s}) - f(Q_m) \right] Q_m \left[-\frac{Q_m \left(Q_m + Q_{fix} \right)}{4qN_D\varepsilon_{si}} - \frac{Q_m}{2C_{ox}} + 2U_T \frac{Q_m + Q_{fix}}{Q_m + 2Q_{fix}} \right] dQ_m.$$
(8.51)

Further, neglecting the logarithmic term in (8.30), $f(Q_m)$ simplifies to:

$$f(Q_m) = -\frac{Q_m^3}{12qN_D\varepsilon_{si}} - Q_m^2 \left(\frac{1}{8C_{si}} + \frac{1}{C_{ox}} \right) + 2U_T Q_m,$$
(8.52)

Replacing (8.52) in (8.51), an explicit solution for the equivalent drain charge is obtained (see Appendix D): The total mobile charge density (Q_G) in depletion mode is calculated by integrating $Q_m(y)$ over the entire channel length. Again, using relation (8.17), Q_G can be expressed as follows:

$$Q_G + WL_G Q_{fix} = -W \int_0^{L_G} Q_m dy = \frac{W^2\mu}{I_{DS}} \int_0^{V_{DS}} Q_m^2 dV_{ch}$$

$$= \frac{W^2\mu}{I_{DS}} \int_{Q_{m,s}}^{Q_{m,d}} \left[\frac{Q_m^2 \left(Q_m + Q_{fix} \right)}{4qN_D\varepsilon_{si}} + \frac{Q_m^2}{2C_{ox}} - 2U_T \left(Q_m - Q_{fix} + \frac{2Q_{fix}^2}{Q_m + 2Q_{fix}} \right) \right] dQ_m$$

$$= \frac{W^2\mu}{I_{DS}} \left[\frac{Q_m^4}{16qN_D\varepsilon_{si}} + \frac{Q_m^3 Q_{fix}}{12qN_D\varepsilon_{si}} + \frac{Q_m^3}{6C_{ox}} \right.$$

$$\left. \left. -U_T Q_m^2 + 2U_T Q_m Q_{fix} + 4U_T Q_{fix}^2 \ln \left(\frac{1}{Q_m + 2Q_{fix}} \right) \right] \right|_{Q_{m,s}}^{Q_{m,d}}$$
(8.53)

Charge conservation gives the equivalent charge at the source:

$$\overline{Q}_{m,s} = -Q_G - \overline{Q}_{m,d} - WL_G Q_{fix}.$$
(8.54)

Applying relation (8.44) to the mobile charge densities at source and drain derived in depletion mode ($\overline{Q}_{m,s}$ and $\overline{Q}_{m,d}$) gives the transcapacitances.

8.3.2   Channel in Accumulation Mode

When the channel is in accumulation from source to drain i.e., when $V_{GS} - V_{DS} - \Delta\phi_{ms} > U_T \ln(N_D/n_i)$, relation (8.27) retains only one term:

$$I_{DS} = -\frac{W\mu}{y}\left(\int_0^{V_{DS}} Q_m|_{acc}\, dV_{ch}\right). \tag{8.55}$$

Then, following the same derivation as for the totally depleted channel, the channel current takes a simple form:

$$I_{DS} = \frac{W\mu}{L_G}\left[F(Q_{m,d}) - F(Q_{m,s})\right]. \tag{8.56}$$

Similarly, the link between y and $F(Q_{m,y})$ gives:

$$y = L_G\frac{F(Q_{m,s}) - F(Q_m)}{F(Q_{m,s}) - F(Q_{m,d})}, \tag{8.57}$$

$$\frac{y}{L_G} = \frac{W\mu}{I_{DS}L_G}\left[F(Q_m) - F(Q_{m,s})\right], \tag{8.58}$$

where $F(Q_m)$ is obtained from (8.29). Next, merging (8.13) with (8.58) and (8.17) gives the equivalent charge densities at the drain node:

$$\overline{Q}_{m,d} = W\int_0^{L_G}\left(\frac{y}{L_G}\right)Q_m\,dy = -W\int_0^{V_{DS}}\frac{F(Q_{m,s}) - F(Q_m)}{F(Q_{m,s}) - F(Q_{m,d})}\left(\frac{W\mu}{I_{DS}}\right)Q_m^2\,dV_{ch}$$

$$= \eta\int_{Q_{m,s}}^{Q_{m,d}}\left[F(Q_{m,s}) - F(Q_m)\right]Q_m\left[-\frac{Q_m}{2C_{ox}} + 2U_T Q_m\frac{Q_m + Q_{fix}}{8U_T Q_{fix}C_{si} + \left(Q_m + Q_{fix}\right)^2}\right]dQ_m. \tag{8.59}$$

As discussed previously, neglecting the "arctan" term in (8.29), $F(Q_m)$ simplifies to:

$$F(Q_m) = -\frac{1}{4C_{ox}}Q_m^2 + 2U_T\left\{Q_m - \frac{Q_{fix}}{2}\ln\left[8U_T Q_{fix}C_{si} + \left(Q_{fix} + Q_m\right)^2\right]\right\}. \tag{8.60}$$

Replacing (8.60) in (8.59) results in an analytical solution of the global drain charge density in accumulation mode. Note that neglecting the logarithmic term in (8.60) introduces negligible error. In this case, in accumulation mode an explicit relation for $\overline{Q}_{m,d}$ is available (see Appendix E). Moreover, the total charge density at the gate node (Q_G) is expressed as follows:

$$Q_G + WL_G Q_{fix} = -W \int_0^{L_G} Q_m dy = \frac{W^2 \mu}{I_{DS}} \int_0^{L_G} Q_m^2 dV_{ch}$$

$$= -\frac{W^2 \mu}{I_{DS}} \int_0^{L_G} \left[-\frac{Q_m^2}{2C_{ox}} + 2U_T Q_m^2 \frac{Q_m + Q_{fix}}{8U_T Q_{fix} C_{si} + \left(Q_m + Q_{fix}\right)^2} \right] dQ_m$$

$$= \gamma \left(\frac{Q_m^3}{6C_{ox}} - Q_m^2 U_T - \alpha \log \left(Q_m^2 + 2Q_m Q_{fix} + Q_{fix}^2 + 8C_{si} U_T Q_{fix} \right) + 2Q_m Q_{fix} U_T \right)$$

$$- 8\gamma \sqrt{2C_{si} Q_{fix}^3 U_T^3} \arctan \left(\frac{8\sqrt{2C_{si} Q_{fix}^5 U_T^3} + 8Q_m \sqrt{2C_{si} Q_{fix}^3 U_T^3}}{2Q_{fix}^3 U_T - 2Q_{fix}\alpha + 16C_{si} Q_{fix}^2 U_T^2} \right) \Bigg|_{Q_{m,s}}^{Q_{m,d}},$$

$$\tag{8.61}$$

where α and γ are given by $Q_{fix}^2 U_T - 8C_{si} Q_{fix} U_T^2$ and $W^2 \mu / I_{DS}$, respectively. This gives an explicit relationship for the global source charge density ($\overline{Q}_{m,s}$) through (8.54).

Again, differentiating $\overline{Q}_{m,s}$ and $\overline{Q}_{m,d}$, analytical expressions of the transcapacitance matrix elements are obtained in terms of the equivalent mobile charge densities. The next step consists of evaluating $\partial Q_{m,d}/\partial V_i$ and $\partial Q_{m,s}/\partial V_i$.

8.4 Analytical Expressions for the Local Charge Derivatives

Evaluation of the transcapacitances defined from the generic relationship (8.16) requires calculating the derivatives of the drain and source local charge densities with respect to the applied potentials. In principle, this can be obtained by direct differentiation of relations (8.22) and (8.23), depending on whether the channel is in depletion or in accumulation. In particular, at flat-band i.e., when $V_{GS} - V_{ch} = \Delta \phi_{ms} + U_T \ln (N_D / n_i)$ both derivatives of (8.22) and (8.23) give:

$$\frac{\partial Q_m}{\partial (V_{GS} - V_{ch})} \bigg|_{FB} = -2C_{ox}. \tag{8.62}$$

However, this asymptotic value is only reached in accumulation, although the charge densities at source and drain are accurate at flat-band ($Q_{sc} = 0$). For instance, with $T_{sc} = 10\,nm$ and $N_D = 1 \times 10^{19}\,cm^{-3}$ the correct value would be close to $-1.55 \times C_{ox}$. This requires additional developments that are discussed as follows.

In order to simplify the analysis, generic potentials V^* and $\Psi^*_{s/0}$ thus defined so that $V^* = V_{GS} - V_{ch}$ and $\Psi^*_{s/0} = \Psi_{s/0} - V_{ch}$, where V_{ch} is the local Fermi potential and $\Psi_{s/0}$ is the surface or center potential. It can be shown that these quantities are indeed entering in all charge-based relationships. For given generic and surface potentials, the total charge density in the channel becomes:

$$2\left(V^* - \Psi_s^*\right)C_{ox} = -Q_{sc}. \tag{8.63}$$

According to (8.63), the derivative of the charge density with respect to the generic potential is:

$$\frac{\partial Q_m}{\partial V^*} = \frac{\partial Q_{sc}}{\partial V^*} = -2C_{ox}\left(1 - \frac{\partial \Psi_s}{\partial V^*}\right). \tag{8.64}$$

This equation is for the gate capacitor.

On the other hand, the semiconductor links the charge density to the surface and center potentials:

$$Q_{sc}^2 = 4\varepsilon_{si}^2 E_s^2 = 8\varepsilon_{si}qn_i U_T\left[\exp\left(\frac{\Psi_s^*}{U_T}\right) - \exp\left(\frac{\Psi_0^*}{U_T}\right) - \frac{N_D}{n_i U_T}\left(\Psi_s^* - \Psi_0^*\right)\right]. \tag{8.65}$$

This is used to build another relationship for $\dfrac{\partial Q_m}{\partial V^*}$:

$$\frac{\partial Q_m}{\partial V^*} = \frac{\partial Q_{sc}}{\partial V^*} = \frac{4\varepsilon_{si}qn_i}{Q_{sc}}\left[\exp\left(\frac{\Psi_s^*}{U_T}\right)\frac{\partial \Psi_s^*}{\partial V^*} - \exp\left(\frac{\Psi_0^*}{U_T}\right)\frac{\partial \Psi_0^*}{\partial V^*} - \frac{N_D}{n_i}\left(\frac{\partial \Psi_s^*}{\partial V^*} - \frac{\partial \Psi_0^*}{\partial V^*}\right)\right]. \tag{8.66}$$

As discussed in Chapters 3 and 4, considering the first four terms of a MacLaurin series around the center of the channel (under symmetric operation), and introducing the Lambert function, the center potential Ψ_0^* becomes:

$$\Psi_0^* = \Psi_s^* + KN_D - U_T\mathrm{LambertW}\left[\frac{Kn_i}{U_T}\exp\left(\frac{\Psi_s^* + KN_D}{U_T}\right)\right], \tag{8.67}$$

where $K = qT_{sc}^2/8\varepsilon_{si}$. In addition, based on (8.67), there is a link between the derivatives of the surface and center potentials with respect to V_{GS}:

$$\frac{\partial \Psi_s^*}{\partial V^*} = \frac{\partial \Psi_0^*}{\partial V^*}\left[1 + K\frac{n_i}{U_T}\exp\left(\frac{\Psi_0^*}{U_T}\right)\right]. \tag{8.68}$$

Next, combining (8.64) with (8.66) and (8.68), the derivative of the center potential with respect to V^* is readily obtained:

$$\left(\frac{\partial \Psi_0^*}{\partial V^*}\right)^{-1} = 1 + \eta\exp\left(\frac{\Psi_0^*}{U_T}\right) - \frac{2\varepsilon_{si}qn_i}{Q_{sc}C_{ox}}$$
$$\times\left[\exp\left(\frac{\Psi_s^*}{U_T}\right) - \exp\left(\frac{\Psi_0^*}{U_T}\right) + \eta\exp\left(\frac{\Psi_s^* + \Psi_0^*}{U_T}\right) - \eta\frac{N_D}{n_i}\exp\left(\frac{\Psi_0^*}{U_T}\right)\right], \tag{8.69}$$

with $\eta = \dfrac{Kn_i}{U_T}$. Last, combining (8.64), (8.68), and (8.69) the derivatives of mobile charge density with respect to gate, source/drain potentials satisfy:

$$\frac{\partial Q_{m,s}}{\partial V^*} = \frac{\partial Q_{m,d}}{\partial V^*} = -2C_{ox}\left\{1 - \frac{\partial \Psi_0^*}{\partial V^*}\left[1 + \eta\exp\left(\frac{\Psi_0^*}{U_T}\right)\right]\right\}. \tag{8.70}$$

According to the charge-based model in Chapter 3, center and surface potentials in (8.70) can be calculated with (8.63) and (8.67), and therefore all derivatives of the charge densities with respect to the gate, source/drain potentials are now available.

The special case of flat-band condition merit some discussion

At flat-band, the entire channel is neutral and therefore the center and surface potentials become equal:

$$\Psi_{s,FB} = \Psi_{0,FB} = V_{GS,FB} - \Delta\phi_{ms} \approx V_{ch} + U_T \ln\left(\frac{N_D}{n_i}\right). \tag{8.71}$$

In this case, replacing (8.71) in (8.69) and (8.70), the derivative of charge density with respect to V^* can be calculated accurately at flat-band, which was not possible with (8.22) or (8.23).

8.5 Simulations and Discussion

Simulations for specific transcapacitances (in absolute values) versus the gate voltage for $V_{DS} = 0\ V$ and $V_{DS} = 1\ V$ are done for two n-type junctionless double-gate MOSFET, namely $N_D = 1 \times 10^{19}\ cm^{-3}$, $T_{sc} = 20\ nm$ for device A and $N_D = 2 \times 10^{19}\ cm^{-3}$, $T_{sc} = 10\ nm$ for device B. Other devices parameter have the same values i.e., $L_G = 1\ \mu m$, $W = 1\ \mu m$, and $t_{ox} = 1.5\ nm$. Transcapacitances are plotted in Figures 8.2 and 8.3 for device A and Figures 8.4 and 8.5 for device B. Good matching is seen between the model and numerical simulations (TCAD) in all regions of operation. Compared to IM FETs, transcapacitances exhibit a nonmonotonous trend in depletion mode, a peculiarity rooted in the quadratic term in the Q–V relationship (see relation (8.22)). However, this behavior is minimized when the doping

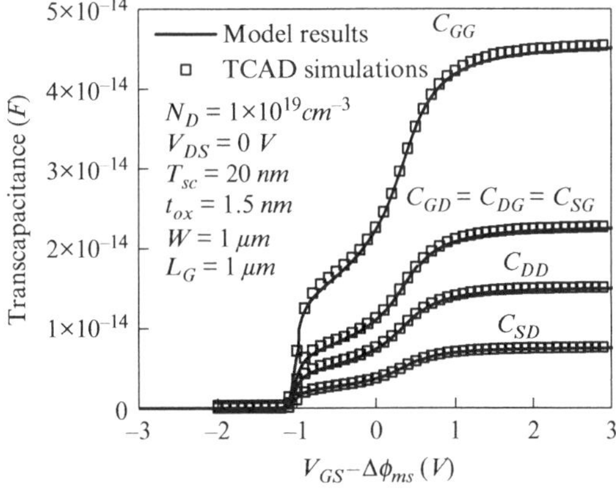

Figure 8.2 Dependence of the transcapacitance versus the gate voltage ($V_{GS} - \Delta\phi_{ms}$) for $V_{DS} = 0\ V$. The lines and symbols are related to the analytical model and the TCAD simulations, respectively ($N_D = 1 \times 10^{19}\ cm^{-3}$, $T_{sc} = 20\ nm$, $L_G = 1\ \mu m$, $W = 1\ \mu m$, and $t_{ox} = 1.5\ nm$). Note that $\Delta\phi_{ms}$ is the difference between the metal work function and an intrinsic reference semiconductor. Reprinted from [118] with permission.

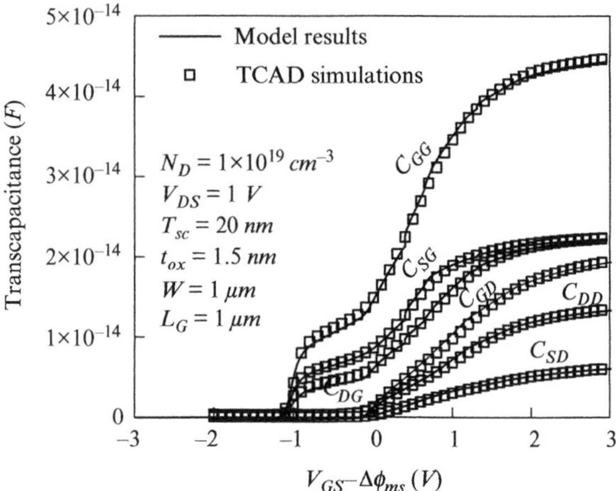

Figure 8.3 Dependence of the transcapacitance versus the gate voltage ($V_{GS} - \Delta\phi_{ms}$) for $V_{DS} = 1\ V$. The lines and symbols are related to the analytical model and the TCAD simulations, respectively ($N_D = 1 \times 10^{19}\ cm^{-3}$, $T_{sc} = 20\ nm$, $L_G = 1\ \mu m$, $W = 1\ \mu m$, and $t_{ox} = 1.5\ nm$). Note that $\Delta\phi_{ms}$ is the difference between the metal work function and an intrinsic reference semiconductor. Reprinted from [118] with permission.

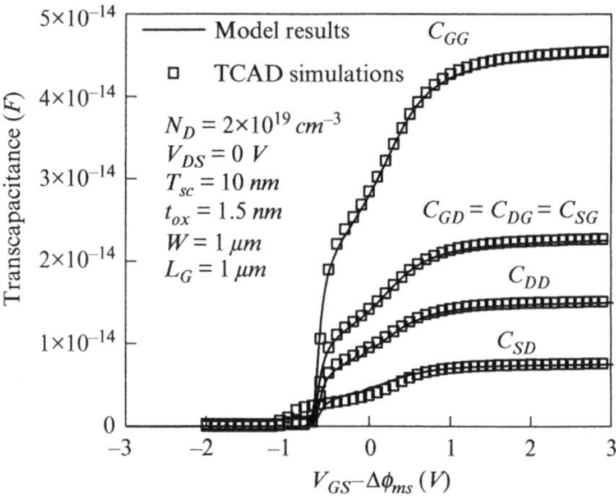

Figure 8.4 Dependence of the transcapacitance versus the gate voltage ($V_{GS} - \Delta\phi_{ms}$) for $V_{DS} = 0\ V$. The lines and symbols are related to the analytical model and the TCAD simulations, respectively ($N_D = 2 \times 10^{19}\ cm^{-3}$, $T_{sc} = 10\ nm$, $L_G = 1\ \mu m$, $W = 1\ \mu m$, and $t_{ox} = 1.5\ nm$). $\Delta\phi_{ms} = W_{ms} - U_T \ln(N_D/n_i)$ where W_{ms} is the metal semiconductor work function difference. Reprinted from [118] with permission.

and thickness/radius are below some values for which an approximate solution can be obtained. This will be detailed in Section 8.8.

Last, Figure 8.6 plots the gate capacitance with respect to effective V_{GS} for $V_{DS} = 0\ V$ for a very highly doped channel ($N_D = 5 \times 10^{19}\ cm^{-3}$) with $10\ nm$ silicon thickness. In this case, an increase in C_{GG} is predicted from numerical simulations for negative gate potentials. As discussed in Chapter 4 and in [144, 146], depending on the

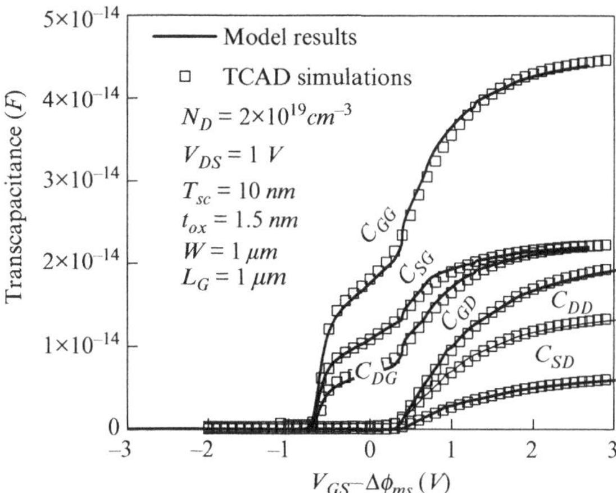

Figure 8.5 Dependence of the transcapacitance versus the gate voltage ($V_{GS} - \Delta\phi_{ms}$) for $V_{DS} = 1\,V$. The lines and symbols are related to the analytical model and the TCAD simulations, respectively ($N_D = 2 \times 10^{19}\,cm^{-3}$, $T_{sc} = 10\,nm$, $L_G = 1\,\mu m$, $W = 1\,\mu m$, and $t_{ox} = 1.5\,nm$). Note that $\Delta\phi_{ms}$ is the difference between the metal work function and an intrinsic reference semiconductor given by $\Delta\phi_{ms} = W_{ms} - U_T \ln(N_D/n_i)$ where W_{ms} is the metal semiconductor work function difference. Reprinted from [118] with permission.

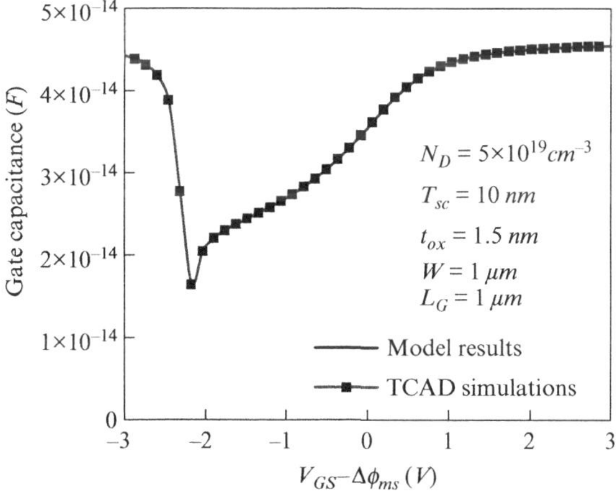

Figure 8.6 Dependence of the gate capacitance versus the gate voltage ($V_{GS} - \Delta\phi_{ms}$) for $V_{DS} = 0\,V$ in a heavily doped channel confirming that a hole-inversion layer can also build up at the channel interface. The lines and symbols are related to the analytical model and the TCAD simulations, respectively ($N_D = 5 \times 10^{19}\,cm^{-3}$, $T_{sc} = 10\,nm$, $L_G = 1\,\mu m$, $W = 1\,\mu m$, and $t_{ox} = 1.5\,nm$). Reprinted from [118] with permission.

technological parameters i.e., doping density and silicon thickness, a hole-inversion layer can build up at the channel interface. This inherent inverted hole layer can be modulated by the gate potential, implying that the gate capacitance starts increasing below some critical gate potential, a feature not included in relation (8.22).

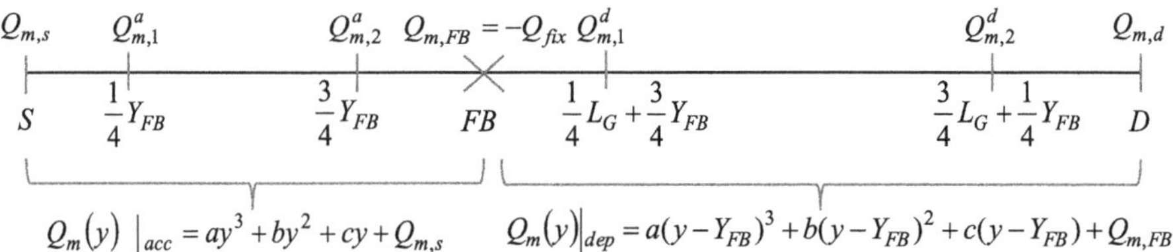

Figure 8.7 Approximation of mobile charge density through the channel by two cubic splines, one in the accumulation region and one in the depletion region.

8.6   Cubic Approximation of $Q_m(y)$ and Transcapacitances

The way transcapacitances have been derived is adequate from a modeling point of view but still it involves calculations that are not efficient for an implementation in electrical simulators. Therefore, approximate – but more explicit – solutions are of interest for circuit applications. Instead of swapping the space variable y with the channel potential V_{ch} as in Sections 8.2 and 8.3, the y dependence of Q_m is approximated with cubic splines, one valid for accumulation and the other for depletion. Likewise, in Section 8.2, a special treatment is carried out for each channel state.

8.6.1   Hybrid Channel

When the channel splits in accumulated and depleted regions (see Section 8.2 and Figure 8.7), each of the regions is interpolated with a third-order spline. Four values of Q_m are sufficient to define unambiguously the polynomial expression. In addition to the source ($y = 0$) and the flat-band positions ($y = Y_{FB}$), two internal nodes are proposed on the accumulated side of the channel (toward the source), $y_1 = 0.25\,Y_{FB}$ and $y_2 = 0.75\,Y_{FB}$, with the corresponding mobile charge densities $Q^a_{m,1} = Q_m(y_1)$ and $Q^a_{m,2} = Q_m(y_2)$, respectively. Similarly, in the depleted part of the channel (toward the drain), in addition to the flat-band and drain coordinates ($y = Y_{FB}$ and $y = L_G$), two additional nodes are defined, $y_3 = 0.25L_G + 0.75\,Y_{FB}$ and $y_4 = 0.25\,Y_{FB} + 0.75L_G$, with $Q^d_{m,1} = Q_m(y_3)$ and $Q^d_{m,2} = Q_m(y_4)$.

To obtain the internal charge densities, the locus of the flat-band position is still missing in case the channel is in the hybrid mode. The coordinate where flat-band occurs Y_{FB} is obtained from (8.28) while imposing $y = Y_{FB}$ and $Q_{m,y} = Q_{m,FB}$ in the accumulated side of the channel:

$$Y_{FB} = \frac{W\mu}{I_{DS}} \left[F\left(Q_{m,FB} = -Q_{fix}\right) - F\left(Q_{m,s}\right) \right]. \tag{8.72}$$

Similarly, when imposing $y = L_G - Y_{FB}$ and $Q_{m,y} = Q_{m,FB}$ in (8.28) in the depleted part of the channel, Y_{FB} can also be expressed as:

$$Y_{FB} = L_G - \frac{W\mu}{I_{DS}} \left[f(Q_{m,d}) - f\left(Q_{m,FB} = -Q_{fix}\right) \right]. \tag{8.73}$$

Once the flat-band position is known, relation (8.28) is applied in each domain to find $F(Q_m)$ (in accumulation) and $f(Q_m)$ (in depletion), at a given channel coordinate y:

$$F(Q_m) \stackrel{acc}{=} \frac{y}{Y_{FB}} F(Q_{m,FB}) + \left(1 - \frac{y}{Y_{FB}}\right) F(Q_{m,s}) \quad \text{when} \quad 0 < y < Y_{FB}, \qquad (8.74)$$

$$f(Q_m) \stackrel{dep}{=} \frac{y - Y_{FB}}{L_G - Y_{FB}} f(Q_{m,d}) + \left(1 - \frac{y - Y_{FB}}{L_G - Y_{FB}}\right) f(Q_{m,FB}) \quad \text{when} \quad Y_{FB} < y < L_G. \qquad (8.75)$$

The Accumulated Domain

In the accumulated part of the channel, the internal charge densities $Q^a_{m,1}$ and $Q^a_{m,2}$ can be obtained from relation (8.74). Then once the coordinate y_1 and y_2 are known, these give back the mobile charge densities $Q^a_{m,1}$ and $Q^a_{m,2}$ at $y = 0.25\,Y_{FB}$ and $0.75\,Y_{FB}$, respectively, ($0 < y < Y_{FB}$) so that:

$$Q_m(y) \stackrel{acc}{\approx} ay^3 + by^2 + cy + Q_{m,s}, \qquad (8.76)$$

where a, b, and c are given by

$$a = \frac{1}{3} Y_{FB}^{-3} \left(-16 Q_{m,s} + 32 Q^a_{m,1} - 32 Q^a_{m,2} + 16 Q_{m,FB}\right), \qquad (8.77)$$

$$b = \frac{1}{3} Y_{FB}^{-2} \left(32 Q_{m,s} - 56 Q^a_{m,1} + 40 Q^a_{m,2} - 16 Q_{m,FB}\right), \qquad (8.78)$$

$$c = \frac{1}{3} Y_{FB}^{-1} \left(-19 Q_{m,s} + 24 Q^a_{m,1} - 8 Q^a_{m,2} + 3 Q_{m,FB}\right), \qquad (8.79)$$

where $Q_{m,FB} = -Q_{fix}$.

The Depleted Domain

In the depleted part of the channel, once the coordinates y_3 and y_4 are known the internal charge densities $Q^d_{m,1}$ and $Q^d_{m,2}$ can be obtained from relation (8.75). Similarly, the mobile charge density in the depleted domain of the channel ($Y_{FB} < y < L_G$) satisfies:

$$Q_m(y) \stackrel{dep}{\approx} a\left(y - Y_{FB}\right)^3 + b\left(y - Y_{FB}\right)^2 + c\left(y - Y_{FB}\right) + Q_{m,FB}, \qquad (8.80)$$

with the new coefficients calculated from (8.77) to (8.79):

$$a = \frac{1}{3} \left(L_G - Y_{FB}\right)^{-3} \left(-16 Q_{m,FB} + 32 Q^d_{m,1} - 32 Q^d_{m,2} + 16 Q_{m,d}\right), \qquad (8.81)$$

$$b = \frac{1}{3} \left(L_G - Y_{FB}\right)^{-2} \left(32 Q_{m,FB} - 56 Q^d_{m,1} + 40 Q^d_{m,2} - 16 Q_{m,d}\right), \qquad (8.82)$$

$$c = \frac{1}{3} \left(L_G - Y_{FB}\right)^{-1} \left(-19 Q_{m,FB} + 24 Q^d_{m,1} - 8 Q^d_{m,2} + 3 Q_{m,d}\right), \qquad (8.83)$$

where the y dependence of Q_m used to determine $Q^d_{m,1}$ and $Q^d_{m,2}$ is now governed by relation (8.75). Therefore, for each domain in the hybrid channel third-order polynomial functions are defined across the flat-band coordinate.

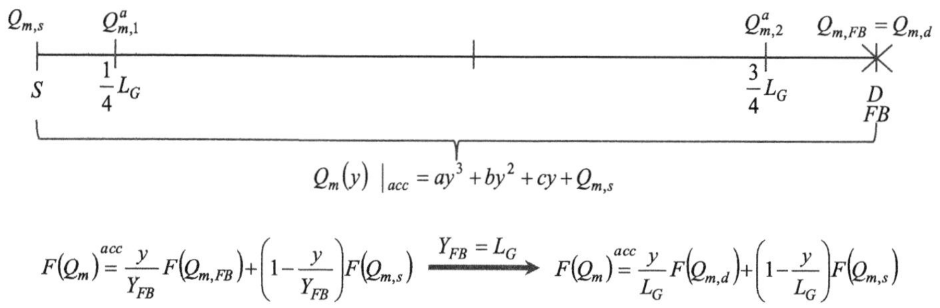

$$Q_m(y)\,\big|_{acc} = ay^3 + by^2 + cy + Q_{m,s}$$

$$F(Q_m) \overset{acc}{=} \frac{y}{Y_{FB}} F(Q_{m,FB}) + \left(1 - \frac{y}{Y_{FB}}\right) F(Q_{m,s}) \quad\xrightarrow{Y_{FB}=L_G}\quad F(Q_m) \overset{acc}{=} \frac{y}{L_G} F(Q_{m,d}) + \left(1 - \frac{y}{L_G}\right) F(Q_{m,s})$$

Figure 8.8 Approximation of mobile charge density through the channel when the whole channel operates in accumulation by a cubic spline.

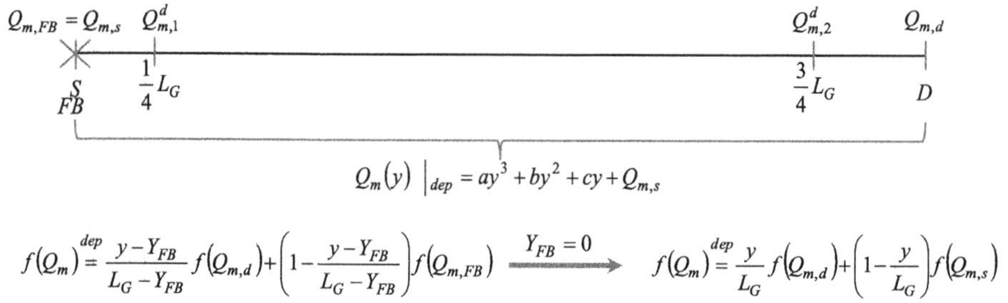

$$Q_m(y)\,\big|_{dep} = ay^3 + by^2 + cy + Q_{m,s}$$

$$f(Q_m) \overset{dep}{=} \frac{y - Y_{FB}}{L_G - Y_{FB}} f(Q_{m,d}) + \left(1 - \frac{y - Y_{FB}}{L_G - Y_{FB}}\right) f(Q_{m,FB}) \quad\xrightarrow{Y_{FB}=0}\quad f(Q_m) \overset{dep}{=} \frac{y}{L_G} f(Q_{m,d}) + \left(1 - \frac{y}{L_G}\right) f(Q_{m,s})$$

Figure 8.9 Approximation of mobile charge density through the channel when the whole channel operates in depletion by a cubic spline.

8.6.2   Uniform Channels

When the whole channel is in accumulation (see Figure 8.8), $V_{GS} - V_{DS} > \Delta\phi_{ms} + U_T \ln(N_D/n_i)$, the link between the mobile charge density and the coordinate has already been expressed in relation (8.74) and is thus rewritten as:

$$F(Q_m) \overset{acc}{=} \frac{y}{L_G} F(Q_{m,d}) + \left(1 - \frac{y}{L_G}\right) F(Q_{m,s}). \tag{8.84}$$

Similarly, when the whole channel is in depletion (see Figure 8.9; i.e., when $V_{GS} < \Delta\phi_{ms} + U_T \ln(N_D/n_i)$) the same kind of polynomial function can be proposed to extract $Q_m(y)$, and the internal points $Q_{m,1}$ and $Q_{m,2}$ are obtained based on relation (8.48) and rewritten as follows:

$$f(Q_m) \overset{dep}{=} \frac{y}{L_G} f(Q_{m,d}) + \left(1 - \frac{y}{L_G}\right) f(Q_{m,s}). \tag{8.85}$$

Therefore, using (8.84) or (8.85), the mobile charge density for accumulation or depletion mode can in principle be expressed as a polynomial function of Q_m by merely changing the coefficients:

$$Q_m(y) = ay^3 + by^2 + cy + Q_{m,s}, \tag{8.86}$$

with a, b, and c given by

$$a = \frac{1}{3} L_G^{-3} \left(-16 Q_{m,s} + 32 Q_{m,1} - 32 Q_{m,2} + 16 Q_{m,d}\right), \tag{8.87}$$

$$b = \frac{1}{3}L_G^{-2}\left(32Q_{m,s} - 56Q_{m,1} + 40Q_{m,2} - 16Q_{m,d}\right), \tag{8.88}$$

$$c = \frac{1}{3}L_G^{-1}\left(-19Q_{m,s} + 24Q_{m,1} - 8Q_{m,2} + 3Q_{m,d}\right). \tag{8.89}$$

In each case, the internal points $Q_{m,1}$ and $Q_{m,2}$ hold for the charge densities in accumulation or depletion and are evaluated at $y = 0.25L_G$ and $0.75L_G$, respectively. For such uniform channels, there is no need to calculate any flat-band position along the channel, which further simplifies the analysis.

Having the y dependence of the mobile charge Q_m, the equivalent charges on the nodes used for transcapacitances follows from relations (8.13), (8.14), and (8.15) for uniform channels (whole channel depleted or accumulated) and (8.20) and (8.21) for hybrid channels:

$$\overline{Q}_{m,s} = WL_G\alpha\left(-6Q_{m,s} - 28Q_{m,1} - 12Q_{m,2} + Q_{m,d}\right), \tag{8.90}$$

$$\overline{Q}_{m,d} = WL_G\alpha\left(Q_{m,s} - 12Q_{m,1} - 28Q_{m,2} - 6Q_{m,d}\right), \tag{8.91}$$

with $\alpha = -11/990$.

8.6.3 Evaluation of Flat-Band Position along the Hybrid Channel

From (8.29), (8.30), (8.74), and (8.75), the mobile charge density can be expressed as a function of y, $Q_{m,s}$, and $Q_{m,d}$, respectively. However, as such, the solutions are not suitable to get the equivalent charge densities at the external nodes, as required for the Ward–Dutton scheme. Figure 8.10 shows the positions of the flat-band transition versus the gate potential as obtained from (8.72) or (8.73). As expected, below a critical gate voltage, the whole channel is in depletion mode. This occurs around $V_{GS} = 0.5\,V$ for the device of interest.

Similarly, above a given gate potential (which now depends on the drain voltage), the whole channel can be in accumulation. In between, the hybrid channel exists: when increasing the gate voltage, the flat-band transition will move gradually from the source to the drain ($y = 0$ to L_G).

8.6.4 Evaluation of the Transcapacitances

Before investigating in detail transcapacitances, it is reasonable to see if the assumption regarding the polynomial approximation of the charge density and the partitioning of the channel is sound.

The polynomial approximation of the mobile charge density is compared with numerical simulations for different values of gate potential with $N_D = 1 \times 10^{19}\,cm^{-3}$, $T_{sc} = 20\,nm$, $L_G = 1\,\mu m$, $W = 1\,\mu m$, and $t_{ox} = 1.5\,nm$. As illustrated in Figure 8.11, the mobile charge densities at the internal points (Q_{m1} and Q_{m2}), which are used to build the polynomial function, are modeled accurately in the depleted and accumulated regions of the channel. Further, the polynomial function reproduces the charge distribution along the channel whatever the channel state.

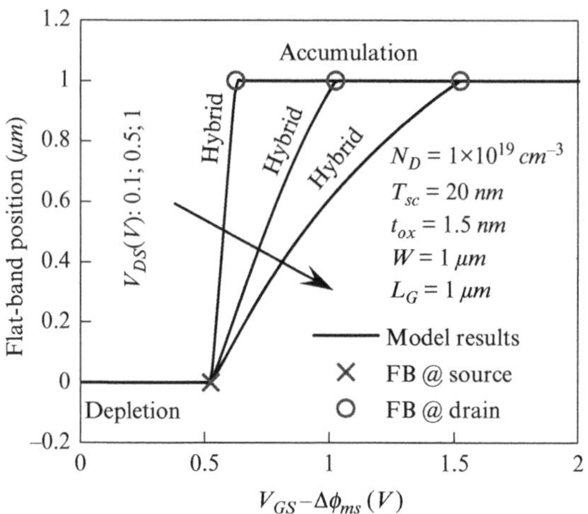

Figure 8.10 Estimation of flat-band position through the channel versus the gate potential ($V_{GS} - \Delta\phi_{ms}$) for different values of V_{DS}, with $N_D = 1 \times 10^{19}\ cm^{-3}$, $T_{sc} = 20\ nm$, $L_G = 1\ \mu m$, $W = 1\ \mu m$, and $t_{ox} = 1.5\ nm$. The potential of the flat-band condition at source and drain is illustrated by the symbols. Note that $\Delta\phi_{ms}$ is the difference between the metal work function and an intrinsic reference semiconductor given by $\Delta\phi_{ms} = W_{ms} - U_T \ln(N_D/n_i)$ where W_{ms} is the metal semiconductor work function difference. Reprinted from [118] with permission.

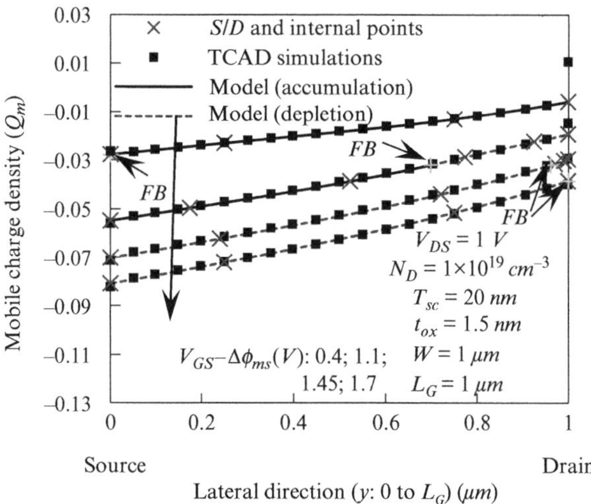

Figure 8.11 Mobile charge density along the channel for different values of gate voltages ($V_{GS} - \Delta\phi_{ms}$) when $V_{DS} = 1\ V$. The dashed lines and black lines are related to the analytical model for depletion and accumulation parts of channel, respectively, and the black symbols are related to the TCAD simulations. The crosses ($\times$) are related to the source, drain, and internal point charge densities ($N_D = 1 \times 10^{19}\ cm^{-3}$, $T_{sc} = 20\ nm$, $L_G = 1\ \mu m$, $W = 1\ \mu m$, and $t_{ox} = 1.5\ nm$). Reprinted from [118] with permission.

From (8.90) and (8.91), in depletion/accumulation the transcapacitances become:

$$C_{SG} = WL_G\alpha\left(\frac{\partial Q_{m,d}}{\partial V_{GS}} - 6\frac{\partial Q_{m,s}}{\partial V_{GS}} - 28\frac{\partial Q_{m,1}}{\partial V_{GS}} - 12\frac{\partial Q_{m,2}}{\partial V_{GS}}\right), \tag{8.92}$$

$$C_{DG} = WL_G\alpha\left(-6\frac{\partial Q_{m,d}}{\partial V_{GS}} + \frac{\partial Q_{m,s}}{\partial V_{GS}} - 12\frac{\partial Q_{m,1}}{\partial V_{GS}} - 28\frac{\partial Q_{m,2}}{\partial V_{GS}}\right). \tag{8.93}$$

In particular, the gate capacitance ($C_{GG} = -C_{SG} - C_{DG}$) is readily obtained:

$$C_{GG} = 5WL_G\alpha\left(\frac{\partial Q_{m,d}}{\partial V_{GS}} + \frac{\partial Q_{m,s}}{\partial V_{GS}} + 8\frac{\partial Q_{m,1}}{\partial V_{GS}} + 8\frac{\partial Q_{m,2}}{\partial V_{GS}}\right). \tag{8.94}$$

Similarly, C_{SD} and C_{DD} are given by:

$$C_{SD} = \frac{\partial \overline{Q}_{m,s}}{\partial V_{DS}} = WL_G\alpha\left(\frac{\partial Q_{m,d}}{\partial V_{DS}} - 28\frac{\partial Q_{m,1}}{\partial V_{DS}} - 12\frac{\partial Q_{m,2}}{\partial V_{DS}}\right), \tag{8.95}$$

$$C_{DD} = \frac{\partial \overline{Q}_{m,d}}{\partial V_{DS}} = WL_G\alpha\left(-6\frac{\partial Q_{m,d}}{\partial V_{DS}} - 12\frac{\partial Q_{m,1}}{\partial V_{DS}} - 28\frac{\partial Q_{m,2}}{\partial V_{DS}}\right), \tag{8.96}$$

$$C_{GD} = 5WL_G\alpha\left(\frac{\partial Q_{m,d}}{\partial V_{DS}} + 8\frac{\partial Q_{m,1}}{\partial V_{DS}} + 8\frac{\partial Q_{m,2}}{\partial V_{DS}}\right). \tag{8.97}$$

Partial Derivatives of the Nodes Charges

Concerning the terms involving $Q_{m,1}$ and Q_{m2}, relations (8.85) and (8.84) lead to:

$$\frac{\partial Q_{m,1}}{\partial V_{GS}} = \frac{\dfrac{1}{4}\dfrac{\partial Q_{m,d}}{\partial V_{GS}}\left.\dfrac{\partial i}{\partial Q_m}\right|_{Q_{m,d}} + \dfrac{3}{4}\dfrac{\partial Q_{m,s}}{\partial V_{GS}}\left.\dfrac{\partial i}{\partial Q_m}\right|_{Q_{m,s}}}{\left.\dfrac{\partial i}{\partial Q_m}\right|_{Q_{m,1}}}, \tag{8.98}$$

$$\frac{\partial Q_{m,2}}{\partial V_{GS}} = \frac{\dfrac{3}{4}\dfrac{\partial Q_{m,d}}{\partial V_{GS}}\left.\dfrac{\partial i}{\partial Q_m}\right|_{Q_{m,d}} + \dfrac{1}{4}\dfrac{\partial Q_{m,s}}{\partial V_{GS}}\left.\dfrac{\partial i}{\partial Q_m}\right|_{Q_{m,s}}}{\left.\dfrac{\partial i}{\partial Q_m}\right|_{Q_{m,2}}}, \tag{8.99}$$

$$\frac{\partial Q_{m,1}}{\partial V_{DS}} = \frac{1}{4}\frac{\dfrac{\partial Q_{m,d}}{\partial V_{DS}}\left.\dfrac{\partial i}{\partial Q_m}\right|_{Q_{m,d}}}{\left.\dfrac{\partial i}{\partial Q_m}\right|_{Q_{m,1}}}, \tag{8.100}$$

$$\frac{\partial Q_{m,2}}{\partial V_{DS}} = \frac{3}{4}\frac{\dfrac{\partial Q_{m,d}}{\partial V_{DS}}\left.\dfrac{\partial i}{\partial Q_m}\right|_{Q_{m,d}}}{\left.\dfrac{\partial i}{\partial Q_m}\right|_{Q_{m,2}}}, \tag{8.101}$$

where $\partial i/\partial Q_m$ represents $\partial F/\partial Q_m$ when the whole channel is in accumulation, or $\partial f/\partial Q_m$ when it is in depletion. These elements have been calculated in Section 8.4. This modeling approach gives data very similar to those obtained in Section 8.2, which is why new plots are omitted.

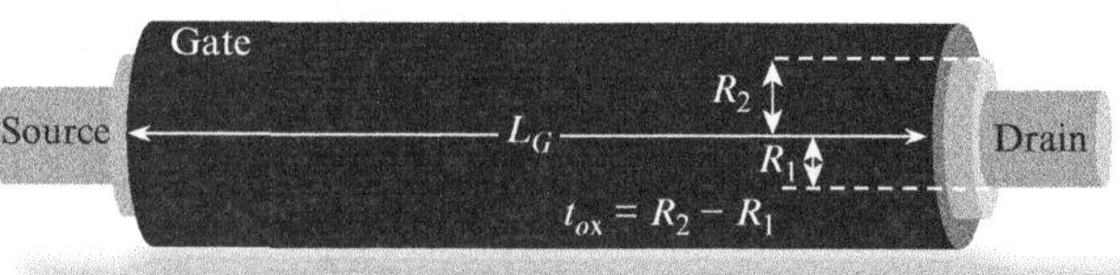

Figure 8.12 Schematic view of an n-type junctionless nanowire FET investigated in this chapter. Reprinted from [134] with permission.

8.7   Transcapacitances in Gate-All-Around Junctionless Nanowire FETs

Following the idea that double-gate and nanowire junctionless FET architectures can share a common model (see Chapter 3) a complete transcapacitance model for the junctionless nanowire FET can also be mapped on the AC core model developed for the junctionless double-gate MOSFET (see Figure 8.12). Note that for channel thicknesses greater than $10\,nm$, quantum mechanical effects can be ignored [154, 155, 158, 159].

8.7.1   Equivalent Parameters Definition

A typical junctionless nanowire FET is shown in Figure 8.12. Here, L_G and $t_{ox} = R_2 - R_1$ represent the gate length and gate-oxide thickness; R_1 and N_D are the semiconductor radius and the dopant density in the channel. As proposed in Chapter 3, the similarities between charge–voltage relationships in double-gate and nanowire junctionless FETs allows us to use the same approach to model transcapacitances in nanowire junctionless FETs [124].

The equivalent model parameters for the double-gate counterpart of a nanowire junctionless FET are given in Table 8.1. Note that the oxide thickness for the equivalent double-gate is given by

$$t_{ox}^{Eq,DG} = \frac{T_{sc}}{2} \ln\left(1 + 2\frac{t_{ox}}{T_{sc}}\right), \tag{8.102}$$

where T_{sc} is given by $2 \times R_1$ and ε_{ox} is the oxide permittivity. As stated in Chapter 3, the main relationships derived for the junctionless double-gate MOSFET case can be used to obtain the mobile charge density at source and drain ($Q_{m,s}$ and $Q_{m,d}$) and also at the specified internal points along the channel (see (8.29), (8.30), (8.74), and (8.75)), which are needed to calculate the channel-charge partitioning. The transcapacitance matrix for the junctionless nanowire FETs are defined as $C_{ij} = \partial \bar{Q}_{m,i}/\partial V_j$, where $\bar{Q}_{m,i}$ hold for the drain, source, and gate and equivalent charge density is defined in (8.13) and (8.14) and (8.15), and V_i is the node potential. Given that the source is used as the reference for the potentials, the matrix is comprised of six transcapacitances only, namely C_{GG}, C_{GD}, C_{DG}, C_{SG}, C_{SD}, and C_{DD}.

Table 8.1. Correspondence between junctionless nanowire physical parameters and equivalent junctionless double-gate FET model parameters

Physical parameters	Nanowire	Equivalent double-gate MOSFET
Radius and thickness	R (radius)	$T_{sc} = 2 \times R$
Oxide thickness	t_{ox}	$\dfrac{T_{sc}}{2} \times \ln\left(1 + 2\dfrac{t_{ox}}{T_{sc}}\right)$
Width	–	$W = \pi \times R$
Doping concentration	N_D	$N_D/2$
Intrinsic carrier concentration	n_i	$n_i/2$

8.7.2 Simulations

Considering an n-type junctionless nanowire FET with $N_D = 2 \times 10^{19}\ cm^{-3}$, $R_1 = 5\ nm$, $L_G = 100\ nm$, and $t_{ox} = 1.5\ nm$, the six transcapacitances (abs. val.) versus the gate-voltage dependence obtained from numerical (TCAD) simulations are plotted in Figure 8.13 (dots) for $V_{DS} = 0\ V$ and $V_{DS} = 1\ V$ (Figure 8.13a and b).

The transcapacitances for the double-gate structure calculated with the equivalent double-gate parameters are plotted in full line.

As can be seen the numerical simulations of the equivalent double-gate FET (squares) are almost identical to the simulations of the original nanowire, especially in linear operation (low V_{DS}). For a higher drain voltage, the mismatch is higher but acceptable and the double-gate model provides good estimation of the different capacitive components.

Transcapacitances in a larger nanowire with $R_1 = 10\ nm$, $N_D = 1 \times 10^{19}\ cm^{-3}$, $L_G = 100\ nm$, and $t_{ox} = 1.5\ nm$ are also shown in Figure 8.14a and b for $V_{DS} = 0\ V$ and $V_{DS} = 1\ V$. Here as well the transcapacitance matrix obtained for the junctionless double-gate FET model with the equivalent parameters can simulate C–V in junctionless nanowire FETs.

In principle, full numerical simulations with equivalent parameters should be the ultimate proof that nanowires and double-gate junctionless FETs can share a common set of capacitance relationships. Numerical TCAD simulations for the equivalent 2D double-gate architectures are carried out, ensuring that in addition to the technological parameters the density of states of the semiconductor is also divided by a factor of two.

The results of these full numerical simulations are represented by the squares in Figure 8.14a and b. This confirms that 2D TCAD simulations can be used in place of 3D TCAD simulations to predict nanowire characteristics. Last, as discussed in Chapter 4, depending on the technological parameters i.e., doping density and silicon thickness, a hole-inversion layer can build up at the channel interface. This inverted hole layer can be modulated by the gate potential, meaning that the gate capacitance will increase again when lowering V_{GS} below some potential.

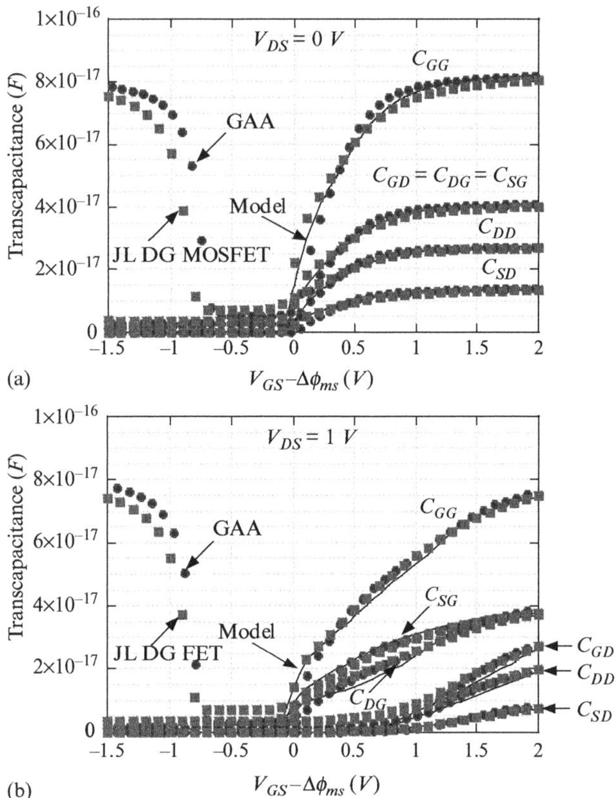

Figure 8.13 Dependence of the transcapacitance versus the gate voltage ($V_{GS} - \Delta\phi_{ms}$) for $V_{DS} = 0$ and $1\,V$ shown in (a) and (b), respectively. The lines and symbols are for the analytical model and TCAD simulations, respectively. The dots and squares are for the TCAD simulations of junctionless gate-all-around (GAA) nanowire FET ($N_D = 2 \times 10^{19}\,cm^{-3}$, $R_1 = 5\,nm$, $L_G = 100\,nm$, and $t_{ox} = 1.5\,nm$) and the equivalent junctionless double-gate FET device ($N_D = 1 \times 10^{19}\,cm^{-3}$, $T_{sc} = 10\,nm$, $L_G = 100\,nm$, $W = 15.7\,nm$, and $t_{ox} = 1.31\,nm$), respectively. Note that $\Delta\phi_{ms}$ is the difference between the metal work function and an intrinsic reference semiconductor. Reprinted from [134] with permission.

Interestingly, even when inversion occurs, the equivalence scheme still holds, but the mismatch is higher.

8.8   A Simplified Approach to Transcapacitance Modeling in Junctionless Nanowire FETs

Based on the long-channel charge based-model described in Section 3.3, explicit expressions can be derived for the intrinsic capacitances of the nanowire junctionless field-effect transistor (see [135]). The assumptions discussed in Section 3.3 end up with two charge–voltage relationships where the quadratic term of the mobile charge

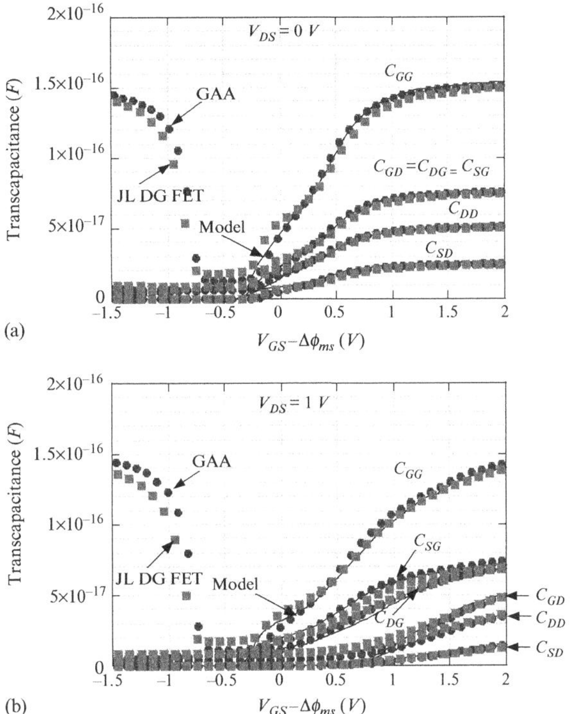

Figure 8.14 Dependence of the transcapacitance versus the gate voltage ($V_{GS} - \Delta\Phi_{ms}$) for $V_{DS} = 0$ and $1\,V$ shown in (a) and (b), respectively. The lines and symbols are for the analytical model and the TCAD simulations, respectively. The dots and squares are for the TCAD simulations of junctionless GAA nanowire FET ($N_D = 1 \times 10^{19}\,cm^{-3}$, $R_1 = 10\,nm$, $L_G = 100\,nm$, and $t_{ox} = 1.5\,nm$) and the equivalent junctionless double-gate FET device ($N_D = 5 \times 10^{18}\,cm^{-3}$, $T_{sc} = 20\,nm$, $L_G = 100\,nm$, $W = 31.4\,nm$, and $t_{ox} = 1.397\,nm$), respectively. Note that $\Delta\phi_{ms}$ is the difference between the metal work function and an intrinsic reference semiconductor. Reprinted from [134] with permission.

density is omitted in depletion mode. As stated in Chapter 3, this limits the model to relatively low doping densities and/or silicon volumes. However, for low doped and thin channels, this approach is acceptable as it results in simpler transcapacitance expressions.

8.8.1   Channel in Depletion Mode

The transcapacitances in depletion mode are obtained from the charge–voltage relations presented in [125] (Section 3.3.1):

$$V_{GS} - \Delta\phi_{ms} - U_T \ln\left(\frac{N_D}{n_i}\right) - V_{ch} + \frac{Q_{fix}}{C_{ox}} = U_T \ln\left(\frac{Q_m}{Q_{cp}}\right) + \frac{Q_m}{C_{ox}}, \tag{8.103}$$

where Q_{cp} is given by $2\varepsilon_{si}U_T/R$.

Next, the total mobile charge density (Q_G) in depletion mode by integrating over the entire channel length is:

$$Q_G = -2\pi R L_G Q_{fix} - 2\pi R \int_0^{L_G} Q_m dy. \tag{8.104}$$

Based on the drift-diffusion, the term dy can be replaced with V_{ch}:

$$I_{DS} dy = -2\pi R \mu Q_m dV_{ch} \Rightarrow dy = -\frac{2\pi R \mu}{I_{DS}} Q_m dV_{ch}, \tag{8.105}$$

and from relation (8.103):

$$dV_{ch} = -U_T \left(\frac{1}{Q_m} + \frac{1}{C_{ox} U_T} \right) dQ_m. \tag{8.106}$$

Combining (8.104), (8.105), and (8.106) results in the total charge density on the gate:

$$Q_G = -2\pi R L_G Q_{fix} - (2\pi R)^2 \frac{\mu}{I_{DS}} U_T \int_{Q_{m,s}}^{Q_{m,d}} Q_m^2 \left(\frac{1}{Q_m} + \frac{1}{C_{ox} U_T} \right) dQ_m$$

$$= -2\pi R L_G Q_{fix} - (2\pi R)^2 \frac{\mu}{I_{DS}} U_T \left(\frac{Q_m^2}{2} + \frac{Q_m^3}{3 C_{ox} U_T} \right) \Bigg|_{Q_{m,s}}^{Q_{m,d}}, \tag{8.107}$$

where $Q_{m,s}$ and $Q_{m,d}$ are the mobile charge densities at source and drain given by (8.103) in depletion mode.

The intrinsic gate capacitance in depletion $C_{GG,dep} = \partial Q_G / \partial V_{GS}$ can be explicitly calculated as soon as the drain current is known, which is obtained after integration of (8.105) from source to drain:

$$I_{DS} = \frac{2\pi R \mu}{L_G} U_T \left(Q_m + \frac{Q_m^2}{2 C_{ox} U_T} \right) \Bigg|_{Q_{m,s}}^{Q_{m,d}} = \frac{2\pi R \mu}{L_G} U_T F(Q_m) \Bigg|_{Q_{m,s}}^{Q_{m,d}}, \tag{8.108}$$

where $F(Q_m)$ is given by

$$F(Q_m) = \left(Q_m + \frac{Q_m^2}{2 C_{ox} U_T} \right). \tag{8.109}$$

Next, integrating (8.105) from source to y leads to

$$I_{DS} = \frac{2\pi R \mu}{y} U_T \left(Q_m + \frac{Q_m^2}{2 C_{ox} U_T} \right) \Bigg|_{Q_{m,s}}^{Q_m} = \frac{2\pi R \mu}{y} U_T F(Q_m) \Bigg|_{Q_{m,s}}^{Q_m}, \tag{8.110}$$

where Q_m is the mobile charge density at y. Comparing (8.108) and (8.110) links y to $Q_m(y)$:

$$\frac{y}{L_G} = \frac{F(Q_m) - F(Q_{m,s})}{F(Q_{m,d}) - F(Q_{m,s})} = 2\pi R \mu U_T \frac{F(Q_m) - F(Q_{m,s})}{I_{DS} L_G}. \tag{8.111}$$

Further, combining (8.105) and (8.108) together with (8.106) gives:

$$\left(\frac{y}{L_G}\right)dy = \frac{(2\pi R \mu U_T)^2}{L_G I_{DS}^2}\left(\frac{1}{Q_m} + \frac{1}{U_T C_{ox}}\right)\left[F(Q_m) - F(Q_{m,s})\right]Q_m dQ_m. \tag{8.112}$$

Following the Ward–Dutton channel charge partitioning scheme, the global source and drain charge densities, as defined in (8.13) and (8.14), can be obtained as follows (note that the width W is replaced with $2\pi R$ for the nanowire):

$$\overline{Q}_{m,d} = -2\pi R \int_0^{L_G} \left(\frac{y}{L_G}\right)Q_m(y)dy$$

$$= \frac{(2\pi R)^3 \mu^2 U_T^2}{L_G I_{DS}^2}\int_{Q_{m,s}}^{Q_{m,d}}\left[F(Q_m) - F(Q_{m,s})\right]Q_m^2\left(\frac{1}{Q_m} + \frac{1}{U_T C_{ox}}\right)dQ_m$$

$$= \frac{(2\pi R)^3 \mu^2 U_T^2}{L_G I_{DS}^2}\int_{Q_{m,s}}^{Q_{m,d}} Q_m^2\left(Q_m - Q_{m,s} + \frac{Q_m^2 - Q_{m,s}^2}{2U_T C_{ox}}\right)\left(\frac{1}{Q_m} + \frac{1}{U_T C_{ox}}\right)dQ_m$$

$$= \frac{(2\pi R)^3 \mu^2 U_T^2}{L_G I_{DS}^2}\left(\frac{Q_m^3}{3} + 3\frac{Q_m^4}{8 C_{ox} U_T} - Q_{m,s}\frac{Q_m^2}{2} - Q_{m,s}\frac{Q_m^3}{3 C_{ox} U_T} + \frac{Q_m^5}{10 C_{ox}^2 U_T^2}\right.$$

$$\left.- Q_{m,s}^2\frac{Q_m^2}{4 C_{ox} U_T} - Q_{m,s}^2\frac{Q_m^3}{6 C_{ox}^2 U_T^2}\right)\Bigg|_{Q_{m,s}}^{Q_{m,d}}, \tag{8.113}$$

$$\overline{Q}_{m,s} = -Q_G - \overline{Q}_{m,d} - 2\pi R L_G Q_{fix}. \tag{8.114}$$

By differentiating $\overline{Q}_{m,s}$, $\overline{Q}_{m,d}$, and Q_G the analytical expressions of transcapacitance matrix elements are obtained in terms of the mobile charge densities at source and drain.

8.8.2 Channel in Accumulation Mode

Relying on the approximate solution described in Section 3.3.2, the mobile charge-density in the accumulation regime of operation is given by

$$V_{GS} - \Delta\phi_{ms} - U_T \ln\left(\frac{N_D}{n_i}\right) - V_{ch} + \frac{Q_{fix}}{C_{ox}} = U_T \ln\left(\frac{Q_m}{Q_{cp}}\left(\frac{Q_m}{Q_{fix}} - 1\right)\right) + \frac{Q_m}{C_{ox}}. \tag{8.115}$$

Assuming $Q_m/Q_{fix} \gg 1$ in accumulation regime, relation (8.115) simplifies into

$$V_{GS} - \Delta\phi_{ms} - U_T \ln\left(\frac{N_D}{n_i}\right) - V_{ch} + \frac{Q_{fix}}{C_{ox}} = U_T \ln\left(\frac{Q_m^2}{Q_{cp} Q_{fix}}\right) + \frac{Q_m}{C_{ox}}. \tag{8.116}$$

In addition, differentiating (8.115) leads to an analytical expression of dV_{ch} as a function of dQ_m:

$$dV_{ch} = -U_T\left(\frac{2}{Q_m} + \frac{1}{C_{ox} U_T}\right)dQ_m. \tag{8.117}$$

Following the same idea as for the depletion mode:

$$Q_G = -2\pi R L_G Q_{fix} - (2\pi R)^2 \frac{\mu}{I_{DS}} U_T \left(Q_m^2 + \frac{Q_m^3}{3 C_{ox} U_T} \right) \Bigg|_{Q_{m,s}}^{Q_{m,d}},$$ (8.118)

and the drain current I_{DS} in accumulation is given by

$$I_{DS} = \frac{2\pi R \mu}{L_G} U_T \left(2Q_m + \frac{Q_m^2}{2 C_{ox} U_T} \right) \Bigg|_{Q_{m,s}}^{Q_{m,d}} = \frac{2\pi R \mu}{L_G} U_T G(Q_m) \Bigg|_{Q_{m,s}}^{Q_{m,d}},$$ (8.119)

where $G(Q_m)$ is given by

$$G(Q_m) = \left(2Q_m + \frac{Q_m^2}{2 C_{ox} U_T} \right),$$ (8.120)

and $Q_{m,s}$ and $Q_{m,d}$ are the charge carrier densities at source and drain from relation (8.115).

The intrinsic capacitance in accumulation $C_{GG,acc} = \partial Q_G / \partial V_{GS}$ is obtained through the derivative of total mobile charge density with respect to gate potential. Similar to Section 8.8.1, integrating the transport equation from source to y gives:

$$I_{DS} = \frac{2\pi R \mu}{y} U_T \left(2Q_m + \frac{Q_m^2}{2 C_{ox} U_T} \right) \Bigg|_{Q_{m,s}}^{Q_m} = \frac{2\pi R \mu}{y} U_T G(Q_m) \Bigg|_{Q_{m,s}}^{Q_m},$$ (8.121)

where Q_m is the mobile charge density at y in the depletion region of operation. Again, imposing the same current in (8.119) and (8.121), an additional link between y and the mobile charge-densities Q_m is obtained:

$$\frac{y}{L_G} = \frac{G(Q_m) - G(Q_{m,s})}{G(Q_{m,d}) - G(Q_{m,s})} = 2\pi R \mu U_T \frac{G(Q_m) - G(Q_{m,s})}{I_{DS} L_G}.$$ (8.122)

Merging (8.105), (8.117), and (8.122) leads to

$$\left(\frac{y}{L_G} \right) dy = \frac{(2\pi R \mu U_T)^2}{L_G I_{DS}^2} \left(\frac{2}{Q_m} + \frac{1}{U_T C_{ox}} \right) [G(Q_m) - G(Q_{m,s})] \, Q_m dQ_m.$$ (8.123)

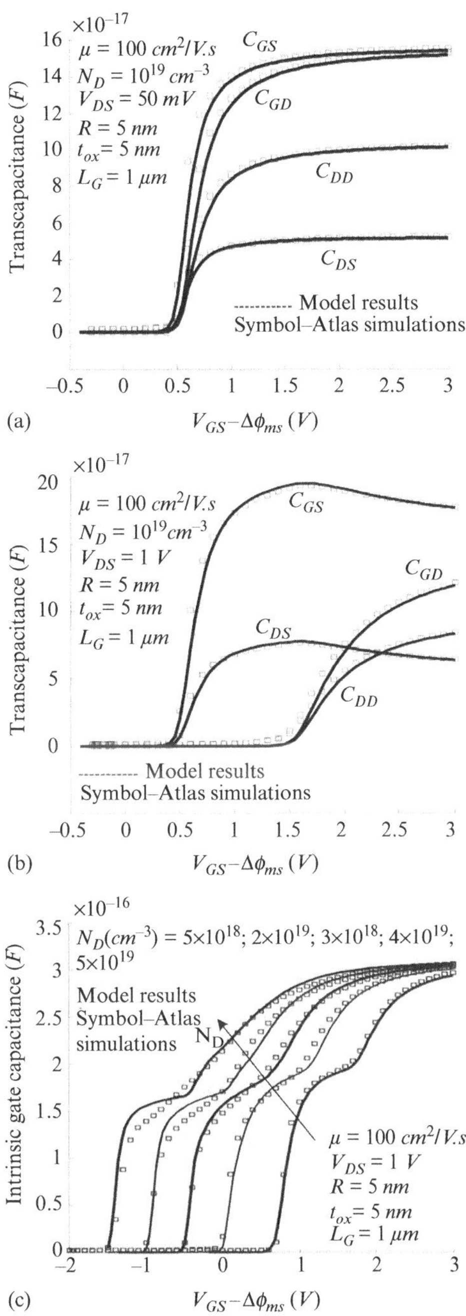

Figure 8.15 Dependence of the transcapacitance versus the gate voltage ($V_{GS} - \Delta\phi_{ms}$) for $V_{DS} = 0$ and $1\ V$ shown in (a) and (b), respectively, and intrinsic gate capacitance (C_{GG}) for $V_{DS} = 1$ for different values of channel-doping concentration illustrated in (c). The lines and symbols are for the analytical model and TCAD simulations, respectively. Note that $\Delta\phi_{ms}$ is the difference between the metal work function and an intrinsic reference semiconductor. Reprinted from [135] with permission.

The Ward–Dutton charge partitioning is used to calculate the equivalent charge on the drain:

$$
\begin{aligned}
\overline{Q}_{m,d} &= 2\pi R \int_0^{L_G} \left(\frac{y}{L_G}\right) Q_m(y)\,dy \\[2mm]
&= \frac{(2\pi R)^3 \mu^2 U_T^2}{L_G I_{DS}^2} \int_{Q_{m,s}}^{Q_{m,d}} [G(Q_m) - G(Q_{m,s})]\, Q_m^2 \left(\frac{2}{Q_m} + \frac{1}{U_T C_{ox}}\right) dQ_m \\[2mm]
&= \frac{(2\pi R)^3 \mu^2 U_T^2}{L_G I_{DS}^2} \int_{Q_{m,s}}^{Q_{m,d}} Q_m^2 \left(2[Q_m - Q_{m,s}] + \frac{Q_m^2 - Q_{m,s}^2}{2 U_T C_{ox}}\right)\left(\frac{2}{Q_m} + \frac{1}{U_T C_{ox}}\right) dQ_m \\[2mm]
&= \frac{(2\pi R)^3 \mu^2 U_T^2}{L_G I_{DS}^2} \left(\frac{4 Q_m^3}{3} + \frac{3 Q_m^4}{4 C_{ox} U_T} - 2 Q_{m,s} Q_m^2 - 2 Q_{m,s}\frac{Q_m^3}{3 C_{ox} U_T} \right. \\[2mm]
&\qquad\qquad \left. \left. + \frac{Q_m^5}{10\,(C_{ox} U_T)^2} - Q_{m,s}^2 \frac{Q_m^2}{2 C_{ox} U_T} - Q_{m,s}^2 \frac{Q_m^3}{6\,(C_{ox} U_T)^2}\right)\right|_{Q_{m,s}}^{Q_{m,d}}
\end{aligned}
$$

$$\tag{8.124}$$

The equivalent charge on the source node follows:

$$
\overline{Q}_{m,s} = -Q_G - \overline{Q}_{m,d} - 2\pi R L_G Q_{fix}. \tag{8.125}
$$

These equivalent charges are used to derive the transcapacitance matrix elements in either depletion or accumulation modes of operation. Figure 8.15 shows some transcapacitances in accumulation and depletion. The total capacitance model is given by merging the capacitances in depletion and in accumulation modes though an hyperbolic tangent function with dedicated fitting parameters [135]. The TCAD simulations in a long-channel device with a channel length of $L_G = 1\,\mu m$, a silicon film radius of $R = 5\,nm$, a gate-oxide thickness of $t_{ox} = 5\,nm$, and a doping density of $N_D = 10^{19}\,cm^{-3}$ provide good agreement with the model for C_{GD}, C_{GS}, C_{DD}, and C_{DS} as functions of the gate voltage and for different values of the drain–source voltage [135].

8.9   Summary

In this chapter, three different approaches were proposed to calculate the transcapacitance elements in junctionless FETs in double-gate and nanowire geometries. The Ward–Dutton partitioning scheme was adopted together with the drift-diffusion transport equations. The different models cover depletion and accumulation modes and are valid in all regions of operation, from linear to saturation. Complete capacitance networks were then obtained and constituted a major step toward a coherent compact model for circuit simulation based on junctionless double-gate MOSFETs.

9 Modeling Asymmetric Operation of Double-Gate Junctionless FETs

Junctionless double-gate MOSFETs can exhibit either symmetric (SDG-MOSFET) or asymmetric (ADG-MOSFET) modes of operation. Symmetric operation presumes that both gates have the same technological parameters and share the same potential $V_{GS} - \Delta\phi_{ms}$, whereas asymmetric operation arises when gates, electrodes, and insulators are not identical and/or are biased at different gate potentials. Most modeling efforts have been devoted to simulating symmetric operation and obtaining information on drain current [80,84,86,90,92,118,146], subthreshold swing, short-channel effects [110,111,160], threshold voltage [110,160,161], neglecting the asymmetric mode [82, 102], which, as a matter of fact, is most likely to take place in real devices.

Indeed, symmetric operation is unlikely since architectures are scarcely symmetric in general. Biasing the gates independently offers more flexibility in terms of circuit operation (tuning the threshold voltage of one gate using the counter gate for instance).

Few researchers have tackled this challenging problem. The model proposed in [82] generalizes the depletion approximation by including the exponential contribution of the Boltzmann statistics through linearization of the charge density in the Poisson–Boltzmann equation. The approach derived in [102] derives the subthreshold current in asymmetric and symmetric junctionless double-gate MOSFETs.

The main issue in modeling asymmetric junctionless double-gate FETs is that under asymmetric operation, the center of the channel is no longer the coordinate where the electric field vanishes. This lack of symmetry makes any analytical handling of the Poisson–Boltzmann equation challenging. Exploiting the analysis developed in Chapter 3 for symmetric operation, this chapter proposes to model the asymmetric operation in junctionless double-gate MOSFETs as a combination of two virtual symmetric devices.

9.1 General Considerations in Asymmetric Junctionless Double-Gate FETs

A typical n-type doped long-channel ($L_G = 100\,nm$) junctionless asymmetric double-gate MOSFET (JL ADG MOSFET) is illustrated in Figure 9.1a. Here, L_G is the gate length, T_{sc} and N_D are the semiconductor thickness and channel doping density, t_{ox1} and t_{ox2} are the oxide thicknesses for each gate, and $\Delta\phi_{ms1}$ and $\Delta\phi_{ms2}$ hold for the

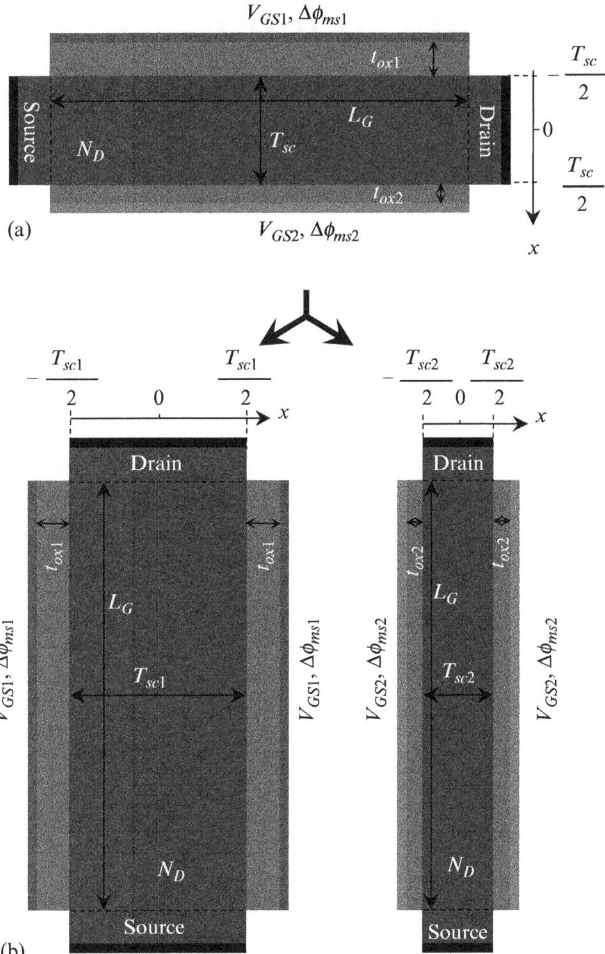

Figure 9.1 (a) n-type junctionless asymmetric double-gate MOSFET and (b) two junctionless symmetric double-gate MOSFETs with equivalent physical parameters and gate potentials in order to obtain the same potential distribution as for the asymmetric case (b). Reprinted from [119] with permission.

difference between the metal work functions and the semiconductor Fermi potential when referenced to the intrinsic Fermi potential, $\Delta\phi_{ms1,2} = W_{ms1,2} - U_T \ln(N_D/n_i)$.

The asymmetry in the potential and in the charge distribution can have different roots. The structure itself can exhibit an intrinsic asymmetry due to the technology, such as a difference in the gate electrode work functions or in the gate capacitance. However, asymmetry can also be obtained through different biasing of the gate electrodes, giving more flexibility to circuit designers.

The Poisson equation combined with nondegenerate Boltzmann statistics gives:

$$\frac{\partial^2 \Psi}{\partial x^2} = \frac{q n_i}{\varepsilon_{si}} \left[\exp\left(\frac{\Psi - V_{ch}}{U_T}\right) - \frac{N_D}{n_i} \right], \tag{9.1}$$

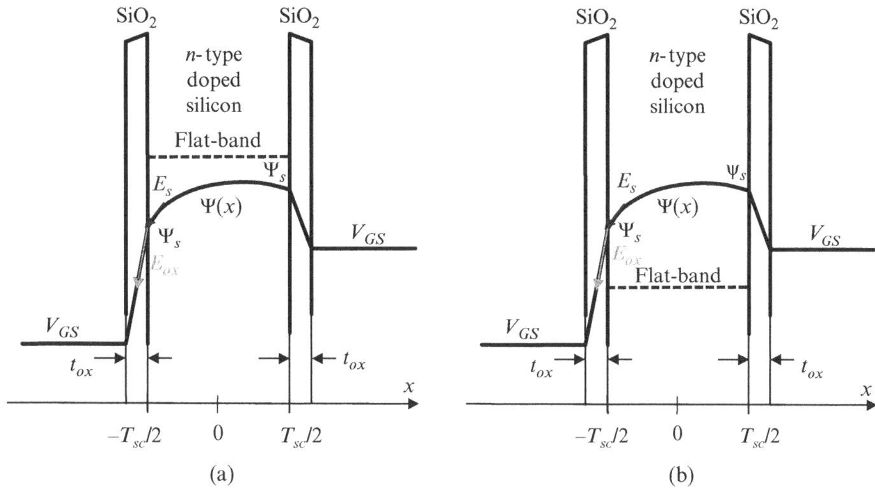

Figure 9.2 (a) Potential distribution between two gates when $\Psi_s < \Psi_{FB}$, which leads to $V_{GS} < V_{GS,FB}$. (b) Potential distribution across the channel when V_{GS} is imposed to be less than $V_{GS,FB}$. In this case Ψ_s must be less than Ψ_{FB} meaning that the situation in (b) cannot happen.

where x is the coordinate across the two gates ($-T_{sc}/2 < x < T_{sc}/2$), U_T is the thermal voltage, n_i and ε_{si} are the silicon intrinsic carrier density and permittivity, and V_{ch} is the shift in quasi-Fermi potential for the electrons due to the channel potential. It should be noted that Ψ and V_{ch} are evaluated with respect to the intrinsic Fermi potential of the semiconductor at the source and is set at 0 V. In an asymmetric configuration, each gate has a well-defined flat-band voltage V_{GSFB1} and V_{GSFB2} such that $V_{GSFB1,2} = V_{ch} + \Delta\phi_{ms1,2} + U_T \ln(N_D/n_i)$ [86, 146], where $\Delta\phi_{ms1,2}$ hold for the difference between the metal work function and the intrinsic Fermi potential so that $\Delta\phi_{ms1,2} = W_{ms1,2} - U_T \ln(N_D/n_i)$. Depending on the applied voltages, different situations can be encountered.

Forbidden modes of operation: Before proceeding, it is useful to note how the surface potential and the gate voltage are linked to each other under different conditions. For instance, when the surface potential is lower than the flat-band potential ($\Psi_s < \Psi_{FB}$), the channel is in depletion and $Q_{sc} > 0$, $\partial^2\Psi(x)/\partial x^2 < 0$ (see the potential distribution in Figure 9.2a). In addition, the continuity of the displacement vector at the semiconductor–insulator interface requires $V_{GS} < \Psi_s$ and consequently $V_{GS} < V_{GS,FB}$.

In contrast, when $V_{GS} < V_{GS,FB}$, if it is assumed that $\Psi_s > \Psi_{FB}$, then the potential distribution across two gates would be the same as in Figure 9.2b. This assumption ($\Psi_s > \Psi_{FB}$) means accumulation mode and therefore $Q_{sc} < 0$. As a result $\partial^2\Psi(x)/\partial x^2 > 0$, in contrast to the potential depicted in Figure 9.2b. Therefore, the condition $V_{GS} < V_{GS,FB}$ is equivalent to $\Psi_s < \Psi_{FB}$.

The channel in depletion: The channel operating in depletion mode is the most common situation and corresponds to $V_{GS1} < V_{GSFB1}$ and $V_{GS2} < V_{GSFB2}$. The potential distribution is shown in Figure 9.3a. Here, the potential is chosen in such a way that there is an extremum i.e., maximum in this case noted x_{ext} so that $\partial \Psi(x)/\partial x|_{x=x_{ext}} = 0$. The extrema can be inside or outside the channel limits (see Figure 9.3a and e, respectively). An extremum between the gates is likely to happen for relatively highly doped channels.

The channel in accumulation: When both gate voltages are above the flat-band voltage i.e., $V_{GS1} > V_{GSFB1}$ and $V_{GS2} > V_{GSFB2}$ the channel between the gates is set to accumulation (see Figure 9.3c).

The channel in mixed-mode: Between these regimes, depletion and accumulation can coexist across the gates (not to be mislead with what was called hybrid operation in symmetric double-gate junctionless FETs). This can happen when the surface potentials verify $\Psi_{s2} < \Psi_{FB} < \Psi_{s1}$, as illustrated in Figure 9.3f.

When modeling asymmetric double-gate junctionless MOSFETs, those different situations require dedicated approaches as will be developed in the following sections.

9.2   Analysis Restricted to Depletion or Accumulation

9.2.1   Potential Extremum Inside the Channel

As stated in Chapter 3, relation (9.1) does not have a closed analytical solution. However, it is still possible to simplify the problem by analyzing the position of the potential extremum (if any).

When the extremum lies inside the channel, the electric fields at each channel/insulator interface are obtained by integrating (9.1) from $-T_{sc}/2$ to x_{ext}, then from x_{ext} to $T_{sc}/2$, while noting that the electric field vanishes at x_{ext}:

$$E_{s1,2}^2 = \frac{2qn_iU_T}{\varepsilon_{si}} \left[\exp\left(\frac{\Psi_{s1,2} - V_{ch}}{U_T} \right) - \exp\left(\frac{\Psi_{ext} - V_{ch}}{U_T} \right) - \frac{N_D}{n_iU_T}(\Psi_{s1,2} - \Psi_{ext}) \right]. \quad (9.2)$$

Here, $E_{s1,2}$ and $\Psi_{s1,2}$ are the surface electric fields and the surface potentials at $-T_{sc}/2$ and $T_{sc}/2$. In the very special case of symmetric operation, $x_{ext} = 0$ and $\Psi_{s1} = \Psi_{s2}$, which gives the charge-based solution developed in Chapter 3. But relation (9.2) can be grasped as the solution of two symmetric devices ($DG1$ and $DG2$ in Figure 9.1b) electrically connected at the extrema. That is, each should have a channel thickness given by

$$T_{sc1} = |T_{sc} + 2x_{ext}|, \quad (9.3)$$

$$T_{sc2} = |T_{sc} - 2x_{ext}|. \quad (9.4)$$

Absolute values make definitions consistent when the extremum lies outside of the channel. This intuitive picture is now analyzed in detail. Figure 9.3a and c exhibits an asymmetric shape in the potential distribution, whereas Figure 9.3b and d illustrates their equivalent symmetric counterparts. Each virtual device has its own equivalent parameters (T_{sc}, t_{ox}, and $\Delta\phi_{ms}$). Since adopting this equivalence results in the same surface potentials as for the asymmetric case, the charge densities can also

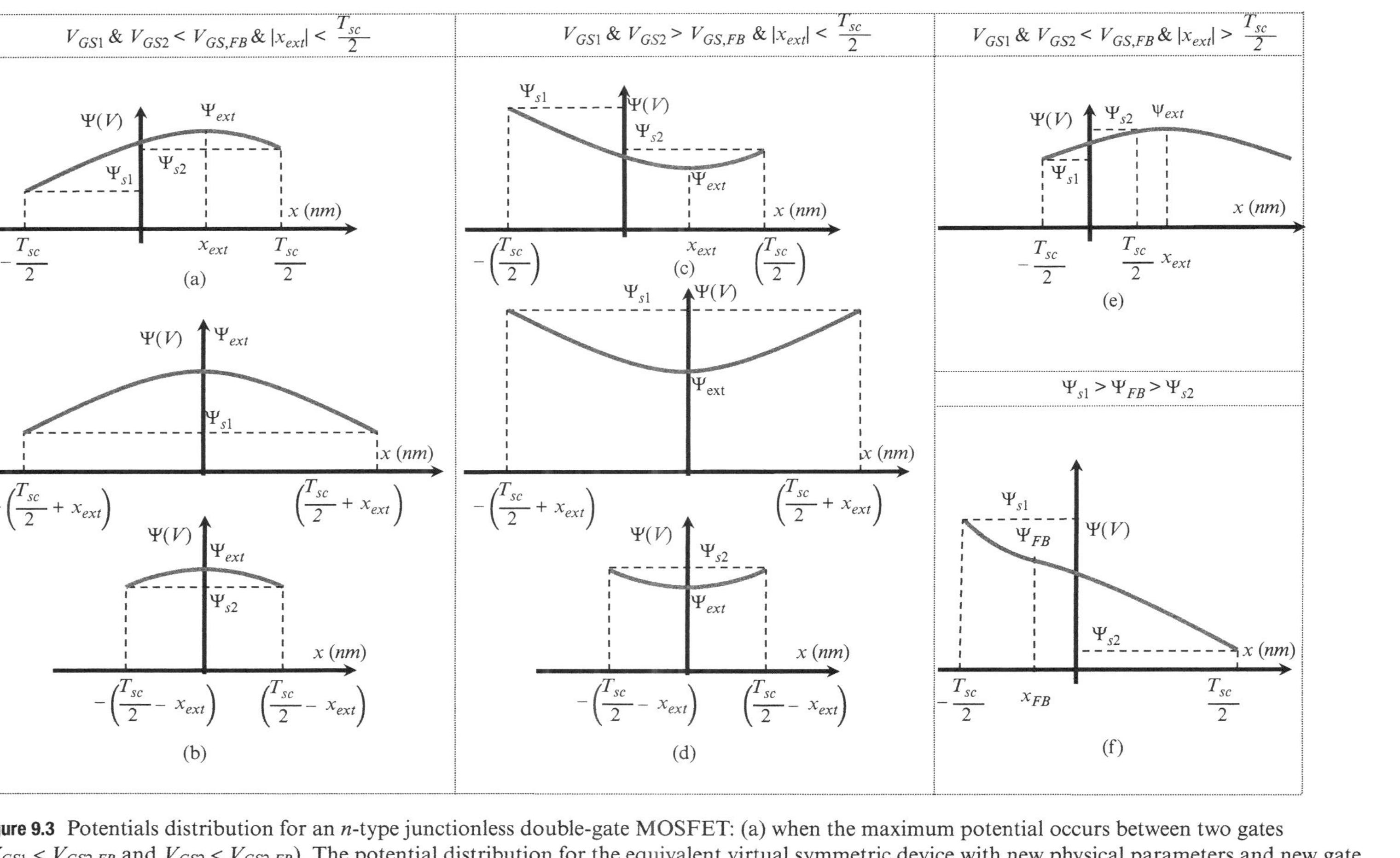

Figure 9.3 Potentials distribution for an *n*-type junctionless double-gate MOSFET: (a) when the maximum potential occurs between two gates ($V_{GS1} < V_{GS2,FB}$ and $V_{GS2} < V_{GS2,FB}$). The potential distribution for the equivalent virtual symmetric device with new physical parameters and new gate potentials is shown in (b). For the case when the gate potentials are higher than $V_{GS1,FB}$ and $V_{GS2,FB}$ (accumulation mode), the potential distribution across the two gates in the asymmetric device and in the equivalent symmetric are as shown in (c) and (d), respectively. A similar situation is shown in (e), but with the extremum outside of the channel. Case (f) is when $\Psi_{s1} > \Psi_{FB} > \Psi_{s2}$ i.e., the device is partly in depletion and partly in accumulation. In this case, there is no extremum between the two gates. Reprinted from [119] with permission.

be accurately calculated. A transformation where an asymmetric device is mapped onto a virtual symmetric one is shown in Figure 9.1.

As in Chapter 3, introducing the second-order MacLaurin series (around $x = 0$) of the potential distribution across two gates gives:

$$\Psi(x) = \Psi_{ext} + \frac{qn_i}{2\varepsilon_{si}} x^2 \left[\exp\left(\frac{\Psi_{ext} - V_{ch}}{U_T} \right) - \frac{N_D}{n_i} \right]. \tag{9.5}$$

Since the extremum is not at $x = 0$ in virtual devices, Taylor expansion of $\Psi(x - x_{ext})$ around x_{ext} is used. From (9.5), the surface potentials are obtained:

$$\Psi_{s1,2} = \Psi_{ext} + \frac{qn_i}{8\varepsilon_{si}} (T_{sc1,2})^2 \left[\exp\left(\frac{\Psi_{ext} - V_{ch}}{U_T} \right) - \frac{N_D}{n_i} \right]. \tag{9.6}$$

Similarly, from the continuity of the displacement vector at the silicon/insulator interface, the surface electric fields are linked to the potential drops across the gate insulators:

$$E_{s1,2}^2 = \left(-\frac{Q_{sc1,2}}{2\varepsilon_{si}} \right)^2 = \left[\frac{\varepsilon_{ox}}{\varepsilon_{si} t_{ox1,2}} (V_{GS1,2} - \Psi_{s1,2} - \Delta\phi_{ms1,2}) \right]^2, \tag{9.7}$$

where $t_{ox1,2}$ and $Q_{sc1,2}$ are the insulator thicknesses for each gate and the total charge densities. For a given set of gates and channel potentials, relations (9.2), (9.3), (9.6), and (9.7) on one side and relations (9.2), (9.4), (9.6), and (9.7) on the other side generate two equations with two unknowns (Ψ_{ext} and x_{ext}). Even though these can be used to extract the charge densities with respect to the applied potentials, numerical calculations are still needed, which is not desirable in compact modeling.

Instead, the charge density in asymmetric double-gate junctionless MOSFETs can be obtained using simple relationships as proposed for the symmetric case (see Chapter 3) that are contained in the two master equations for depletion and accumulation:

$$V_{GS} - V_{ch} - \Delta\phi_{ms} - U_T \ln\left(\frac{N_D}{n_i} \right) \overset{dep}{\approx} -\frac{Q_{sc}^2}{8qN_D\varepsilon_{si}} - \frac{Q_{sc}}{2C_{ox}} + U_T \ln\left[1 - \left(\frac{Q_{sc}}{Q_{fix}} \right)^2 \right], \tag{9.8}$$

$$V_{GS} - V_{ch} - \Delta\phi_{ms} - U_T \ln\left(\frac{N_D}{n_i} \right) \overset{acc}{\approx} -\frac{Q_{sc}}{2C_{ox}} + U_T \ln\left(1 + \frac{Q_{sc}^2}{8qN_D\varepsilon_{si}U_T} \right). \tag{9.9}$$

When the device is divided into two symmetric junctionless FETs, the quantity $V_{GS} - \Delta\phi_{ms}$ is either $V_{GS1} - \Delta\phi_{ms1}$ or $V_{GS2} - \Delta\phi_{ms2}$, where T_{sc} represents T_{sc1} or T_{sc2}. Since only "half" of each device is used (see Figure 9.3), the total charge density is half the sum of the charge density obtained for virtual symmetric devices:

$$Q_{sc}^{asym} = C_{ox1}(\Psi_{s1} + \Delta\phi_{ms1} - V_{GS1}) + C_{ox2}(\Psi_{s2} + \Delta\phi_{ms2} - V_{GS2}). \tag{9.10}$$

Relations (9.8) to (9.10) constitute a set of equations that can be solved a priori. However, whereas the silicon thicknesses T_{sc1} and T_{sc2} are absent in accumulation mode (see relation (9.9)) they still occur in depletion mode (9.8). This peculiarity requires extra analysis.

Given that each virtual device should share the same potential extremum, subtracting the surface potentials given by (9.6) for the two gates gives $\Psi_{s2} - \Psi_{s1}$. Then using (9.7) twice together with (9.3) and (9.4), an important relationship between the extrema coordinate x_{ext} and the potentials is obtained:

$$\Psi_{s2} - \Psi_{s1} = -\frac{qT_{sc}x_{ext}n_i}{\varepsilon_{si}}\left[\exp\left(\frac{\Psi_{ext} - V_{ch}}{U_T}\right) - \frac{N_D}{n_i}\right]$$

$$= (V_{GS2} - \Delta\phi_{ms2}) - (V_{GS1} - \Delta\phi_{ms1}) + \frac{Q_{sc2}}{2C_{ox2}} - \frac{Q_{sc1}}{2C_{ox1}}. \tag{9.11}$$

In addition, the extremum potential can be extracted by calculating $(\Psi_{s2} - \Psi_{ext})/(\Psi_{s1} - \Psi_{ext})$ from (9.6) together with the definitions of the virtual thicknesses obtained in (9.7). It follows that the extremum potential is given by

$$\Psi_{ext} = \frac{(V_{GS2} - \Delta\phi_{ms2}) - \eta\,(V_{GS1} - \Delta\phi_{ms1}) + \dfrac{Q_{sc2}}{2C_{ox2}} - \eta\dfrac{Q_{sc1}}{2C_{ox1}}}{(1 - \eta)}, \tag{9.12}$$

where η is defined by $(T_{sc2}/T_{sc1})^2$. Substituting the value in (9.11) and combining it with (9.8) either with T_{sc1} or T_{sc2}, the problem ends up with three equations and three unknowns i.e., Q_{sc1}, Q_{sc2}, and x_{ext}.

The case where the whole semiconductor layer across the gates is in accumulation is the simplest situation: since the high carrier density at the channel interface acts as a shield, the total charge density obtained from the charge-based model will be almost independent of the virtual silicon thicknesses and the device will essentially operate as two independent channels. This is already shown in relation (9.9). The real thickness T_{sc} can then be used for T_{sc1} and T_{sc2} without affecting the solution.

9.2.2 Extremum Potential Outside the Channel

When the channel is low-doped in regard to its thickness, the extremum potential may not be not inside but instead outside of the channel (see Figure 9.3e; note that such a situation is unlikely in highly doped junctionless FET devices). From relations (9.6) and (9.7), the surface potentials can be calculated:

$$\Psi_{s1} = \Psi_{ext} + \frac{qn_i}{8\varepsilon_{si}}T_{sc1}^2\left[\exp\left(\frac{\Psi_{ext} - V_{ch}}{U_T}\right) - \frac{N_D}{n_i}\right] = V_{GS1} - \Delta\phi_{ms1} + \frac{Q_{sc1}}{2C_{ox1}}, \tag{9.13}$$

$$\Psi_{s2} = \Psi_{ext} + \frac{qn_i}{8\varepsilon_{si}}T_{sc2}^2\left[\exp\left(\frac{\Psi_{ext} - V_{ch}}{U_T}\right) - \frac{N_D}{n_i}\right]$$

$$= V_{GS2} - \Delta\phi_{ms2} - \frac{qn_i}{2C_{ox2}}T_{sc2}\left[\exp\left(\frac{\Psi_{ext} - V_{ch}}{U_T}\right) - \frac{N_D}{n_i}\right]. \tag{9.14}$$

In (9.13), Q_{sc1} is obtained from the charge-based model (9.8) for a given value for T_{sc1}. This in turn gives Ψ_{ext}. Next, differentiating the potential in (9.5) gives the electric field at the channel interface. Once combined with the continuity of the displacement vector, this process will link Ψ_{s2} to V_{GS2} (see relation (9.14)).

Finally, introducing the extremum potential (Ψ_{ext}) in (9.14) gives the surface potential at gate-2 (Ψ_{s2}). Therefore, combining relations (9.8), (9.13), and (9.14) generates three equations with three unknowns i.e., Ψ_{ext}, x_{ext}, and Q_{sc1}. Following this approach, the mobile charge density for each gate is obtained by integrating the related parabolic potential across the gates.

9.2.3   The Iterative Solution

An alternative approach consists of using an initial value for x_{ext} discussed as follows. The case where the extremum potential is between the gates is the most likely. Initially, x_{ext} is obtained from (9.11) with Q_{sc1} and Q_{sc2} calculated with (9.8) while assigning to T_{sc1} and T_{sc2} the real semiconductor thickness i.e., $T_{sc1} = T_{sc2} = T_{sc}$. Next, the value of x_{ext} is used to calculate the new silicon thicknesses T'_{sc1} and T'_{sc2}, which in turn will give the updated charge densities Q'_{sc1} and Q'_{sc2} based on relation (9.8). In practice, good accuracy is obtained after some iterations, and a single loop may be sufficient in many situations. The method has been assessed with TCAD-Synopsys simulations for a device with $T_{sc} = 10\,nm$, $N_D = 1 \times 10^{19}\,cm^{-3}$, $L_G = 100\,nm$, and $t_{ox} = 1.5\,nm$ ($V_{ch} = 0\,V$).

9.2.4   Simulations

Figure 9.4a shows the potential distribution for asymmetric operation in depletion ($V_{GS1} - \Delta\phi_{ms1} = -1\,V$, $V_{GS2} - \Delta\phi_{ms2} = -0.5\,V$) and in accumulation modes ($V_{GS1} - \Delta\phi_{ms1} = 1.5\,V$, $V_{GS2} - \Delta\phi_{ms2} = 0.7\,V$) where a potential extremum is located between the two gates. The surface potentials as well as the minimum potential and position estimated from the iterative scheme are supported by numerical simulations.

Extremum Inside the Channel

When the whole channel (between two gates) operates in depletion i.e., $V_{GS1} - \Delta\phi_{ms1} = -1\,V$, $V_{GS2} - \Delta\phi_{ms2} = -0.5\,V$, the iteration method leads to two virtual junctionless symmetric devices with $T_{sc1} = 6.6\,nm$ and $V_{GS1} - \Delta\phi_{ms1} = -1\,V$, and $T_{sc2} = 13.4\,nm$ with $V_{GS2} - \Delta\phi_{ms2} = -0.5\,V$, respectively. Conversely, when the whole channel across two gates operates in accumulation i.e., $V_{GS1} - \Delta\phi_{ms1} = 1.5\,V$, $V_{GS2} - \Delta\phi_{ms2} = 0.7\,V$ the virtual devices thicknesses are $T_{sc1} = 11\,nm$ with $V_{GS1} - \Delta\phi_{ms1} = 1.5\,V$, and $T_{sc2} = 9\,nm$ with $V_{GS2} - \Delta\phi_{ms2} = 0.7\,V$.

Extremum Outside the Channel

When the extremum lies outside of the channel, the surface and extremum potentials obtained from the charge-based approach relying on relations (9.8), (9.13), and (9.14) are compared with numerical simulations on Figure 9.4b, for a device where $T_{sc} = 10\,nm$, $N_D = 5 \times 10^{18}\,cm^{-3}$, $L_G = 100\,nm$, and $t_{ox} = 1.5\,nm$. Numerical (dashed line) and analytical (solid line and symbols) simulations reveal that the potential distribution across the gates is well predicted with $V_{GS1} - \Delta\phi_{ms1} = -0.7\,V$ and $V_{GS2} - \Delta\phi_{ms2} = 0.4\,V$.

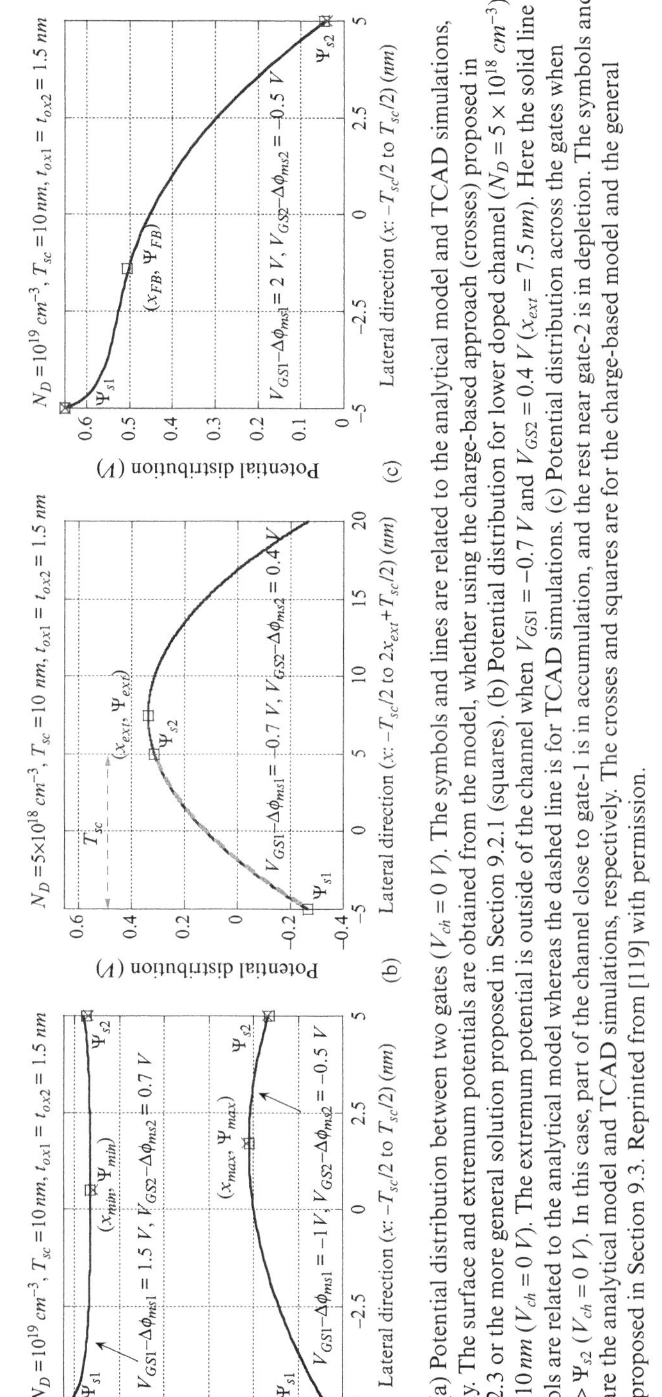

Figure 9.4 (a) Potential distribution between two gates ($V_{ch} = 0\,V$). The symbols and lines are related to the analytical model and TCAD simulations, respectively. The surface and extremum potentials are obtained from the model, whether using the charge-based approach (crosses) proposed in Section 9.2.3 or the more general solution proposed in Section 9.2.1 (squares). (b) Potential distribution for lower doped channel ($N_D = 5 \times 10^{18}\,cm^{-3}$) with $T_{sc} = 10\,nm$ ($V_{ch} = 0\,V$). The extremum potential is outside of the channel when $V_{GS1} = -0.7\,V$ and $V_{GS2} = 0.4\,V$ ($x_{ext} = 7.5\,nm$). Here the solid line and symbols are related to the analytical model whereas the dashed line is for TCAD simulations. (c) Potential distribution across the gates when $\Psi_{s1} > \Psi_{FB} > \Psi_{s2}$ ($V_{ch} = 0\,V$). In this case, part of the channel close to gate-1 is in accumulation, and the rest near gate-2 is in depletion. The symbols and solid line are the analytical model and TCAD simulations, respectively. The crosses and squares are for the charge-based model and the general approach proposed in Section 9.3. Reprinted from [119] with permission.

9.3   Coexistence of Depletion and Accumulation

So far, the analysis based on virtual devices invoked the existence of an extremum potential, which happens when the semiconductor across the gates is either in depletion or accumulation (still the extremum can be inside or outside the channel). This happens as soon as the channel between the gates combines depletion and accumulation at the same time (at a given coordinate between source and drain). When this condition is met, the device cannot be modeled relying on Section 9.2 and another strategy is required.

9.3.1   Modeling Depleted–Accumulated Channels

The case where depletion and accumulation coexists between the gates is shown in Figure 9.3f. This happens when $\Psi_{s2} < \Psi_{FB}$ and $\Psi_{s1} > \Psi_{FB}$. Since the surface potential changes from depletion at drain to accumulation at source, there is a particular coordinate (x_{FB}) where it equals the flat-band potential i.e., $\Psi(x_{FB}) = \Psi_{FB} = V_{ch} + U_T \ln(N_D/n_i)$. In addition, since at flat-band the local charge density is zero, this particular locus also satisfies $\partial^2 \Psi(x)/\partial x^2 \big|_{x=x_{FB}} = 0$. Integrating the relation (9.1) from $x_1 = -T_{sc}/2$ to x_{FB}, then from x_{FB} to $x_2 = T_{sc}/2$, the surface electric fields $E_{s1,2}$ can be given by

$$E_{s1,2}^2 - E_{FB}^2 = \frac{2qn_i U_T}{\varepsilon_{si}} \left\{ \exp\left(\frac{\Psi_{s1,2} - V_{ch}}{U_T} \right) - \frac{N_D}{n_i} - \frac{N_D}{n_i U_T} \left[\Psi_{s1,2} - V_{ch} - U_T \ln\left(\frac{N_D}{n_i} \right) \right] \right\}.$$

(9.15)

Note that the electric field E_{FB} at flat-band comes into play, whereas the electric field at the extremum was zero in the symmetric cases. Following the same method as in Chapter 4, the Laplacian in (9.1) is replaced by considering the first four terms (instead of two for a symmetric operation) of a Taylor series around the flat-band position (x_{FB}). After some manipulation, the electric field at flat-band E_{FB} and the surface potentials are linked:

$$\Psi_{s1,2} = V_{ch} + U_T \ln\left(\frac{N_D}{n_i} \right) + E_{FB} T_{sc1,2} \left[1 + \frac{qN_D T_{sc1,2}^2}{6\varepsilon_{si} U_T} \right],$$

(9.16)

where the virtual semiconductor thicknesses $T_{sc1,2}$ are given by

$$T_{sc1} = T_{sc}/2 + x_{FB},$$

(9.17)

$$T_{sc2} = T_{sc}/2 - x_{FB}.$$

(9.18)

Combining (9.7), (9.15), and (9.16) with (9.17) for gate-1, and (9.7), (9.15), and (9.16) with (9.18) for gate-2 gives three equations with three unknown i.e., Ψ_{s1}, Ψ_{s2}, and x_{FB} for given gates and channel potentials. Therefore, this approach can be used to calculate charges (using $E_{s1,2}$) and potentials through a system of equations. Further simplification is possible as detailed in the next section where a different approach is discussed.

9.3.2 Simplifying Assumptions

When depletion and accumulation occur between the gates, virtual symmetric devices are meaningless concepts since there no extremum. However, in some situations one can simplify the analysis that mitigates this conclusion. First, accumulation in symmetric operation supports that the semiconductor thickness has a negligible impact on the charge–voltage relationship, see relation (9.9), which targets a virtual junctionless FET in accumulation. Next, the potential shape happens to be almost "flat" near the accumulated layer. These elements show that the rest of the channel reverts to a symmetric junctionless FET in depletion and that relation (9.8) can be exploited (in depletion) while using $2 \times T_{sc}$ for the silicon thickness.

Note Q_{sc1} and Q_{sc2} as the total charge densities for the equivalent symmetric devices obtained from (9.8) and (9.9) (with $2 \times T_{sc}$) and the mobile charge density for the asymmetric depleted–accumulated mode is given by

$$Q_m^{asym} = \frac{1}{2}(Q_{sc1,acc}^{sym} + Q_{sc2,dep}^{sym}) - qN_D T_{sc}. \tag{9.19}$$

The soundness of this analysis is now analyzed in detail. The surface and the extremum potentials obtained with relations (9.15) to (9.18) and the equivalent virtual devices based on relations (9.8) and (9.9) are compared to numerical TCAD simulations in Figure 9.4c (squares and crosses, respectively).

9.3.3 Assessment of Continuity at the Transition Coordinates

As discussed, when gate potentials are below/above the flat-band voltages the extremum potential is inside or outside the channel. Transition between these modes must satisfy continuity conditions. Assuming that the extremum potential i.e., where the electric field vanishes is at the Si/SiO$_2$ interface for the gate-2 implies that $\Psi_{s2} = \Psi_{ext} = V_{GS2} - \Delta\phi_{ms2}$ and $T_{sc1} = 2 \times T_{sc}$ for the virtual device. From relation (9.6) the surface potential at gate-1 is given by

$$\Psi_{s1} = V_{GS2} - \Delta\phi_{ms2} + \frac{qn_i}{2\varepsilon_{si}} T_{sc}^2 \left[\exp\left(\frac{V_{GS2} - \Delta\phi_{ms2} - V_{ch}}{U_T} \right) - \frac{N_D}{n_i} \right]. \tag{9.20}$$

Next, imposing Ψ_{s1} in (9.2) and combining with (9.7) gives V_{GS1}^*, which guarantees that the extremum potential is at gate-2. Here, it should be noted that if $V_{GS2} < V_{GS,FB2}$ and $V_{GS1} < V_{GS1}^*$, then the maximum potential is outside of the channel. If $V_{GS1}^* < V_{GS1} < V_{GS,FB1}$, the maximum potential occurs between the gates.

However, the challenge is when $V_{GS2} > V_{GS,FB2}$, which builds an accumulation layer. According to (9.9) the total charge density at gate-2 is not a function of T_{sc}, and so the virtual device (DG2) is electrically isolated from gate-1. As a result, the minimum potential will not lay outside of the channel (if the minimum had to stay between the gates, we would need $V_{GS1,2} > V_{GS,FB1,2}$).

Consequently, a maximum potential can take place inside or outside of the channel, but a minimum potential is expected to appear only between the gates.

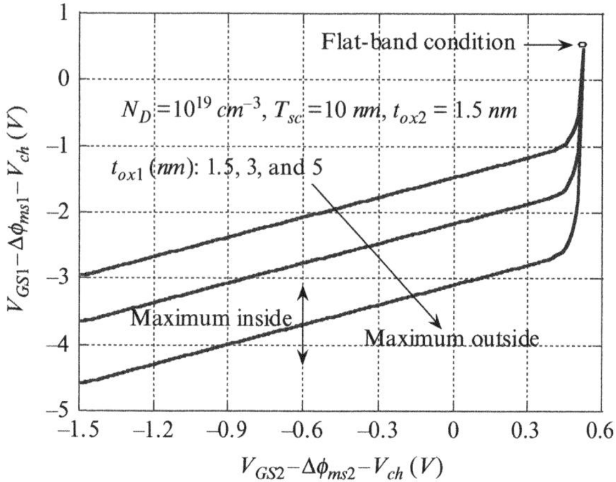

Figure 9.5 Transition of the extremum potential inside or outside of the channel arises when gates are not equivalent and/or biased at different potentials (note that $\Delta\phi_{ms}$ is the difference between the metal work function and an intrinsic reference semiconductor given by $\Delta\phi_{ms} = W_{ms} - U_T \ln(N_D/n_i)$ where W_{ms} is the metal semiconductor work function difference.) Reprinted from [119] with permission.

Finally, it is worth noting that crossing the flat-band between the gates results in coexistence of depletion and accumulation, which leads to an inflection point of potential profile across two gates ($\partial^2 \Psi / \partial x^2 = 0$). To make it clearer, the frontier between extremum inside/outside of the channel is plotted as a function of the gate voltages for different values of t_{ox1} in Figure 9.5. It is very instructive to see how this transition follows an almost linear dependence with respect to the gate voltages, unless we approach the flat-band condition.

9.4   Derivation of the Current

The charge density in accumulated, depleted, and mixed channels has been derived for a given channel potential. However, this quantity varies continuously from source to drain, which means that the channel changes gradually and so does the virtual symmetric junctionless FET thickness. The current can no longer be mapped on a single pair of equivalent devices; instead, it must be calculated by numerical integration along the channel. Adopting drift-diffusion transport, the current is given by

$$I_{DS} = -W\mu Q_m(V_{ch})\frac{dV_{ch}}{dy}, \tag{9.21}$$

where μ is the carrier mobility (assumed constant along the channel), y is the coordinate along the channel, W is the silicon width, and V_{ch} is the shift in quasi-Fermi potential. A simple finite difference based on channel segmentation into n elements, where the local channel potential is assumed constant and given by $\Delta V_{ch} = V_{DS}/n$ and the mobile charge density in each cell is given by $Q_m(V_{ch})$, leads to

$$\Delta y^i I_{DS} = -W\mu Q_m(V_{ch}^i)\Delta V_{ch}, \tag{9.22}$$

where i represents the number of each cell and $Q_m(V^i_{ch})$ corresponds to the mobile charge density obtained from the virtual symmetric devices. After summation over i, the drain current is

$$I_{DS} = -\mu \frac{W}{L_G} \frac{V_{DS}}{n} \sum_{i=1}^{n} Q_m(V^i_{ch}). \tag{9.23}$$

9.5 Simulations

9.5.1 Potential-Induced Asymmetry

Figure 9.6a displays the mobile charge density and the charge density for each of the gates in a $10\,nm$ silicon channel doped at $1 \times 10^{19}\,cm^{-3}$ for different values of the effective gate potentials $V^* = V_{GS} - \Delta\phi_{ms} - V_{ch}$. The agreement with the numerical simulations indicates that the concept of the virtual device obeying the charge-based equations (9.8) and (9.9) is sound.

The situations corresponding to depletion, accumulation, and mixed states are illustrated in Figure 9.6a in linear and logarithmic scales. Even when considering a thicker device with $T_{sc} = 15\,nm$ doped at $5 \times 10^{18}\,cm^{-3}$, the virtual symmetric device concept merged with the symmetric charge-based model gives good results for different combinations of gates potentials (see Figure 9.6b).

9.5.2 Structural Asymmetry

In practice, due to the inherent fabrication process, double-gate devices have different gate insulator thicknesses and/or different work functions. Such technological asymmetry is already included in the analytical expressions (see relation (9.7)).

The asymmetry originating from the mismatch in the gate capacitances is accurately predicted (see Figure 9.6c). The variation of mobile charge density with respect to effective gate potentials is plotted in linear and logarithmic scales for $t_{ox1} = 1.5\,nm$ and $t_{ox2} = 10\,nm$. This asymmetry originates from the buried oxide layer in an SOI-based junctionless MOSFET such as reported in [82]. As can be seen, the mobile charge density is modified by the effective back-gate potential (V_{GS2}).

All possible conditions corresponding to $|x_{ext}| > T_{sc}/2$, $|x_{ext}| < T_{sc}/2$, and $|x_{FB}| < T_{sc}/2$ are illustrated in Figure 9.7 for $N_D = 5 \times 10^{18}\,cm^{-3}$ and $10\,nm$ silicon thickness. Also, the position of extremum potential versus effective potential is well captured by the model.

Figure 9.8 shows the drain current versus the gate potential at low ($V_{DS} = 0.1\,V$) and high ($V_{DS} = 1\,V$) drain potential for $10\,nm$ silicon thickness doped at $1 \times 10^{19}\,cm^{-3}$ ($n = 5$) and for different values of V_{GS2}. Both in linear and logarithmic scales, the agreement between model and the numerical simulations from deep depletion to accumulation is satisfactory.

9.5.3 Approximate Expression for the Current

It can be shown that when the device is working in full depletion (mobile charge density negligible with respect to the doping), the position of extremum potential does

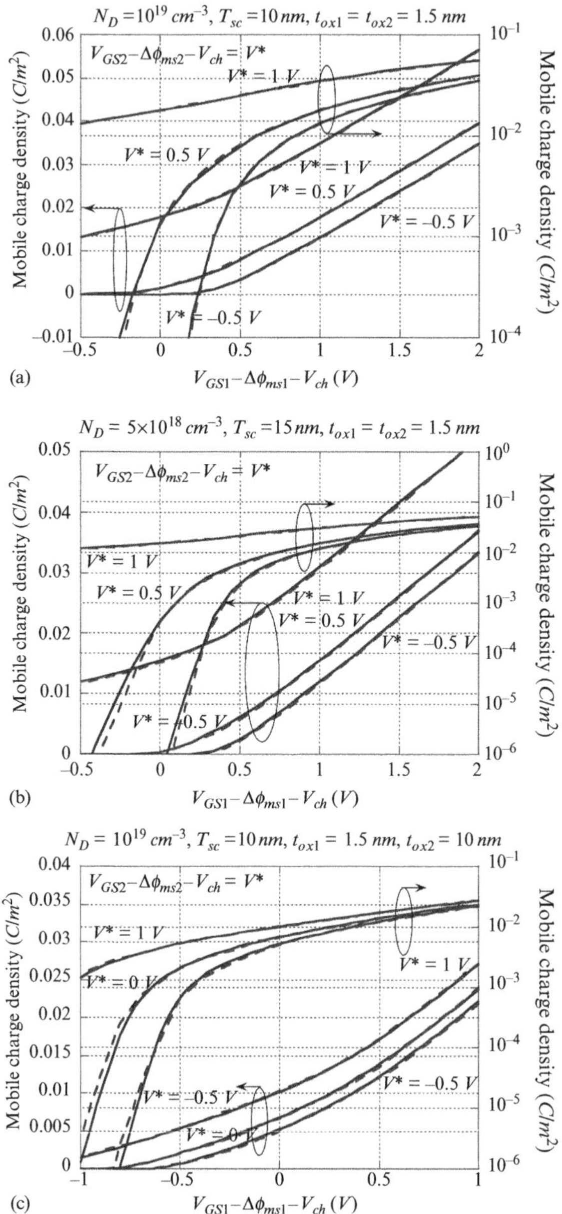

Figure 9.6 Mobile charge density (C/m^2) with respect to the effective gate potentials in linear and logarithmic scales for different junctionless asymmetric double-gate MOSFET devices. (a) $N_D = 10^{19}\ cm^{-3}$, $T_{sc} = 10\ nm$, $t_{ox1} = t_{ox2} = 1.5\ nm$, (b) $N_D = 5 \times 10^{18}\ cm^{-3}$, $T_{sc} = 15\ nm$, $t_{ox1} = t_{ox2} = 1.5\ nm$, and (c) $N_D = 10^{19}\ cm^{-3}$, $T_{sc} = 10\ nm$, $t_{ox1} = 1.5\ nm$, $t_{ox2} = 10\ nm$. The dashed lines and solid lines are related to the charge-based model and TCAD simulations, respectively (note that $\Delta\phi_{ms}$ is the difference between the metal work function and an intrinsic reference semiconductor given). Reprinted from [119] with permission.

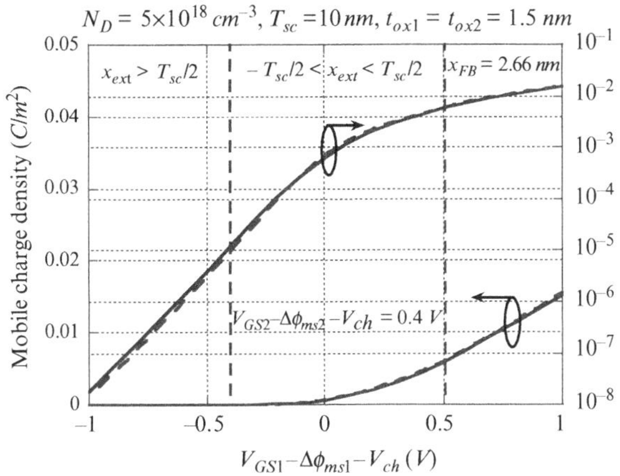

Figure 9.7 Mobile charge density (C/m^2) with respect to the effective gate potential in linear and logarithmic scales. The dashed lines and solid lines are related to the charge-based model and TCAD simulations, respectively (note that $\Delta\phi_{ms}$ is the difference between the metal work function and an intrinsic semiconductor reference). Reprinted from [119] with permission.

not change significantly through the channel with respect to the channel potential, meaning that discretization of the channel is no longer needed to obtain the current.

Indeed, the mobile charge density at source and drain for each virtual device is twice the "real" drain current, which represents an essential simplification for AC analysis of junctionless FETs [118, 134].

9.5.4   Limitations

As discussed in Chapter 4 for symmetric devices, when the gate potential is lowered below a critical value (for an n-type channel), an inversion layer is created at the channel interface. For this reason, this asymmetric model is not suitable to predict the charge density in junctionless double-gate MOSFETs under inversion. When the gate potential is set to "$-1\,V$," a hole density as large as $8 \times 10^{19}\,cm^{-3}$ can be seen in the TCAD simulations, which cannot be predicted from the current work.

9.6   Summary

In this chapter, asymmetric operation was addressed in terms of virtual symmetric devices. Based on the analysis of the potential profile, each equivalent symmetric device was characterized by a silicon thickness that was a function of the gate and pseudo-Fermi voltages. This approach is a great simplification with respect to the analytical solution of asymmetric potential and reuses the charge-based model developed for symmetric junctionless FETs.

The electrically induced as well the inherent technological asymmetry aspects were accurately taken into account. This approach is particularly suitable for modeling junctionless FETs with different gate-oxide thicknesses in Ultra Thin Body

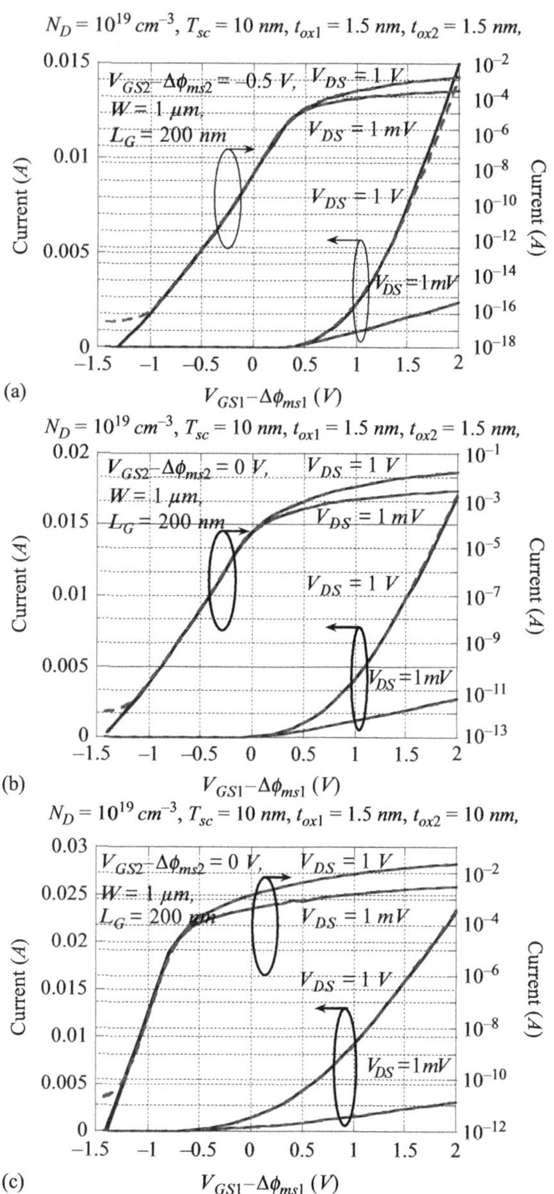

Figure 9.8 Drain current versus gate potential in a linear and saturated regime in a $10\,nm$ silicon thickness junctionless ADG MOSFET doped at $1 \times 10^{19}\,cm^{-3}$ for different values of drain potential. The solid lines and dashed lines are related to the model and TCAD simulations, respectively. The asymmetry in gate-oxide thickness and gate potential is shown in (c). Reprinted from [119] with permission.

Silicon On Insulator (UTBSOI) technology. Additionally, this serves as the basis for a complete capacitance network for asymmetric junctionless MOSFETs based on developments that will be detailed in Chapter 8 [118, 134].

10 Modeling Noise Behavior in Junctionless FETs

In addition to modeling DC and AC characteristics of junctionless devices [37,86,111,118,119,122,124,134,146,162], modeling thermal noise is another fundamental parameter in circuit design. Relying on the charge-based relationships presented in Chapter 3 together with the transcapacitance approach developed in Chapter 8, explicit models for thermal noise, induced gate noise (IGN), and cross-correlation noise in junctionless double-gate MOSFETs in all regions of operation are derived in this chapter and compared with inversion-mode FETs.

10.1 Thermal-Noise Modeling

Noting that $S_{\Delta V_{ch}}$ is the thermal-noise power spectral density (PSD) of the channel voltage, the drain-current thermal noise $dS_{\Delta I_{DS}}$ is obtained from

$$dS_{\Delta I_{DS}} = g_{ch}^2 \times dS_{\Delta V_{ch}}, \tag{10.1}$$

where $g_{ch} = dI_{DS}/dV_{ch}$ is the local channel conductance and V_{ch} is the shift in quasi-Fermi potential due to the channel potential. Relying on drift-diffusion transport, g_{ch} and $dS_{\Delta V_{ch}}$ are given by $g_{ch} = -\mu Q_m W / L_G$ and $dS_{\Delta V_{ch}} = -4k_B T dy / W \mu Q_m$.

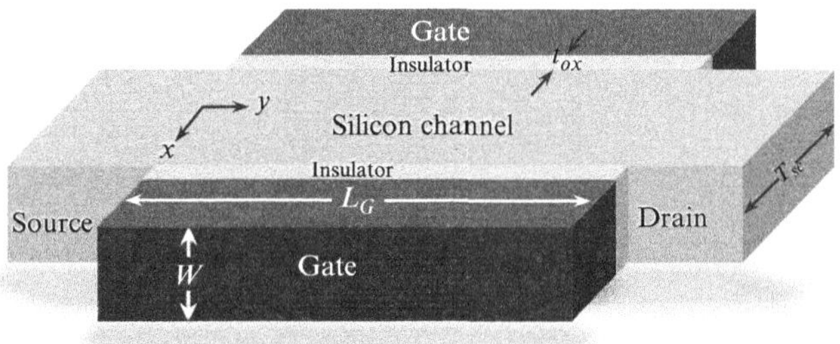

Figure 10.1 Schematic view of an n-type junctionless double-gate MOSFET.

The drain-current PSD gives

$$S_{\Delta I_{DS}} = -\frac{1}{L_G^2} \int_0^{L_G} 4k_B T \mu W Q_m(y) dy. \qquad (10.2)$$

This expression is very general and holds for junctionless and inversion-mode MOSFETs. Alternatively, relying on charge neutrality

$$Q_G + W \int_0^{L_G} Q_m dy + W L_G Q_{fix} = 0. \qquad (10.3)$$

Relation (10.2) can be rewritten in terms of the global charge density on the gate Q_G:

$$S_{\Delta I_{DS}} = -\frac{1}{L_G^2} 4k_B T \mu W \int_0^{L_G} Q_m(y) dy = \frac{1}{L_G^2} 4k_B T \mu \left(Q_G + W L_G Q_{fix} \right). \qquad (10.4)$$

As for transcapacitance, the channel can be in a depletion, accumulation, or hybrid state. For each case, an analytical expression for Q_G is obtained as follows:

$$Q_G = -W L_G Q_{fix} - W \int_0^{L_G} Q_m dy = -W L_G Q_{fix} - W \int_{Q_{m,s}}^{Q_{m,FB}} Q_m |_{acc} \, dy - W \int_{Q_{m,FB}}^{Q_{m,d}} Q_m |_{dep} \, dy$$

$$= -W L_G Q_{fix} + \frac{W^2 \mu}{I_{DS}} \int_0^{V_{ch,FB}} Q_m^2 |_{acc} \, dV_{ch} + \frac{W^2 \mu}{I_{DS}} \int_{V_{ch,FB}}^{V_{DS}} Q_m^2 |_{dep} \, dV_{ch}$$

$$= -W L_G Q_{fix} + \frac{W^2 \mu}{I_{DS}} \left(I_{acc} + I_{dep} \right), \qquad (10.5)$$

where

$$I_{acc} = \int_0^{V_{ch,FB}} Q_m^2(y) |_{acc} \, dV_{ch}$$

$$= -\int_{Q_{m,s}}^{-Q_{fix}} Q_m \left[-\frac{Q_m}{2C_{ox}} + 2U_T Q_m \frac{Q_m + Q_{fix}}{8U_T Q_{fix} C_{si} + \left(Q_m + Q_{fix} \right)^2} \right] dQ_m$$

$$= \frac{Q_m^3}{6C_{ox}} - Q_m^2 U_T - \alpha \log \left(Q_m^2 + 2Q_m Q_{fix} + Q_{fix}^2 + 8C_{si} U_T Q_{fix} \right) + 2Q_m Q_{fix} U_T$$

$$- 8 \sqrt{2C_{si} Q_{fix}^3 U_T^3} \arctan \left(\frac{8 \sqrt{2C_{si} Q_{fix}^5 U_T^3} + 8Q_m \sqrt{2C_{si} Q_{fix}^3 U_T^3}}{2Q_{fix}^3 U_T - 2Q_{fix}\alpha + 16C_{si} Q_{fix}^2 U_T^2} \right) \Bigg|_{Q_{m,s}}^{-Q_{fix}}, \qquad (10.6)$$

$$I_{dep} = \int_{V_{ch,FB}}^{V_{DS}} Q_m^2(y)\big|_{dep}\, dV_{ch}$$

$$= -\int_{-Q_{fix}}^{Q_{m,d}} Q_m \left[-\frac{Q_m\left(Q_m + Q_{fix}\right)}{4qN_D\varepsilon_{si}} - \frac{Q_m}{2C_{ox}} + 2U_T\frac{Q_m + Q_{fix}}{Q_m + 2Q_{fix}} \right] dQ_m$$

$$= \frac{W^2\mu}{I_{DS}}\left[\frac{Q_m^4}{16qN_D\varepsilon_{si}} + \frac{Q_m^3 Q_{fix}}{12qN_D\varepsilon_{si}} + \frac{Q_m^3}{6C_{ox}} \right.$$

$$\left. - U_T Q_m^2 + 2U_T Q_m Q_{fix} + 4U_T Q_{fix}^2 \ln\left(\frac{1}{Q_m + 2Q_{fix}}\right) \right]\Bigg|_{-Q_{fix}}^{Q_{m,d}} , \tag{10.7}$$

with $\alpha = Q_{fix}^2 U_T - 8C_{si}Q_{fix}U_T^2$.

Therefore, combining (10.5) and (10.4) gives

$$S_{\Delta I_{DS}} = \frac{W^3\mu^2}{I_{DS}L_G^2} 4k_B T\left(I_{acc} + I_{dep}\right). \tag{10.8}$$

The PSD of the channel thermal noise for an n-type junctionless double-gate MOS-FET with $N_D = 1 \times 10^{19}\ cm^{-3}$, $T_{sc} = 20\ nm$, $L_G = 1\ \mu m$, $W = 1\ \mu m$, and $t_{ox} = 1.5\ nm$ is plotted in Figure 10.2 and compared to an inversion-mode FET with the same physical parameters and $N_A = 1 \times 10^{14}\ cm^{-3}$. Figure 10.2a represents the PSD of the (thermal) current noise obtained with respect to the effective gate voltage for different values of the drain potential ($V_{DS} = 10\ mV$ and $V_{DS} = 1\ V$). Note that in order to illustrate the channel thermal noise with respect to the drain current and g_m/I_{DS}, V_{DS} was set to a low, but nonzero, value) and channel-doping concentrations ($N_D = 7 \times 10^{18}\ cm^{-3}$, $N_D = 1 \times 10^{19}\ cm^{-3}$). The lines and symbols hold for the analytical model and numerical TCAD simulations, respectively.

The thermal noise predicted by the charge-based model is accurate in all regions of operation. In addition, since fundamental differences exist between a junctionless and inversion-mode double-gate MOSFET i.e., the charge–voltage dependence is different [37], comparing the PSD between these devices is not relevant when using the gate voltage as a reference. The PSD of the channel thermal noise is analyzed with respect to the drain current and to the transconductance-to-current ratio g_m/I_{DS}, a useful metric for circuit design.

Figure 10.2b and c shows the $S_{\Delta I_{DS}}$ versus the current and g_m/I_{DS} for a junctionless double-gate MOSFET and inversion-mode double-gate MOSFET (with the same technological parameters, except for the doping). Interestingly, when adopting this representation, it can be seen that *for a given drain voltage* there is almost no difference for the thermal noise spectral density between the junctionless and inversion-mode FETs *biased at the same current*. Thus, while these devices are different, the thermal noise in junctionless FETs can be predicted from the noise models of inversion-mode MOSFETs provided a representation in terms of the current is used. In fact, this observation is unsurprising since the mobile charge density needed

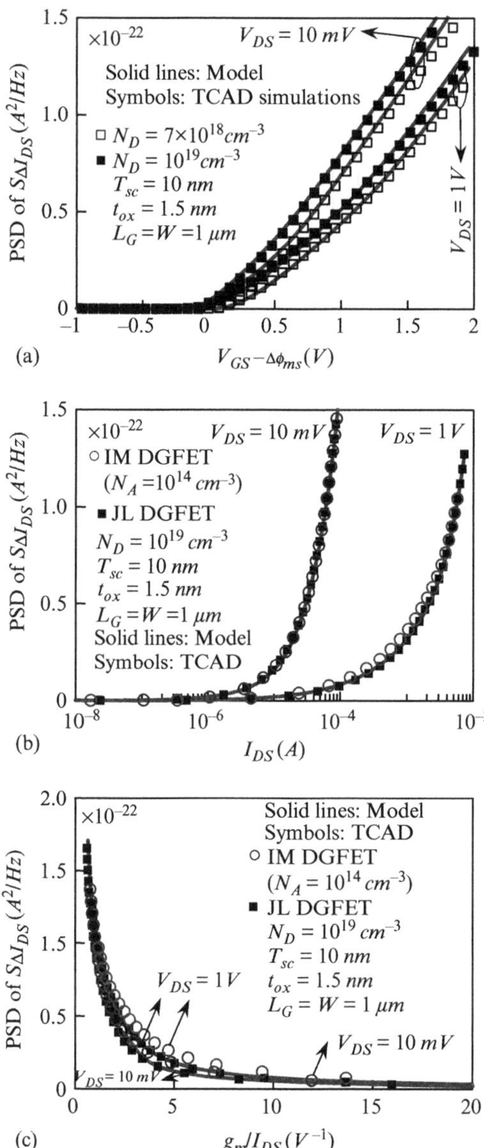

Figure 10.2 Thermal-noise PSD of the drain-current fluctuation ($S_{\Delta I_{DS}}$) (A^2/Hz) versus (a) effective gate voltage, (b) drain current, and (c) g_m/I_{DS} for different values of drain potential ($V_{DS} = 10\,mV$ and $1\,V$) and channel-doping concentration ($N_D = 7 \times 10^{18}\,cm^{-3}$ and $1 \times 10^{19}\,cm^{-3}$) in junctionless double-gate MOSFETs and inversion-mode double-gate MOSFETs ($N_A = 10^{14}\,cm^{-3}$). Reprinted from [163] with permission.

to carry a given current is almost the same in inversion-mode and junctionless FETs (with the same width and length). Indeed, assuming that the quasi-Fermi potential along the channel in junctionless and inversion-mode double-gate MOSFETs have almost the same derivatives, for a given drain current, the mobile charge densities will be the same as well. Then, the channel thermal noise PSD in junctionless and inversion-mode MOSFETs can be expected to be almost equal.

10.2   Induced Gate Noise in Junctionless FET

In radio frequency (RF) applications, random potential fluctuations resulting from the channel noise are strongly coupled to the gate terminal through the gate capacitance, causing the so-called IGN ($S_{\Delta_{IG}}$) [164–167]. In principle, IGN in junctionless MOSFETs can be modeled by taking advantage of the explicit solution for the total mobile charge density.

The PSD of IGN versus drain current in junctionless and inversion-mode FETs at $100\,MHz$ is obtained from numerical TCAD simulations and shown in Figure 10.3 for different values of the drain potential. Again, the IGN is plotted versus the drain current is used. The IGN is lower in junctionless FETs than in inversion-mode FETs, whatever the operating point, despite the PSD of the channel thermal noise being almost the same.

Thus, while the channel noise looks almost identical for inversion-mode and junctionless devices operating at the same current, this is not the case for the IGN. In fact, IGN is affected not only by the mobile charge density profile, but also by the equivalent gate capacitance, which can be quite different since in junctionless FETs the depletion region is modulated and acts as a variable capacitance.

The PSD of the IGN is shown in Figure 10.4 with respect to the effective gate potential at $V_{DS} = 10\,mV$ and $V_{DS} = 1\,V$ and for different technological parameters.

Interestingly, numerical TCAD simulations show that IGN is almost independent of the drain voltage. Good estimation can then be obtained from an expression derived for low drain-to-source voltage i.e., $V_{DS} = 10\,mV$, where an explicit solution exists. In this case, only the local IGN at $V_{DS} = 0\,V$ should be considered.

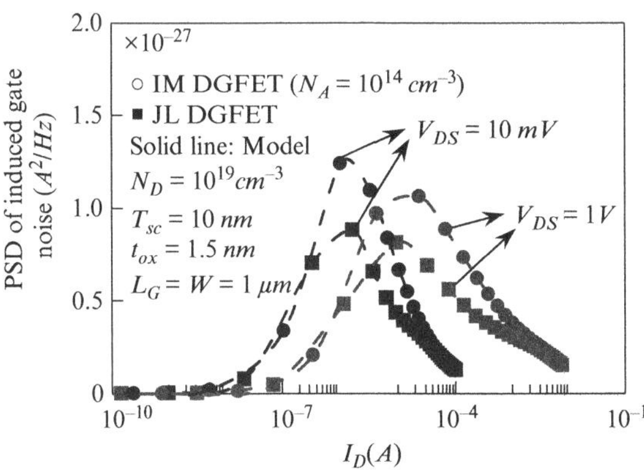

Figure 10.3 The PSD of induced gate noise ($S_{\Delta I_G}$) (A^2/Hz) versus drain current for different drain potential values ($V_{DS} = 10\,mV$ and 1 V), (frequency = $100\,MHz$). Reprinted from [163] with permission.

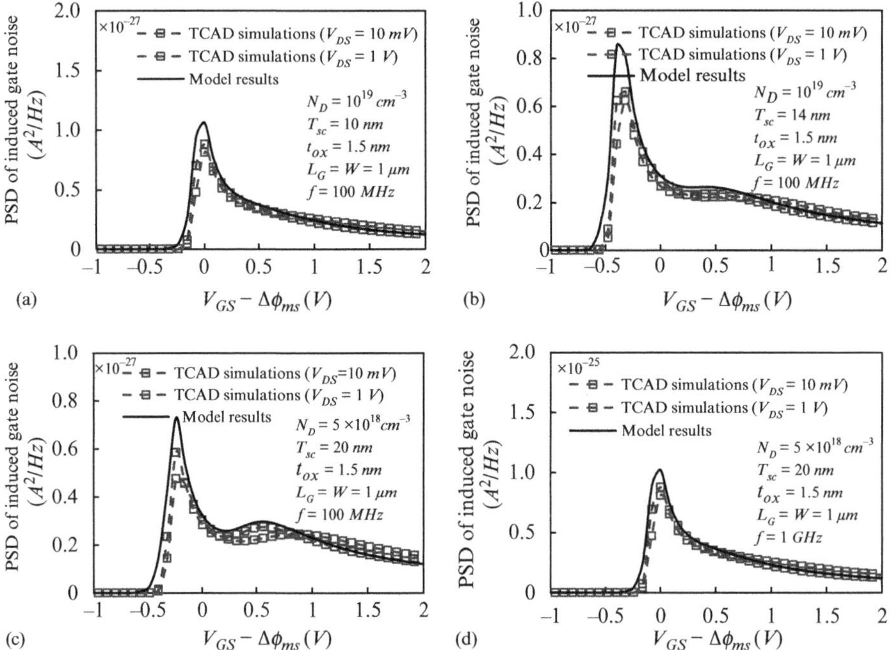

Figure 10.4 The PSD of IGN ($S_{\Delta I_G}$) (A^2/Hz) versus effective gate voltage for different physical parameter values, (a–c) $V_{DS} = 10\,mV$, $V_{DS} = 1\,V$, frequency = 100 MHz and (d) 1 GHz. The solid lines and symbols correspond to the proposed model and TCAD simulation results for IGN. Reprinted from [163] with permission.

Following this idea, the IGN PSD gives:

$$S_{\Delta I_G} = -\omega^2 C_{sc}^2 S_{\Delta V_{ch}} = -\omega^2 C_{sc}^2 \int_0^{L_G} \left(-\frac{4KT}{\mu W Q_m(y)} \right) dy = \frac{4KT\omega^2 C_{sc}^2 L_G}{\mu W Q_m}. \tag{10.9}$$

Here, ω is the angular frequency and $C_{SC} = 2C_{GG}$, where C_{SC} and Q_m do not depend on the coordinate since $V_{DS} = 10\,mV$. Instead, the charge-based relationships in Chapter 3 can be used.

Relation (10.9) can be a first step toward IGN modeling in junctionless MOS-FETs operating in linear or saturation regimes.

Last, it should be noted that modulation of the depletion region results in less capacitance coupling (C_{GG}) in junctionless with respect to inversion mode FETs, as can be seen in Figure 10.3. When the junctionless MOSFET operates in accumulation, as for instance for high drain currents, the IGN becomes comparable for both devices. IGN at low frequency is proportional to ω^2 because it is generated through a capacitor coupling $C_{SC} = 2C_{GG}$. However, as can be seen in charge-based MOS modeling, at high frequency IGN becomes smaller than expected from the first-order approximation. It still increases, but only as ω^2 instead of ω^2 due to nonquasistatic effects. Note that similar behavior was observed by L.-J. Pu and Y. Tsividis.

10.3 Cross-Correlation Noise in Junctionless FETs

Since channel thermal noise and IGN are generated from the same source of noise, it is possible to estimate the correlation between them. Relying on [167], the cross-correlation PSD can be obtained from relation (10) in [167]:

$$S_{\Delta I_{DS}\Delta I_G} = -j\frac{4KT\omega W^4\mu^3}{I_{DS}^3 L_G^2} \times \int_0^{V_{DS}} Q_m^2(V_{ch}) \int_0^{V_{DS}} Q_m(V'_{ch})\left[Q_G(V'_{ch}) - Q_G(V_{ch})\right] dV'_{ch} dV_{ch},$$

(10.10)

where V_{ch} and Q_G denote the electron quasi-Fermi potential and gate charge density, respectively (it should be noted that the correlation spectrum density is a made-up quantity). An analytical expression is obtained for (10.10) when assuming that the mobile charge density varies linearly with respect to the Fermi potential:

$$Q_m(V_{ch}) = \frac{Q_{m,d} - Q_{m,s}}{V_{DS}} V_{ch} + Q_{m,s}.$$

(10.11)

Here, $Q_{m,d}$ and $Q_{m,s}$ are the mobile charge densities at drain and source (relation (10.11) becomes less accurate as the drain-to-source potential is increased). Merging (10.10) and (10.11) leads to an analytical model for the PSD of cross-correlation noise:

$$S_{\Delta I_{DS}\Delta I_G} = j\frac{KT\omega W^4\mu^3 V_{DS}^2}{6 I_{DS}^3 L_G^2}(Q_{m,d} - Q_{m,s})^4$$

$$\times \left[\frac{1}{6} + \frac{Q_{m,s}}{Q_{m,d} - Q_{m,S}} + \left(\frac{Q_{m,s}}{Q_{m,d} - Q_{m,s}}\right)^2\right].$$

(10.12)

Figure 10.5 compares the cross-correlation noise obtained from numerical TCAD simulations and the model for the n-type junctionless double-gate MOSFET with $N_D = 10^{19}\,cm^{-3}$, $T_{sc} = 10\,nm$, and $15\,nm$, $L_G = 200\,nm$, and $1\,\mu m$, $W = 1\,\mu m$ and $t_{ox} = 1.5\,nm$. The cross-correlation noise is accurate for relatively small drain-to-source potentials i.e., below $100\,mV$; see Figure 10.5a, b, e, and f. When increasing V_{DS}, the charge linearization becomes less valid and relation (10.12) fails to simulate the cross-correlation noise below the flat-band (see Figure 10.5c and d). Nevertheless, above the flat-band relation (10.12) works properly. Although, for higher V_{DS}, the linear approximation of mobile charge density is no longer valid in the depleted region, the agreement between the numerical and analytical calculations is acceptable in accumulation and hybrid domains.

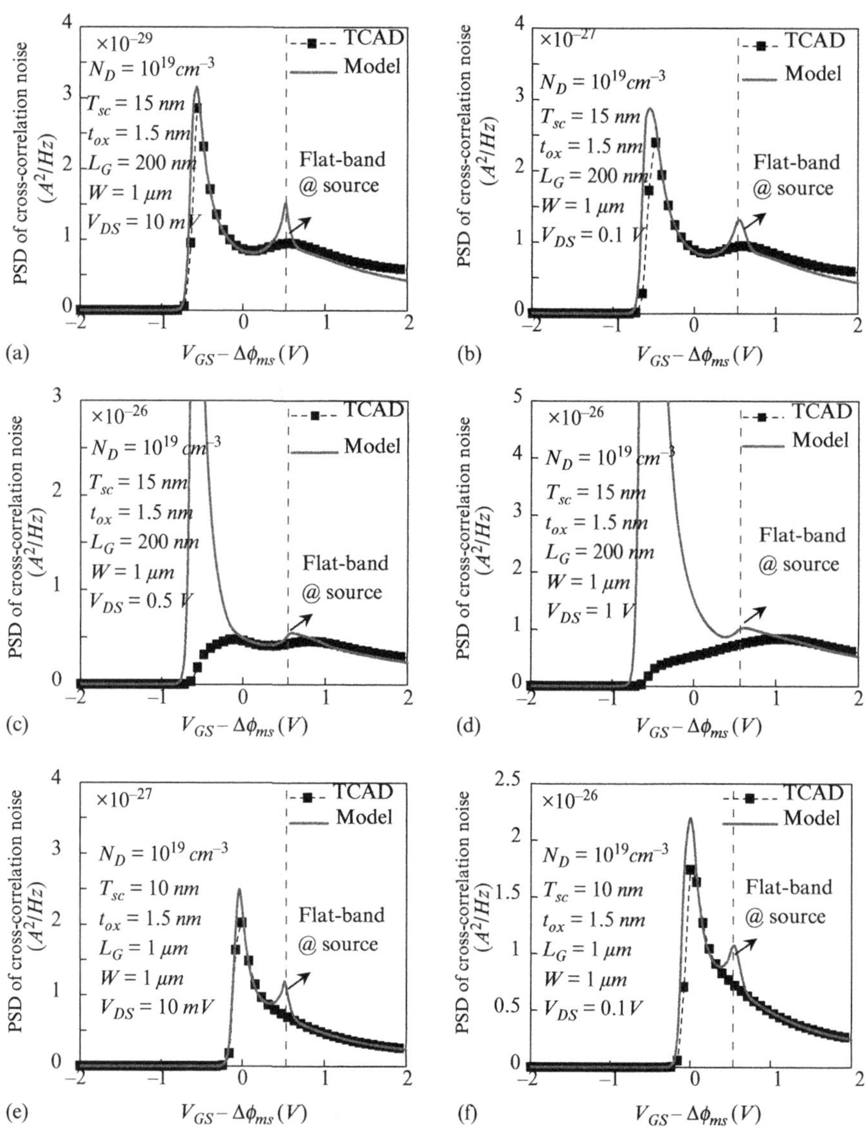

Figure 10.5 The cross-correlation noise is accurate for relatively small drain-to-source potentials (a, b, e, and f) and less valid for high values of drain-to-source potentials (c and d). The PSD of cross-correlation noise (A^2/Hz) versus effective gate voltage for different drain potential values (frequency = 100 MHz). Noting that $\Delta\phi_{ms}$ is difference between the metal work function and an intrinsic reference semiconductor given by $\Delta\phi_{ms} = W_{ms} - U_T \ln(N_D/n_i)$ where W_{ms} is the metal semiconductor work function difference. Reprinted from [163] with permission.

10.4　Summary

The analytical approach to noise analysis (thermal noise, induced gate noise, and cross-correlation noise) in junctionless double-gate FETs relying on a charge-based model in this chapter showed that junctionless and inversion-mode double-gate MOSFETs have comparable channel thermal noise PSD at a given current. For induced gate noise the junctionless FET performed better than its inversion-mode counterpart. In addition, assuming linear approximation of the mobile charge density with the channel potential, an analytical cross-correlation noise for junctionless double-gate MOSFETs was obtained.

11 Carrier Mobility Extraction Methodology in JL and Inversion-Mode FETs

The free-carrier mobility in FETs is one of the main parameters that impacts device performance. Although different techniques have been developed to extract the free-carrier mobility in MOSFETs [168–174], this is still an issue for junctionless MOSFETs since the current–voltage relationships are different, especially in depletion mode [175–180]. For instance, in inversion-mode FETs the low-field mobility is usually extracted either from the standard Y function ($I_D/\sqrt{g_m}$) introduced by Ghibaudo in [174], or from the split-CV method [181], which in addition to the $I_D - V_G$ characteristics requires CV measurements. The Y-function implicitly assumes that the mobility follows the well-known homographic function $\mu = \mu_0/[1 + \theta(V_{GS} - V_T)]$, where μ_0 is the low-field mobility, V_T is the threshold voltage, V_{GS} is the gate voltage, and θ is the mobility reduction factor.

The split-CV this technique is generic as it does not depend on the mobility, but it becomes inappropriate for short and narrow devices since the channel capacitance to be measured includes important extrinsic contributions [175–178, 181].

This chapter presents a generic method compatible with carrier mobility extraction in junctionless transistors and provides an extension to conventional inversion-mode MOSFETs.

11.1 *Y* Function and Mobility Extraction in Junctionless FETs

The free-carrier mobility obtained from the Y function introduced by Ghibaudo [174] presumes that the carrier mobility follows the well-known homographic function:

$$\mu(V_{GS}) = \frac{\mu_0}{1 + \theta(V_{GS} - V_T)},$$

(11.1)

where μ_0 is the low-field mobility, V_T is the charge threshold voltage, V_{GS} is the gate-source voltage, and θ is the mobility reduction coefficient. Introducing this mobility and assuming drift-diffusion transport, in linear mode (low V_{DS}) the drain current above the threshold is approximated by

$$I_{DS} = \frac{WC_{ox}}{L_G} \frac{\mu_0}{1 + \theta(V_{GS} - V_T)}(V_{GS} - V_T)V_{DS},$$

(11.2)

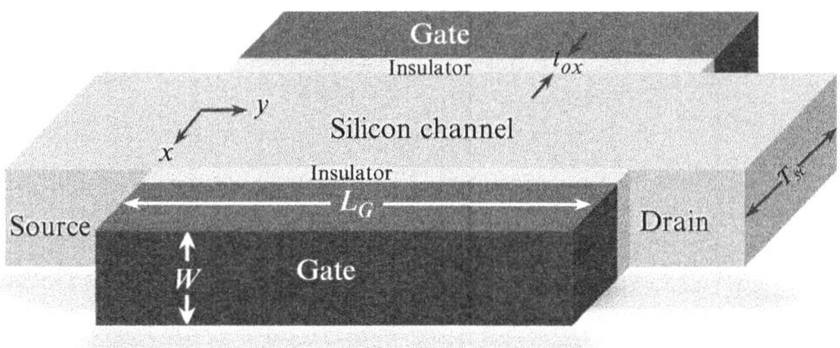

Figure 11.1 Schematic view of an *n*-type junctionless double-gate MOSFET.

where the symbols have their usual meaning. Dividing the current by the square root of the transconductance leads to a linear function with respect to ($V_{GS} - V_T$):

$$\frac{I_{DS}}{\sqrt{g_m}} = \sqrt{\frac{WC_{ox}}{L_G}\mu_0 V_{DS}}(V_{GS} - V_T). \tag{11.3}$$

In strong inversion and linear mode, the linear extrapolation of $I_{DS}/\sqrt{g_m}$ gives the charge threshold voltage (V_T) from the *x*-axis intersection, while the slope is related to the low-field mobility (μ_0) (see [174]). It should be noted that the charge threshold voltage should not be confused with the extrapolated threshold voltage. Finally, knowing the charge threshold voltage, the mobility reduction factor (θ) is also obtained:

$$\theta = \frac{I_{DS}}{g_m(V_{GS} - V_T)^2} - \frac{1}{(V_{GS} - V_T)}. \tag{11.4}$$

However, this method still presumes that relation (11.1) is satisfied.

A long-channel ($L_G \geq 100\,nm$) *n*-type doped symmetric junctionless double-gate MOSFET (Figure 11.1) biased in accumulation is investigated to assess the limit of validity of the *Y* function by introducing two electron mobility dependences such as those in Figure 11.2a and b. The device parameters are $N_D = 10^{19}\,cm^{-3}$, $T_{sc} = 10\,nm$, $L_G = 1\,\mu m$, $W = 1\,\mu m$, and $t_{ox} = 1.5\,nm$. Figure 11.2c and d shows the ratio $I_{DS}/\sqrt{g_m}$ versus the effective gate voltage for each mobility model (homographic and nonhomographic) with the device biased in linear mode, $V_{DS} = 50\,mV$.

As expected, $I_{DS}/\sqrt{g_m}$ is no longer linear, nor in depletion or accumulation. Even when the carrier mobility follows the homographic function (see Figure 11.2c), in double-gate junctionless FETs some uncertainty remains. Across flat-band, two different slopes are observed, and thus different charge threshold voltages (V_T) are obtained from the *Y* function. Next, the electron mobility extracted from the *Y* function is shown in Figure 11.2a and b assuming homographic and nonhomographic functions. Figure 11.2a and b shows the mismatch between the carrier mobility obtained with the *Y* function either assuming homographic or nonhomographic dependences. The relative error for the low-field mobility μ_0 exceeds by 26 percent the expected value in the junctionless double-gate MOSFET.

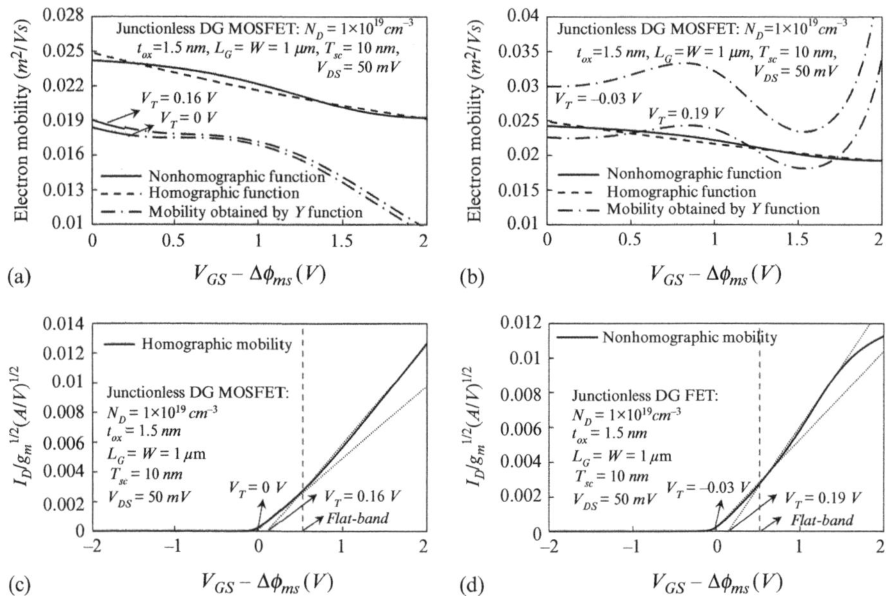

Figure 11.2 Electron mobility given by nonhomographic (tangent hyperbolic) and homographic models and obtained by the Y function with respect to the effective gate potential (a and b). The $I_{DS}/(\sqrt{g_m})$ versus the gate potential obtained by the Y function for two different mobility models (c and d). Note that $\Delta\phi_{ms}$ is the difference between the metal work function and an intrinsic reference semiconductor. Reprinted from [182] with permission.

In conclusion, despite the Y function giving accurate results in inversion-mode FETs, its validity remains questionable in junctionless FETs.

11.2 Model-Independent Mobility-Extraction Method in Junctionless FETs

To overcome some of the limitations inherent to the Y function, an alternative approach to measure free-carrier mobility in junctionless FET channels is developed based on drift-diffusion model.

11.2.1 General Treatment

The drift-diffusion based current is given by

$$I_{DS} = -W\mu(V_{GS})Q_m \frac{dV_{ch}}{dy}, \tag{11.5}$$

where V_{ch} is the quasi-Fermi channel potential. Assuming the quasi-Fermi potential constant along the channel, which is satisfied for small drain-to-source potential values, relation (11.5) becomes:

$$I_{DS} = -\frac{W}{L_G}\mu(V_{GS})Q_m V_{DS},\tag{11.6}$$

where Q_m is the mean value of the mobile charge densities at source (Q_{ms}) and drain (Q_{md}) i.e., $Q_m = (Q_{ms} + Q_{md})/2$. Applying relation (11.6) twice for two small values of drain voltage (V_{DS1} and V_{DS2}) and subtracting the drain currents ($I_{DS1} - I_{DS2}$) gives:

$$I_{DS1} - I_{DS2} = -\frac{W}{L_G}\mu(V_{GS})(Q_{m1} V_{DS1} - Q_{m2} V_{DS2}),\tag{11.7}$$

or equivalently:

$$I_{DS1} - I_{DS2} = -\frac{W}{L_G}\mu(V_{GS})[Q_{m1}(V_{DS1} - V_{DS2}) + V_{DS2}(Q_{m1} - Q_{m2})],\tag{11.8}$$

where Q_{m1} and Q_{m2} are the mean values of the mobile charge densities along the channel at V_{DS1} and V_{DS2}, respectively. The channel conductance can be approximated by dividing relation (11.8) by ($V_{DS1} - V_{DS2}$):

$$g_{ds} = -\frac{W}{L_G}\mu(V_{GS})\left[Q_{m1} + V_{DS2}\frac{\partial Q_m}{\partial V_{DS}}\right].\tag{11.9}$$

Using relations (11.9) and (11.6), after some manipulation the derivative of the mobile charge density with respect to V_{DS} is obtained:

$$g_{ds} = \frac{I_{DS1}}{V_{DS1}} + \frac{I_{DS2}}{Q_{m2}}\left(\frac{\partial Q_m}{\partial V_{DS}}\right) \Rightarrow \frac{1}{Q_{m2}}\left(\frac{\partial Q_m}{\partial V_{DS}}\right) = \frac{g_{ds}}{I_{DS2}} - \left(\frac{I_{DS1}}{I_{DS2}}\right)\frac{1}{V_{DS1}}.\tag{11.10}$$

Note that the RHS of (11.10) is a measurable quantity as it involves currents and voltages only, whereas the RHS and LHS terms of (11.10) are given in Figure 11.3 as a function of V_{GS}, confirming that relation (11.10) holds to a great extent (the humps come from the convergence limits in the numerical simulations when the device operates in deep depletion). Next, using the identity $\partial Q_m/\partial V_{GS} = -2\partial Q_m/\partial V_{DS}$ valid at low V_{DS} (see Appendix C) and integrating (11.10) over V_{GS} from V_{G1} to v_g gives the ratio of the averaged mobile charge densities corresponding to V_{G1} and v_g:

$$\int_{V_{G1}}^{v_g} \frac{1}{Q_{m2}}\left(\frac{\partial Q_m}{\partial V_{GS}}\right)dV_{GS} = \ln\left[\frac{Q_m(v_g)}{Q_m(V_{G1})}\right] = -2\int_{V_{G1}}^{v_g}\left(\frac{g_{ds}}{I_{DS2}} - \frac{I_{DS1}}{I_{DS2}}\frac{1}{V_{DS1}}\right)dV_{GS}.\tag{11.11}$$

As stated in (11.11), the ratio of the averaged mobile densities for two values of the gate potential is known as soon as the drain current and the drain conductance are known with respect to the gate potential (experimentally for instance). So far, the derivation and relations remain quite general and no assumption is made concerning the nature of the field-effect device.

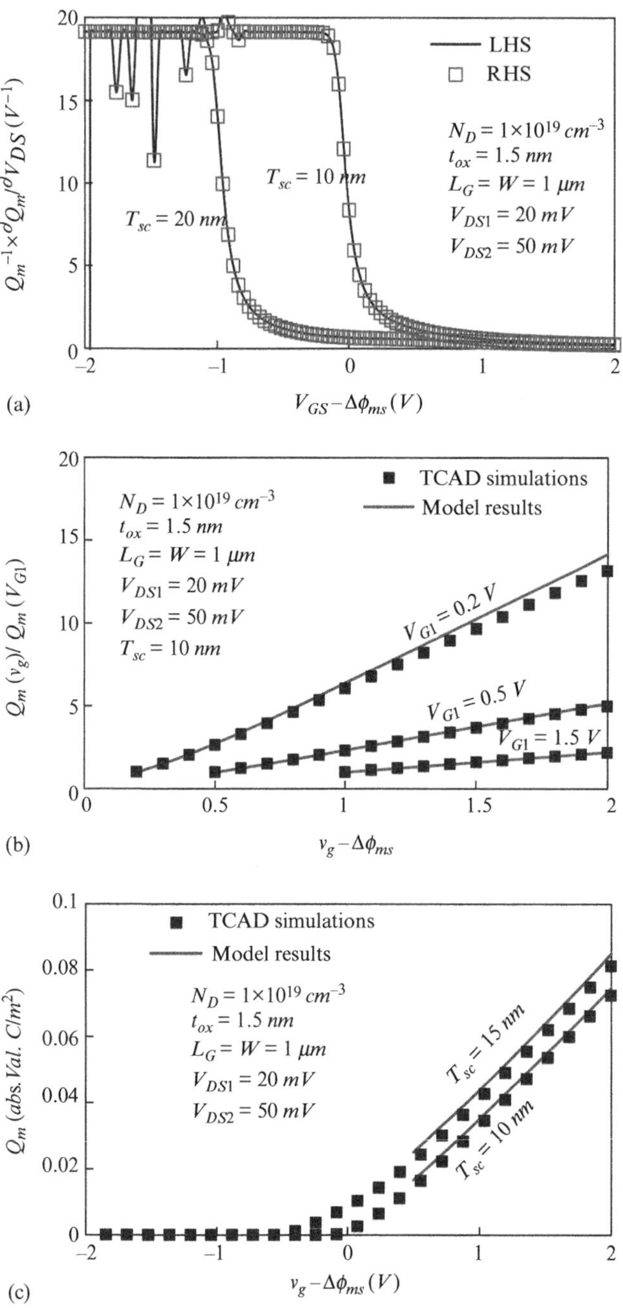

Figure 11.3 (a) Comparison between RHS and LHS of relation (11.10) in an n-type junctionless double-gate MOSFET for two values of silicon thickness. (b) Ratio of mobile charge densities at V_{G1} and v_g for different values of V_{G1}. (c) The mobile charge density with respect to v_g for 10 and 15 nm silicon thicknesses obtained by (11.16). Symbols and lines are for the TCAD simulations and the model, respectively. Reprinted from [182] with permission.

11.2.2   Mobility Extraction in Double-Gate Junctionless FETs

The method is used to extract mobility in double-gate junctionless FETs. Figure 11.3b shows that the ratio of the average mobile charge density and the RHS integral in (11.11) with respect to v_g for different values of V_{G1} is accurately predicted by the model.

Relying on the charge-based model developed in Chapter 3, in accumulation the total charge density (Q_{sc}) is given by (see relation 3.43)):

$$V_{GS} - V_{ch} - \Delta\phi_{ms} - U_T \ln\left(\frac{N_D}{n_i}\right) \overset{acc}{\approx} -\frac{Q_{sc}}{2C_{ox}} + U_T \ln\left(1 + \frac{Q_{sc}^2}{8qN_D\varepsilon_{si}U_T}\right). \qquad (11.12)$$

The derivative of the total charge density with respect to V_{GS} becomes:

$$\frac{\partial Q_{sc}}{\partial V_{GS}} \overset{acc}{\approx} \left(-\frac{1}{2C_{ox}} + U_T\frac{\dfrac{2Q_{sc}}{8qN_D\varepsilon_{si}U_T}}{1 + \dfrac{Q_{sc}^2}{8qN_D\varepsilon_{si}U_T}}\right)^{-1}. \qquad (11.13)$$

Assuming accumulation mode $\dfrac{Q_{sc}^2}{8qN_D\varepsilon_{si}U_T} \gg 1$ and (11.13) simplifies to:

$$\frac{\partial Q_{sc}}{\partial V_{GS}} \overset{acc}{\approx} \left(-\frac{1}{2C_{ox}} + \frac{2U_T}{Q_{sc}}\right)^{-1}. \qquad (11.14)$$

In addition, if Q_{sc} is approximated by the value it takes above flat-band i.e., $Q_{sc} = -qN_D T_{sc}$ and the derivative of the mobile charge density with respect to V_{GS} becomes:

$$\frac{\partial Q_m}{\partial V_{GS}} \overset{acc}{\approx} -\left(\frac{1}{2C_{ox}} + \frac{2U_T}{qN_D T_{sc}}\right)^{-1} = -\left(\frac{1}{2C_{equ}}\right)^{-1} = \frac{Q_m(v_g) - Q_m(V_{G1})}{v_g - V_{G1}}. \qquad (11.15)$$

It is worth noting that relation (11.15) is almost the same as for inversion double-gate FETs biased in the strong inversion regime of operation, whose slope is about $2C_{ox}$. Therefore *in accumulation* a seemingly equivalent gate capacitance $2C_{equ}$ can be seen. Its definition involves the doping density, the channel thickness, and more unexpectedly the temperature.

Merging (11.11) and (11.15) leads to explicit expressions for mobile charge densities at V_{G1} and v_g:

$$Q_m(V_{G1}) = \frac{(V_{G1} - v_g)\left(\dfrac{1}{2C_{ox}} + \dfrac{2U_T}{qN_D T_{sc}}\right)^{-1}}{\exp\left[-2\displaystyle\int_{V_{G1}}^{v_g}\left(\frac{g_{ds}}{I_{DS2}} - \frac{I_{DS1}}{I_{DS2}}\frac{1}{V_{DS1}}\right)dV_{GS}\right] - 1}, \qquad (11.16)$$

$$Q_m(v_g) = \frac{(V_{G1} - v_g)\left(\dfrac{1}{2C_{ox}} + \dfrac{2U_T}{qN_D T_{sc}}\right)^{-1}}{1 - \exp\left[2\displaystyle\int_{V_{G1}}^{v_g}\left(\frac{g_{ds}}{I_{DS2}} - \frac{I_{DS1}}{I_{DS2}}\frac{1}{V_{DS1}}\right)dV_{GS}\right]}. \qquad (11.17)$$

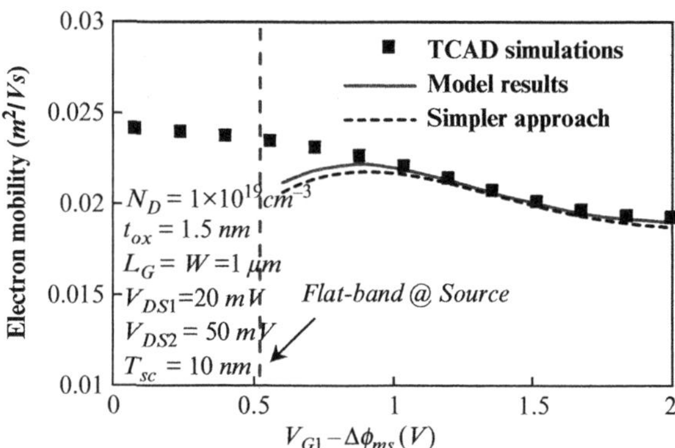

Figure 11.4 Electron mobility given by (11.18) and (11.20) for an n-type junctionless double-gate MOSFET (symbols and lines denote the TCAD simulations and proposed model, respectively). Reprinted from [182] with permission.

To evaluate the validity of (11.17), the mobile charge density obtained with this method is compared to numerical simulations for 10 and 15 nm silicon thicknesses in Figure 11.3. Having rebuilt the mobile charge density from simulations/measurements, the mobility is extracted for each value of the gate voltage by combining (11.16) and (11.6):

$$\mu(V_{G1}) = -\frac{L_G}{W}\frac{I_{DS1}}{V_{DS1}}\frac{\exp\left[-2\int_{V_{G1}}^{v_g}\left(\frac{g_{ds}}{I_{DS2}} - \frac{I_{DS1}}{I_{DS2}}\frac{1}{V_{DS1}}\right)dV_{GS}\right] - 1}{V_{G1} - v_g}\left(\frac{1}{2C_{ox}} + \frac{2U_T}{qN_D T_{sc}}\right). \tag{11.18}$$

The prediction of the analytical approach is illustrated in Figure 11.4 where numerical simulations have been used as virtual data. The mobility is quite accurate in accumulation. This is attributed to the slope of the mobile charge density versus V_{GS}, which is almost constant in accumulation (above flat-band) as required from relation (11.15). However, as expected from the assumption leading to relation (11.15), near flat-band the linear dependence of the mobile charge with V_{GS} is no longer satisfied, and a mismatch is visible.

11.2.3   A Simplified Approach

Relation (11.18) still requires integration over V_{GS}. Now, assuming a linear approximation of the mobile charge density with respect to V_{GS} as in (11.15) greatly simplifies the carrier mobility extraction in accumulation. Combining relations (11.10) and (11.15), then recalling that $\partial Q_m/\partial V_{GS} = -2\partial Q_m/\partial V_{DS}$ (see Appendix C), the mobile charge density is obtained without integrating over V_{GS}:

$$Q_{m2} = \frac{1}{2}\left(\frac{1}{2C_{ox}} + \frac{2U_T}{qN_D T_{sc}}\right)^{-1}\left[\frac{g_{ds}}{I_{DS2}} - \left(\frac{I_{DS1}}{I_{DS2}}\right)\frac{1}{V_{DS1}}\right]^{-1}. \tag{11.19}$$

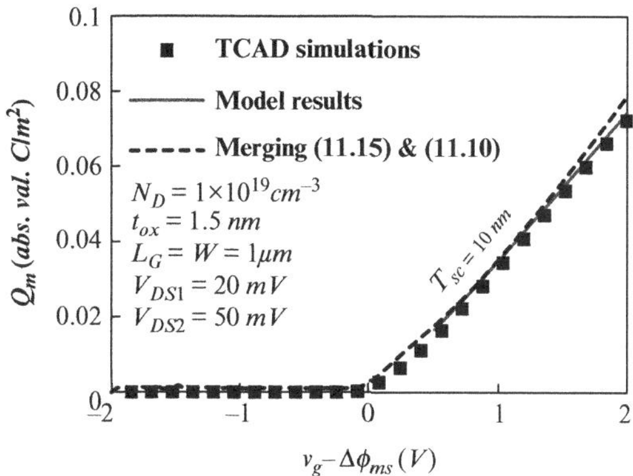

Figure 11.5 The mobile charge density with respect to v_g for $10\,nm$ silicon thickness obtained by (11.17) and (11.19) in junctionless double-gate MOSFETs (the dashed and solid lines are related to the model and the symbols are for the TCAD simulations). Reprinted from [182] with permission.

Therefore, introducing (11.19) into (11.6), a simplified relationship for the carrier mobility is obtained:

$$\mu(V_{G1}) = -\frac{L_G}{W}\frac{2}{V_{DS2}}\left(\frac{1}{2C_{ox}} + \frac{2U_T}{qN_D T_{sc}}\right)(g_{ds} - g_{ds0})$$

$$\approx -\frac{2L_G}{W}\left(\frac{1}{2C_{ox}} + \frac{2U_T}{qN_D T_{sc}}\right)\frac{\partial g_{ds}}{\partial V_{DS}} \approx -\frac{2L_G}{W}\left(\frac{1}{2C_{ox}} + \frac{2U_T}{qN_D T_{sc}}\right)\frac{\partial^2 I_{DS}}{\partial V_{DS}^2},$$

$$(11.20)$$

where g_{ds0} holds for I_{DS1}/V_{DS1}.

The electron mobility obtained from TCAD simulations is plotted with respect to V_{G1} in Figure 11.4, then compared to (11.18) and (11.20). The carrier mobility estimated from (11.20) remains accurate in accumulation and does not introduce any substantial errors with regard to the more complete expression (11.18). Then, assuming linear approximation of the charge density with respect to V_{GS} makes the model simpler and this could be used in place of (11.18). As illustrated in Figure 11.5, the mobile charge density obtained from (11.20) does not introduce large errors compared to (11.17), even though some discrepancy is seen near and below flat-band for reasons discussed previously.

11.3 Extending the Method to Inversion-Mode FETs

As seen in relations (11.18) and (11.20), the expression of mobility involves the gate potential dependence of the mobile charge above flat-band. It consists of two terms, one of which depends on the doping density. Assuming a low-doped channel, relation (11.12) gives " $-2C_{ox}$ " as expected for inversion-mode double-gate FETs

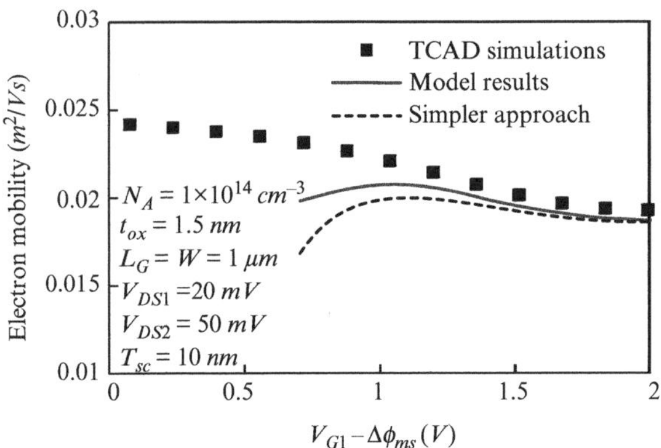

Figure 11.6 Electron mobility given by (11.21) and (11.22) for an inversion-mode double-gate MOSFET (the symbols and lines denote the TCAD simulations and proposed model, respectively). Reprinted from [182] with permission.

(and "$-C_{ox}$" for planar MOSFETs). Therefore, following the same derivation, the electron mobility in inversion-mode double-gate FET simplifies to:

$$\mu(V_{G1}) = \frac{L_G}{W}\frac{I_{DS1}}{V_{DS1}}\frac{\exp\left[-2\int_{V_{G1}}^{v_g}\left(\frac{g_{ds}}{I_{DS2}} - \frac{I_{DS1}}{I_{DS2}}\frac{1}{V_{DS1}}\right)dV_{GS}\right] - 1}{2C_{ox}(v_g - V_{G1})}. \tag{11.21}$$

Alternatively, using the same simplified approach, the mobility is also given by

$$\mu(V_{G1}) = -\frac{L_G}{W}\frac{1}{V_{DS2}C_{ox}}\left(g_{ds} - \frac{I_{DS1}}{V_{DS1}}\right) \approx -\frac{L_G}{WC_{ox}}\frac{\partial^2 I_{DS}}{\partial V_{DS}^2}. \tag{11.22}$$

The two mobility dependences obtained from (11.21) and (11.22) evaluated against TCAD simulations in Figure 11.6 show that the method is also applicable to inversion-mode FETs above the threshold, where the slope of the mobile charge density is almost linear and close to $-2C_{ox}$ for double-gate FETs. Note that as soon as the transistor departs from the strong inversion regime (here when $V_{GS} - \Delta\phi_{ms}$ is lower than 1 volt) the discrepancy between real and extracted mobilities increases considerably since the linear dependence of the mobile charge density versus V_{GS} becomes inaccurate.

11.4   Limitations

Regarding the limitations of the method, two aspects have to be considered. First, the mismatch between the mobility extracted from the model and the values predefined in TCAD simulations becomes larger when V_{GS} is close to the flat-band voltage. Although the charge densities at source and drain for depletion and accumulation can be obtained accurately from the charge-based model in Chapter 3, at flat-band the derivatives of the charge densities with respect to the potential (where $Q_{sc} = 0$) give $2C_{ox}$. This asymptotic value is only reached in strong accumulation, whereas the correct value is in fact close to $1.55C_{ox}$ for $T_{sc} = 10\,nm$ and $N_D = 10^{19}\,cm^{-3}$ (a discussion on this point is available in Chapter 8), which explains the larger mismatch for mobility around the flat-band condition.

A second limitation comes from the long-channel assumption. The approach is suitable for long-channel devices with small values of drain-to-source potentials i.e., with no velocity saturation.

11.5 Summary

Inherent limitations of the so-called Y function method used to perform mobility extraction in junctionless devices was discussed in detail in this chapter. In the presence of different slopes in I–V characteristics, as in junctionless FETs, the Y function may become inaccurate and generate errors if used without care.

An alternative and more generic approach was proposed based on the derivative of the mobile charge density with respect to the gate voltage. This approach relies on current and transconductance measurements and is applicable to junctionless FETs and inversion-mode FETs biased in accumulation and in strong inversion, respectively. Whereas the Y function method assumes a predefined mobility law, the method proposed here overcomes some of these limitations and can be used independently of the kind of field-effect device and gate-dependent mobility law provided an almost linear dependence can be stated between the mobile charge density and the gate voltage. This method can also likely be applied to nanowire FETs using the equivalent double-gate MOSFET model parameters discussed in Chapter 3.

12 Revisiting the Junction FET: A Junctionless FET with an ∞ Gate Capacitance

The junction field-effect transistor (JFET, also called the JUGFET or junction unipolar gate FET) is the first field-effect device produced before being supplanted by the MOSFET. Even today though its use is limited, the JFET is still used in specific low-noise applications and in new applications in power electronics with new architectures and new materials (e.g., SiC). Just like the junctionless FET biased below the flat-band, the JFET operates in depletion and the current flows in the volume, not at an interface. However, unlike the junctionless FET, the JFET has no gate-insulator layer i.e., there is no semiconductor-insulator interface. This could result in low-frequency noise, which mainly originates from interface traps, and is a clear advantage. Similarly, surface-roughness scattering [183], which degrades mobility, is alleviated to a great extent.

Despite the long use of JFETs in discrete electronics, physics-based compact models today still adopt a conservative approach and deal with the so-called full-depletion approximation where the deep-depletion regime is essentially modeled empirically. Today, FET compact models [183–186] use regional approximations bridged with smoothing functions. For instance, Sansen and Das [184] adopted a MOSFET-like approach to interpolate the subthreshold, quadratic, and linear modes of operation. Semiempirical log-exponential functions have also been introduced to ensure a smooth transition across the pinch-off domain [184], a technique also used in [187, 188].

This chapter discusses a unified charge-based model for double-gate JFETs addressing fundamental DC, AC, and noise characteristics. The central element considers the JFET as a junctionless FET with a negligible insulating layer i.e., a junctionless FET with an *infinite gate capacitance*. As for the double-gate junctionless FET, the physics-based model for the JFET will cover all regions of operation, with the exception of accumulation, which is not compatible with the device architecture (forward-biasing of the gate-to-channel *pn* junctions).

12.1 Principle Operation of the JFET

Like the junctionless FET, the JFET has no source and drain *pn* junction and the channel consists of a highly doped semiconductor layer. However, a major difference exists between these two devices: whereas the junctionless FET has an insulating

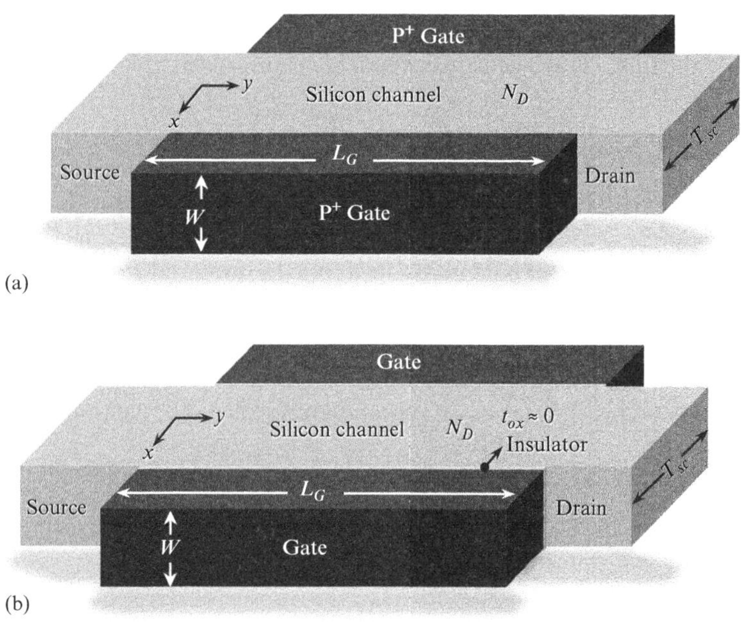

Figure 12.1 Schematic view of (a) an n-type double-gate JFET and (b) its equivalent (an n-type double-gate junctionless FET with infinite oxide capacitance).

layer between the channel and the gate electrode(s), in JFETs this role is played by a reversed biased gate-to-channel *pn* junction i.e., the gate is highly doped to avoid depletion into the gate.

In the case of an n-channel JFET, when a negative V_{GS} bias is imposed, the depletion layer in the channel modulates the conductance. However, unlike a junctionless FET, a JFET is usually in an *on*-state at $V_{GS} = 0$. Mapping the charge-based model developed in Chapter 3 to the double-gate junctionless FET for double-gate JFETs is the objective of this chapter.

12.2 Charge-based Modeling of Double-Gate JFETs

Figure 12.1 shows a schematic of (a) an n-type doped long-channel double-gate junctionless FET and (b) JFET. To develop a physics-based model for mobile charge density in double-gate JFETs, an analytical expression for the charge density in junctionless FETs with an **infinite gate capacitance** ($t_{ox} \approx 0$) is used.

12.2.1 Charge–Voltage and Pinch-Off Voltage in Double-Gate JFETs

When the whole channel operates in depletion, the gate-to-source voltage must satisfy $V_{GS} < V_{bi} = U_T \ln\left(N_D N_G / n_i^2\right)$ where N_G is the p-type gate doping and V_{bi} is the gate-channel built-in potential. Since the JFET can be viewed as a JLFET with infinite gate capacitance i.e., there is no insulating layer as such, from relation (3.46) the total charge density is implicitly expressed as

$$V_{GS} - V_{ch} - V_{bi} \overset{dep}{\approx} -\frac{Q_{sc}^2}{8qN_D\varepsilon_{si}} + U_T \ln\left[1 - \left(\frac{Q_{sc}}{Q_{fix}}\right)^2\right], \tag{12.1}$$

where Q_{sc} is the total charge density comprised of mobile Q_m and fixed $Q_{fix} = qN_DT_{sc}$ charge densities, with other symbols having their usual meaning (the source is still the reference for potentials). Defining the threshold voltage as the gate voltage where $Q_m = 0$ (at $V_{DS} = 0$) and neglecting the logarithmic term in (12.1) gives:

$$V_{TH} = V_{FB} - V_P, \tag{12.2}$$

where $V_{FB} = V_{ch} + V_{bi}$. V_P is commonly called the pinch-off voltage and is defined as

$$V_P = \frac{qT_{sc}^2 N_D}{8\varepsilon_{si}} = \frac{Q_{fix}}{8C_{si}}, \tag{12.3}$$

where $C_{si} = \varepsilon_{si}/T_{sc}$.

12.2.2   Channel Current in Double-Gate JFETs

The total electron current density is obtained by adding the diffusion and drift contributions. In this case, the current is given by

$$I_{DS} = -W\mu Q_m \frac{dV_{ch}}{dx} = -\frac{W}{L_G}\int_S^D \mu Q_m dV_{ch} = \frac{W}{L_G}\mu\int_S^D (qN_DT_{sc} - Q_{sc})dV_{ch}$$

$$= \frac{W}{L_G}\mu qN_DT_{sc}V_{DS} - \frac{W}{L_G}\mu\int_S^D Q_{sc}dV_{ch}, \tag{12.4}$$

where μ is the carrier mobility (assumed constant along the channel), W, L_G are the width and gate length of the channel, respectively, and $Q_{sc}dV_{ch}$ and $Q_m dV_{ch}$ are expressed by the following relationships obtained by (see Chapter 3) (C_{ox} is infinite)

$$Q_{sc}dV_{ch} \overset{dep}{\approx} \left[\frac{Q_{sc}^2}{4qN_D\varepsilon_{si}} + 2U_T\frac{\left(\dfrac{Q_{sc}}{qN_DT_{sc}}\right)^2}{1 - \left(\dfrac{Q_{sc}}{qN_DT_{sc}}\right)^2}\right] dQ_{sc}, \tag{12.5}$$

$$Q_m dV_{ch} \overset{dep}{\approx} \left[\frac{Q_m(Q_m + Q_{fix})}{4qN_D\varepsilon_{si}} - 2U_T\frac{Q_m + Q_{fix}}{Q_m + 2Q_{fix}}\right] dQ_m. \tag{12.6}$$

Similar to the junctionless FETs in depletion mode (see Chapter 3), the total drain current is obtained from

$$I_{DS} = \frac{W\mu}{y}f(Q_m)\Big|_S^y = \frac{W\mu}{L_G}f(Q_m)\Big|_S^D = \frac{W\mu}{L_G}[f(Q_{m,d}) - f(Q_{m,s})], \tag{12.7}$$

where

$$f(Q_m) = -\frac{Q_m^3}{12qN_D\varepsilon_{si}} - \frac{Q_m^2}{8C_{si}} + 2U_T Q_m - 2U_T Q_{fix} \ln\left(Q_m + 2Q_{fix}\right). \tag{12.8}$$

12.2.3 Simulations

The device of interest is an n-type JFET with $L_G = 20\,\mu m$, $W = 1\,\mu m$, where the length is chosen so that short-channel effects can be ignored. A constant carrier mobility is used (note that the constant mobility used in the model is a restriction that can be circumvented using a first- or second-order mobility model).

Figure 12.2a–d depicts the channel current versus the gate voltage for a $500\,nm$ silicon thickness doped at $N_D = 5 \times 10^{16}\,cm^{-3}$ biased in saturation ($V_{DS} = 1\,V$, Figure 12.2a and b) and linear ($V_{DS} = 10\,mV$, Figure 12.2c and d) modes. The agreement between the numerical calculations and the model is good, over more than ten orders of magnitude.

Similarly, in Figure 12.2e–h the $600\,nm$ silicon thickness device with $N_D = 5 \times 10^{16}\,cm^{-3}$ is well modeled in linear and saturation regimes. This thicker channel exhibits significant quadratic current-voltage characteristics at low V_{DS}, which is evidence of the depletion taking place in the channel.

12.3 Modeling Small Signals in JFETs

Small-signal characteristics (gate transconductance and output conductance) help highlight how different the transfer characteristics between junctionless FETs and JFETs are in double-gate JFETs.

12.3.1 Transconductance

Concerning the gate transconductance (g_m), the derivative of (12.7) with respect to V_{GS} is expressed by

$$g_m \frac{L_G}{W\mu} = \frac{\partial}{\partial V_{GS}}\left[f(Q_{m,d}) - f(Q_{m,s})\right] = \frac{\partial f(Q_{m,d})}{\partial Q_{m,d}}\frac{\partial Q_{m,d}}{\partial V_{GS}} - \frac{\partial f(Q_{m,s})}{\partial Q_{m,s}}\frac{\partial Q_{m,s}}{\partial V_{GS}}, \tag{12.9}$$

where $Q_{m,d}$ and $Q_{m,s}$ are the mobile charge density at the drain and source terminals, respectively. Relying on relation (12.8), the gate transconductance can be written as

$$g_m = \frac{W\mu}{L_G}\left[-\frac{Q_m^2}{4qN_D\varepsilon_{si}} - \frac{Q_m}{4C_{si}} - 2U_T\frac{Q_m + Q_{fix}}{Q_m + 2Q_{fix}}\right]\Bigg|_S^D \frac{\partial Q_{m,s}}{\partial V_{GS}}, \tag{12.10}$$

where the term of $\partial Q_{m,s}/\partial V_{GS}$ can be readily obtained from (12.1):

$$\frac{\partial Q_m}{\partial V_{GS}} = \left[\frac{Q_m + Q_{fix}}{4qN_D\varepsilon_{si}} + 2U_T\frac{Q_m + Q_{fix}}{Q_m\left(Q_m + 2Q_{fix}\right)}\right]^{-1}. \tag{12.11}$$

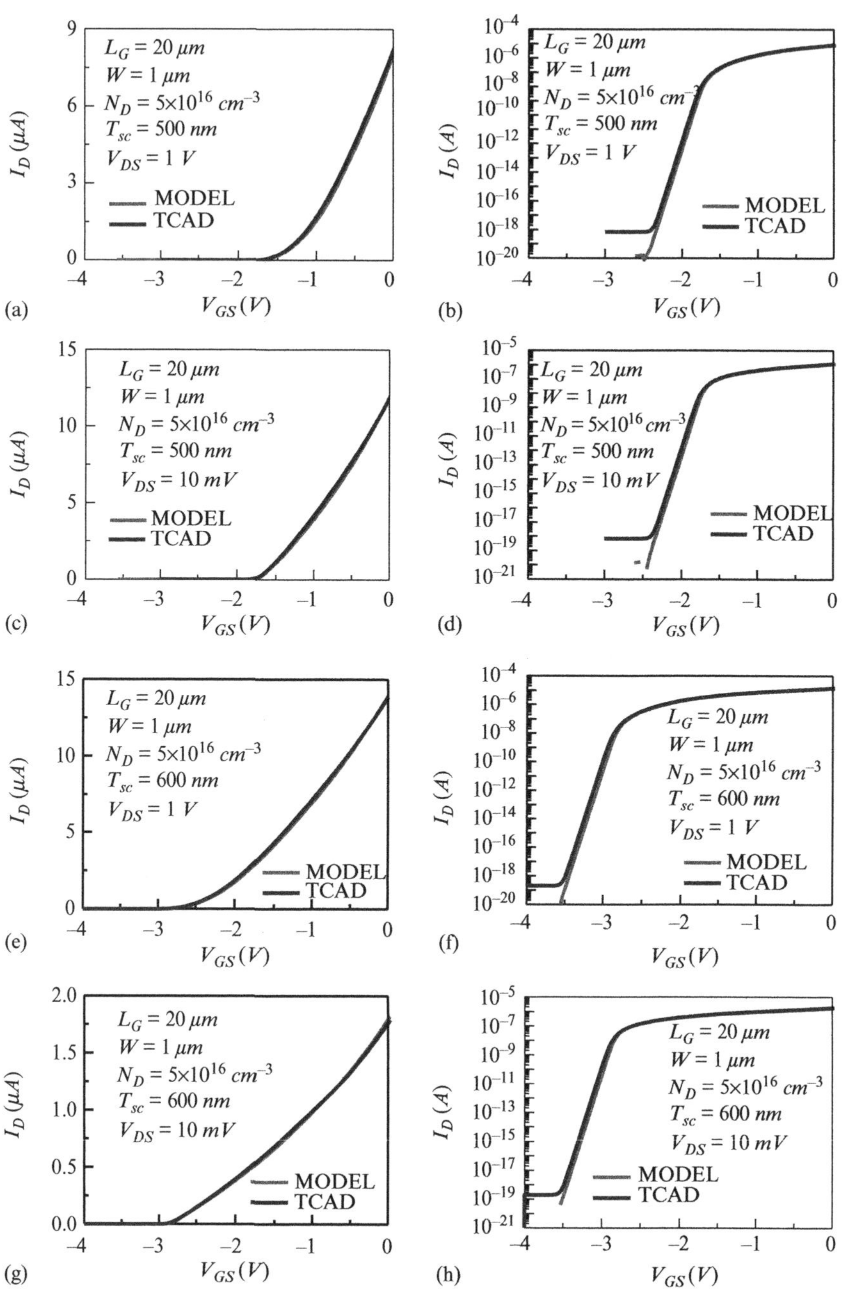

Figure 12.2 Drain current for a 500 *nm* silicon thickness biased in saturation (a and b) and linear (c and d) modes and also for a 600 *nm* silicon thickness biased in saturation (e and f) and linear (g and h) modes. Drain-to-source current versus gate voltage in linear and saturation modes of operation in both linear and logarithmic scales ($L_G = 20\,\mu m$, $W = 1\,\mu m$, $N_D = 5 \times 10^{16}\,cm^{-3}$, $T_{sc} = 500\,nm$, and $L_G = 20\,\mu m$, $W = 1\,\mu m$, $N_D = 5 \times 10^{16}\,cm^{-3}$, $T_{sc} = 600\,nm$).

Neglecting the logarithmic term in (12.1) when the device is working in subthreshold regime, the charge's subthreshold slope can be approximated by

$$\frac{\partial Q_m}{\partial V_{GS}} = \frac{4 Q_{fix} C_{si}}{Q_{sc}}.$$ (12.12)

To express the drain and source transconductances in terms of mobile charges, $f(Q_m)$ is differentiated with respect of V_{DS}, then using (12.7) gives:

$$
\begin{aligned}
g_{ds} &= \frac{W\mu}{L_G} \frac{\partial}{\partial V_{DS}} \left[f(Q_{m,d}) - f(Q_{m,s}) \right] = \frac{W\mu}{L_G} \frac{\partial f(Q_{m,d})}{\partial Q_{m,d}} \frac{\partial Q_{m,d}}{\partial V_{DS}} \\
&= \frac{W\mu}{L_G} \left(-\frac{Q_{m,d}^2}{4 q N_D \varepsilon_{si}} - \frac{Q_{m,d}}{4 C_{si}} - 2 U_T \frac{Q_{m,d} + Q_{fix}}{Q_{m,d} + 2 Q_{fix}} \right) \frac{\partial Q_{m,d}}{\partial V_{DS}}.
\end{aligned}
$$ (12.13)

12.3.2   Transcapacitances in JFET

Modeling transcapacitance in JFETs is essential for circuit design. As discussed in Section 8.3, in a special case of a junctionless FET when the whole channel operates in depletion, the global charge densities at the source $\overline{Q}_{m,s}$ and drain $\overline{Q}_{m,d}$ nodes are obtained from:

$$\overline{Q}_{m,d} = W \int_0^{L_G} \left(\frac{y}{L_G} \right) Q_m(y)\,|_{dep}\, dy = -\frac{W^2}{I_{DS}} \mu \int_0^{V_{DS}} \left(\frac{y}{L_G} \right) Q_m^2(y)\,|_{dep}\, dV_{ch},$$ (12.14)

$$\overline{Q}_{m,s} = W \int_0^{L_G} \left(1 - \frac{y}{L_G} \right) Q_m(y)\,|_{dep}\, dy = -\frac{W^2}{I_{DS}} \mu \int_0^{V_{DS}} \left(1 - \frac{y}{L_G} \right) Q_m^2(y)\,|_{dep}\, dV_{ch}.$$ (12.15)

From the continuity of the current (see relation (12.7)), the term of y/L_G is explicitly expressed as a function of the mobile charge density at y and as a function of the mobile charge densities at source and drain terminals:

$$\frac{y}{L_G} = \frac{f(Q_{m,s}) - f(Q_m)}{f(Q_{m,s}) - f(Q_{m,d})} = \frac{W\mu}{I_{DS} L_G} \left[f(Q_m) - f(Q_{m,s}) \right].$$ (12.16)

Rewriting the equivalent charge at the drain given by (12.14) and inserting the expressions (valid in depletion) for y/L_G and $Q_m dV_{ch}$ as given by (12.16) and (12.6) leads to:

$$
\begin{aligned}
\overline{Q}_{m,d} &= W \int_0^{L_G} \left(\frac{y}{L_G} \right) Q_m dy \\
&= \eta \int_{Q_{m,s}}^{Q_{m,d}} \left[f(Q_{m,s}) - f(Q_m) \right] Q_m \left[-\frac{Q_m (Q_m + Q_{fix})}{4 q N_D \varepsilon_{si}} + 2 U_T \frac{Q_m + Q_{fix}}{Q_m + 2 Q_{fix}} \right] dQ_m,
\end{aligned}
$$ (12.17)

where

$$\eta = -\frac{W^3\mu^2}{L_G I_{DS}^2}.$$

(12.18)

Further, neglecting the logarithmic term in (12.8), $f(Q_m)$ simplifies to

$$f(Q_m) = -\frac{Q_m^3}{12qN_D\varepsilon_{si}} - \frac{Q_m^2}{8C_{si}} + 2U_T Q_m.$$

(12.19)

Replacing (12.19) in (12.17), an explicit solution for the equivalent drain charge is finally obtained and can be found in Appendix D. Next, charge conservation imposes:

$$\overline{Q}_{m,s} + Q_G + \overline{Q}_{m,d} + WL_G Q_{fix} = 0.$$

(12.20)

Again, using $dy = -WQ_m\mu/I_{DS}dV_{ch}$, Q_G can be expressed as follows:

$$Q_G + WL_G Q_{fix} = -\overline{Q}_{m,s} - \overline{Q}_{m,d} = -W\int_0^{L_G} Q_m dy = \frac{W^2\mu}{I_{DS}}\int_0^{V_{DS}} Q_m^2 dV_{ch}$$

$$= \frac{W^2\mu}{I_{DS}}\int_{Q_{m,s}}^{Q_{m,d}}\left[\frac{Q_m^2\left(Q_m + Q_{fix}\right)}{4qN_D\varepsilon_{si}} - 2U_T\left(Q_m - Q_{fix} + \frac{2Q_{fix}^2}{Q_m + 2Q_{fix}}\right)\right]dQ_m$$

$$= \frac{W^2\mu}{I_{DS}}\left[\frac{Q_m^4}{16qN_D\varepsilon_{si}} + \frac{Q_m^3 Q_{fix}}{12qN_D\varepsilon_{si}} - U_T Q_m^2\right.$$

$$\left. + 2U_T Q_m Q_{fix} + 4U_T Q_{fix}^2 \ln\left(\frac{1}{Q_m + 2Q_{fix}}\right)\right]\Bigg|_{Q_{m,s}}^{Q_{m,d}}.$$

(12.21)

Following the same idea, the global charge density at the source terminal results from Q_G and $\overline{Q}_{m,d}$ [156]:

$$\overline{Q}_{m,s} = W\int_0^{L_G}\left(1 - \frac{y}{L_G}\right)Q_m(y)dy = W\int_0^{L_G} Q_m(y)dy - W\int_0^{L_G}\left(\frac{y}{L_G}\right)Q_m(y)dy$$

$$= -Q_G - Q_{fix} - \overline{Q}_{m,d}.$$

(12.22)

Then the transcapacitance matrix elements can be defined according to derivative of the global mobile charge at node i with respect to the applied potential at node j ($C_{ij} = \partial\overline{Q}_{m,i}/\partial V_j$):

$$C_{GG} = \left(\frac{\partial\overline{Q}_{m,s}}{\partial V_{GS}} + \frac{\partial\overline{Q}_{m,d}}{\partial V_{GS}}\right) = -(C_{SG} + C_{DG}),$$

(12.23)

$$C_{GD} = \left(\frac{\partial\overline{Q}_{m,s}}{\partial V_{DS}} + \frac{\partial\overline{Q}_{m,d}}{\partial V_{DS}}\right) = -(C_{SD} + C_{DD}),$$

(12.24)

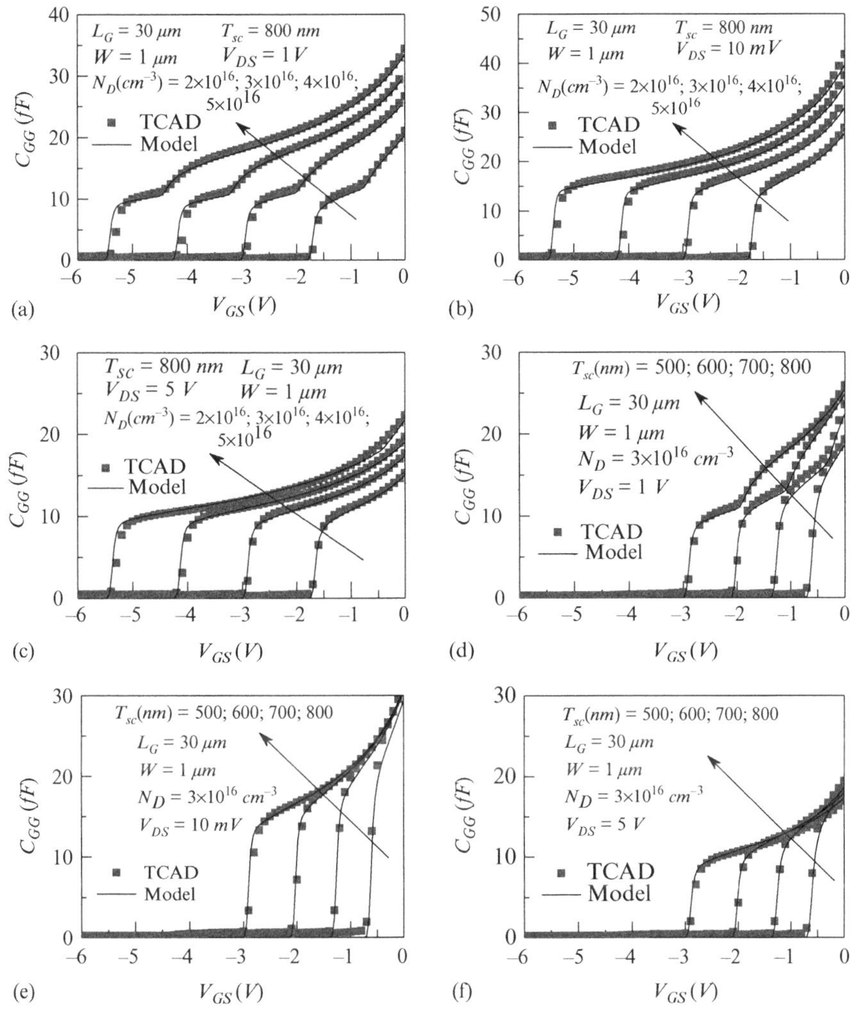

Figure 12.3 The intrinsic gate capacitance versus the gate voltage For a 800 *nm* silicon thickness doped at different doping concentrations (a–c) and also for a silicon doped at $N_D = 3 \times 10^{16} cm^{-3}$ for different thicknesses (d–f). Dependence of the intrinsic gate capacitance (C_{GG}) versus the gate voltage (V_{GS}) and drain-to-source potential (V_{DS}) for different physical parameters of a JFET (N_D and T_{sc}). The lines and symbols are related to the analytical model and TCAD simulations, respectively ($N_D = 2 \times 10^{16}$ cm^{-3} to 5×10^{16} cm^{-3}, $T_{sc} = 500$ to $800\,nm$, $L_G = 30\,\mu m$, $W = 1\,\mu m$).

12.3.3 Simulations

The intrinsic gate capacitance versus gate voltage of a junction double-gate MOSFET obtained from 2D TCAD simulations and the model are plotted on Figure 12.3. Figure 12.3a–c shows the dependence of intrinsic gate capacitance versus gate voltage for $V_{DS} = 1\,V$, $V_{DS} = 10\,mV$, and $V_{DS} = 5\,V$. The doping concentration varies from $2 \times 10^{16}\,cm^{-3}$ to $5 \times 10^{16}\,cm^{-3}$ for a 800 *nm* silicon channel

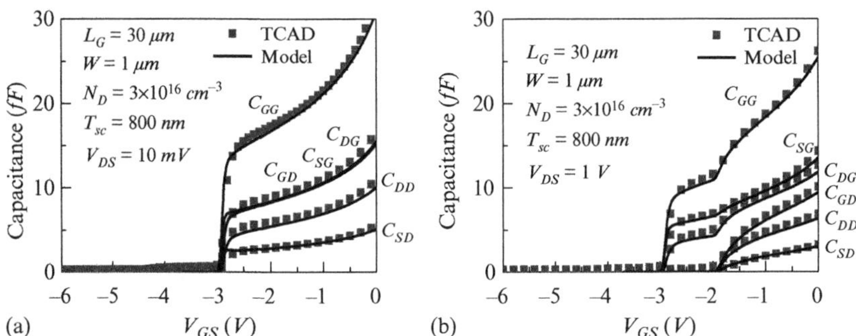

Figure 12.4 Dependence of the transcapacitance matrix elements versus the gate voltage (V_{GS}) and the drain-to-source potential ($V_{DS} = 10\,mV$ and $1\,V$). The lines and symbols are related to the analytical model and TCAD simulations, respectively ($N_D = 3 \times 10^{16}\,cm^{-3}$, $T_{sc} = 800\,nm$, $L_G = 30\,\mu m$, $W = 1\,\mu m$).

thickness JFET with $W = 1\,\mu m$ and $L_G = 30\,\mu m$. The model is accurate in linear and in saturation, from deep depletion to flat-band regimes.

Figure 12.3d–f predicts the intrinsic gate capacitance behavior versus the gate potential for $V_{DS} = 1\,V$, $V_{DS} = 10\,mV$, and $V_{DS} = 5\,V$. The channel silicon thickness varies from $500\,nm$ to $800\,nm$ while the doping concentration in the channel is fixed and equal to $3 \times 10^{16}\,cm^{-3}$ with $W = 1\,\mu m$ and $L_G = 30\,\mu m$. Again, the model predicts very well intrinsic gate capacitance versus gate voltage for different values of silicon thickness and drain-to-source potential.

The transcapacitance versus the gate-source voltage for different drain potentials are shown in Figure 12.4. The agreement between the model developed for junction double-gate FETs and TCAD simulations is good. The lines and symbols hold for the analytical model and TCAD simulations, respectively.

12.4   Summary

The analytical simulations carried out by introducing an infinite gate capacitance in the core model equations of the junctionless FET charge-based model confirm that the JFET can be merely considered as a special case of the junctionless FET. The model relies on physical parameters only, namely the channel thickness, the channel length and width, the channel-doping concentration, as well as the semiconductor material constants. This model covers the operation of the long-channel double-gate JFET among all operating regions, namely from below to above threshold and from linear to saturation operation, without recurring to empirical smoothing functions. As such the model constitutes an ideal basis for a full-circuit simulation model of double-gate, symmetric JFETs.

13 Modeling Junctionless FET with Interface Traps Targeting Biosensor Applications

Often encountered in the literature dealing with label-free biosensing technologies are sensors made of semiconductor nanowires, silicon nanowires in particular. Since semiconductor nanowires are compatible with CMOS technology, they are often considered as ideal devices to detect chemical and biological elements, including DNA, viruses, and cancer markers for instance. The principle of operation and *and* the architecture of these nanowires are very similar to the gate-all-around junctionless FETs, except that there is no solid-state electrode over the gate insulator. These nanowires have no source/drain implant and are highly doped i.e., they are fabricated with minimal technological steps. In addition, they exhibit a high surface-to-volume ratio. These features explain why they are seen as the ultimate devices for label-free detection of ions, nucleic acids, and protein markers to say the least. More generally, nanowire FETs can be configured as highly sensitive biosensors by linking recognition or receptor molecules to their surface [189]. However, despite being used in many "bio" applications, analytical models simulating the intrinsic characteristics of such ungated nanowires are missing. Extending the approach presented in Chapter 3 by including additional features to predict the electrical properties *intrinsic* of semiconductor nanowires when used as biosensors [190] is the goal of this chapter.

13.1 Principle of Semiconductor-based Field-Effect Biosensors

The electrostatic interaction between a solution containing chemical species (ions, biomolecules, etc.) and the gate electrode of a field-effect transistor is a topic of interest in the molecular processes of biology. At the beginning was the so-called ion-sensitive field-effect transistor (ISFET) proposed by Bergveld in the 1970s [191]. This device was aimed at measuring ion activities in electrochemical and biological environments and in particular the action potentials that build up during neuronal activity. An overview of ISFET technology can be found in [192–195] and a typical representation is shown in Figure 13.1 [192]. The principle of operation relies on the modification of the charge density between a liquid and an ion-sensitive gate, which induces variations in the drain current.

When immersing an insulator (silicon dioxide) in an electrolyte, a transition layer is created at the interface between the silicon dioxide and the liquid. This layer

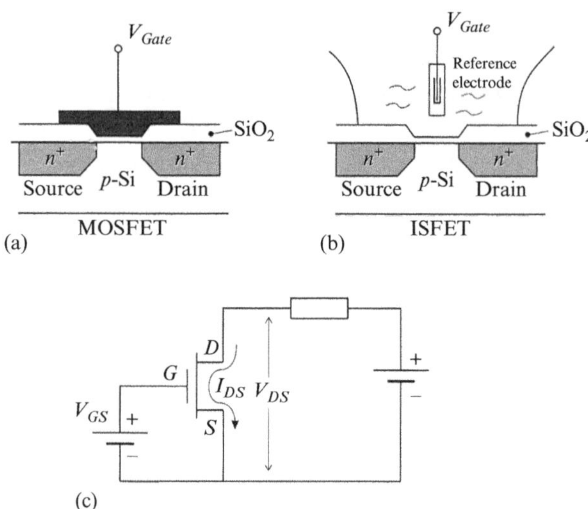

Figure 13.1 Schematic of (a) MOSFET, (b) ISFET, and (c) electronic diagram. Reprinted from [192] with permission.

consists of ions and counter-ions whose arrangement depends on the liquid–solid interface properties and on the applied potential (between the solution and the FET channel). However, unlike in a semiconductor where only electrons or holes are free to move and atoms remain at rest, both negative and positive ions redistribute at the liquid–solid interface, called the electrical double layer (EDL) (see [196, 197] for further reading). This means that provided the potential drop at the liquid–solid interface is known, the induced charge density on the gate insulator (or gate electrode) can serve as the signal input for the nanowire sensor. A very well-known case is the silicon dioxide–electrolyte interface modification upon pH.

Still, two kinds of devices are used to sense ions and biomolecules. One relying on inversion-mode field-effect transistors (IM FET) such as the classical planar MOSFET and related multigate architecture FETs [198–203], while the other relies on the junctionless FET where no *pn* junction is implemented for source and drain. Nanowire for biosensing anticipated the junctionless nanowire FET [204]. From a technology point of view, fabricating unipolar SiNW FETs is more simple than implementing a MOSFET, reason why they have been used as biosensors first. However, whereas experiments confirmed that the conductance of these field-effect devices changed depending on the medium they were in contact with, less attention was given to their principle of operation. In pioneering work [205] stating the proof-of-concept for cancer markers detection using silicon nanowire transistors, it was noted that nanowires operated in accumulation mode, but no quantitative evaluation was offered. Similarly, detection of proteins down to single viruses was also achieved [206] with unprecedented sensitivity. Therefore, upon appropriate surface treatment with specific molecules acting as discriminating elements, semiconductor nanowires can used as sensitive and selective devices for label-free detection of biological macromolecules.

However, these devices have many more crystal and surface defects/contamination than state-of-the-art CMOS transistors, essentially for two reasons. First, if the bare surface of the nanowire exposed to chemicals (liquid and gases) it is likely to be damaged when immersed in harsh environments (in addition to suffering mechanical alterations). Next, these nanowires are fabricated with raw technology that does not use dedicated postprocessing to neutralize fixed-charges and traps. Thus defects that will distort the charge–voltage characteristics are expected, and from a modeling viewpoint, this should be included in compact models. Since relationships governing the potential drop at the liquid–gate–insulator interface are known [196,197] and are not specific to the FET sensor architecture, providing additional features to account for deep levels at the Si/SiO_2 interface and at the surface of the nanowire is required. Here we will extend the model in Chapter 3 to account for surface traps [207] before simulating transfer characteristics of junctionless nanowire targeting biosensor applications.

13.2 Modeling Surface Traps in Junctionless FETs

Introducing interface-trapped charges in compact models has been proposed for inversion-mode MOSFETs (see, e.g., [207–209]). A model to estimate the depletion region induced by the traps at the surface of a silicon nanowire assuming full depletion of the channel has also been reported (see, e.g., [210]). However, these solutions are not appropriate for simulating junctionless FETs because JLFETs are fundamentally different from inversion-mode devices, in part, because the full-depletion approximation is a restrictive condition.

13.2.1 General Considerations for Interface Traps

Traps are classified as donors when they are positively charged upon releasing an electron, or as acceptors when they are negatively charged upon trapping electrons [211]. Traps with energies above the mid-gap exhibit acceptor-like characteristics. These traps are neutral when empty (trap energy above the Fermi level), and become negatively charged when trapping an electron (trap energy below the Fermi level). Interface-charged traps are amphoteric and whether they behave as donors or acceptors depends on their energy in the band-gap [211–213].

Traps with energies lying in the band-gap of the semiconductor [211] are the most critical since their charge varies with the channel and surface potentials and they affect the electrical characteristics of the transistor.

According to Shockley–Read statistics, four processes determine the trap occupancy of a discrete level:

- capture of electrons from the conduction band,
- emission of electrons,

- capture of holes from the valence band,
- emission of holes.

From kinetic rate equations [212] at thermal equilibrium and steady-state conditions, the traps' occupation probability for electrons or holes is $f_A(E_t) = \dfrac{v\sigma_n n_s}{e_n + v\sigma_n n_s}$ and $f_D(E_t) = \dfrac{v\sigma_p p_s}{e_p + v\sigma_p p_s}$, respectively, where E_t is the energy level of the traps; v is the thermal velocity; σ_n and σ_p are the cross-sections of the electrons and holes; n_s and p_s are the free-carrier concentrations; and e_n and e_p are the emission coefficients for the electrons and holes.

The trap-energy level at which the trapped electron has equal probabilities of being thermally excited to the conduction band or recombined with a hole from the valence band can be defined from the equality between the electron emission and the hole-capture rates, leading to the so-called demarcation level E_{s_d} [212]. Referring energies to the Fermi level (E_F) gives $E_{s_d} - E_f = E_i + k_B T \ln\left(\frac{\sigma_p}{4\sigma_n}\right) \approx E_i + 0.04\,eV \approx E_i$, where E_i is the intrinsic Fermi level and $k_B T$ is the thermal energy. This means that electrons trapped at energies below E_{s_d} will soon recombine with holes. Similar calculations are done for trapped holes. In summary, interface traps located above mid-gap will behave as acceptor traps whereas those below mid-gap will be donor traps [211,212]. In practice, only traps located in the band-gap of the semiconductor effectively change their charge state upon biasing.

13.2.2   Modeling Trapped Charges at the Semiconductor/Insulator Interface in Junctionless FETs

Self-Depletion in Ungated Junctionless FETs

A special case of a junctionless architecture where the channel is covered with a bare insulating layer (without any conducting layer) is analyzed in double-gate and nanowire topologies. The channel is n-type doped (N_D) and acceptor-like traps dwell at the semiconductor/oxide interfaces. Since there is no gate electrode, the charge in the channel mirror the traps charge. To simplify the analysis, single-level interface traps located in the band-gap with a degeneracy factor 1 is assumed; a more realistic "U"-shaped distribution will be addressed later (note that the degeneracy factor is expected to equal 2 for electron traps and 4 for holes traps; the equivalent shift is about $30\,mV$ with regard to a hypothetical degeneracy factor of 1). Substitution of $f_A(E_t)$ with the expression of the free-carrier density derived from semiconductor physics, the occupation probability for acceptor traps within the band-gap is:

$$f(E_t) = \frac{1}{1 + \exp\left(\dfrac{E_t - E_f}{k_B T}\right)}, \tag{13.1}$$

where $f(E_t)$ is the trap-occupation probability at energy level E_t. The interface-charge density per unit surface Q_{ss} is obtained by integrating the trap density of states $N_s(E)$ over the band-gap (only mid-gap traps are considered) weighted by the occupation probability $f(E)$ [211, 212]:

$$Q_{ss} = -q \int_{E_V}^{E_C} N_s(E) f(E) dE = -q \int_{E_V}^{E_C} \frac{N_s(E)}{1 + \exp\left(\frac{E - E_f}{k_B T}\right)} dE. \tag{13.2}$$

To be consistent with the charge-based model of Chapter 3, the intrinsic Fermi level evaluated at the surface E_{is} is introduced in relation (13.1) and new variables are defined accordingly ($\Psi_s = -(E_{is} - E_f)/q$ and $E_{t-i} = E_t - E_{is}$). Considering a unique trap energy E_t and a trap density per unit surface N_s, after substituting Ψ_s and E_{t-i} into (13.1), the interface-trapped charge density satisfies:

$$\bar{Q}_{ss} = \frac{-qN_s}{1 + \exp\left(\frac{E_{t-i}}{k_B T}\right) \exp\left(-\frac{\Psi_s - V_{ch}}{U_T}\right)}, \tag{13.3}$$

where $U_T = k_B T/q$ is the thermal voltage and V_{ch} is the channel potential. The presence of acceptor-like interface traps tends to bend the bands upward at the surface of the semiconductor (since they will remove electrons from the channel), even without any gate electrode. Thus, the total charge in the semiconductor and the charge density trapped at the interface (negatively charged since these are acceptors) must compensate for each other (Ψ_s is expected to be negative in relation (13.3)).

This interface-trapped charge density is introduced in the charge-based model for the double-gate topology. The central relationships that link the surface potential (Ψ_s), the central potential (Ψ_0), and the surface electric field (E_s) are as follows:

$$\Psi_s \approx \Psi_0 + \frac{q n_i T_{sc}^2}{8 \varepsilon_{si}} \left[\exp\left(\frac{\Psi_0}{U_T}\right) - \frac{N_D}{n_i} \right], \tag{13.4}$$

$$E_s^2 = \left(\frac{Q_{sc}}{2\varepsilon_{si}}\right)^2 = \frac{2 q n_i U_T}{\varepsilon_{si}} \left[\exp\left(\frac{\Psi_s}{U_T}\right) - \exp\left(\frac{\Psi_0}{U_T}\right) - \frac{N_D}{n_i U_T}(\Psi_s - \Psi_0) \right]. \tag{13.5}$$

In addition, charge neutrality requires neutrality between the charges in the semiconductor and the charges trapped at the interfaces:

$$Q_{sc} = -2\bar{Q}_{ss} = \frac{-2qN_s}{1 + \exp\left(\frac{E_{t-i}}{k_B T}\right) \exp\left(-\frac{q\Psi_s}{k_B T}\right)}. \tag{13.6}$$

Substituting (13.4) and (13.6) into (13.5), the center (Ψ_0) and surface potentials (Ψ_s) are obtained. Not only these quantities depend on the semiconductor doping and thickness, but also on the trap parameters i.e., energy and density. It is worth noting that the usual condition imposed by the gate voltage is substituted with a condition arising from the traps.

For the nanowire geometry, the charge density of the interface traps is calculated with the same set of relationships using the equivalent parameters defined in Table 3.1.

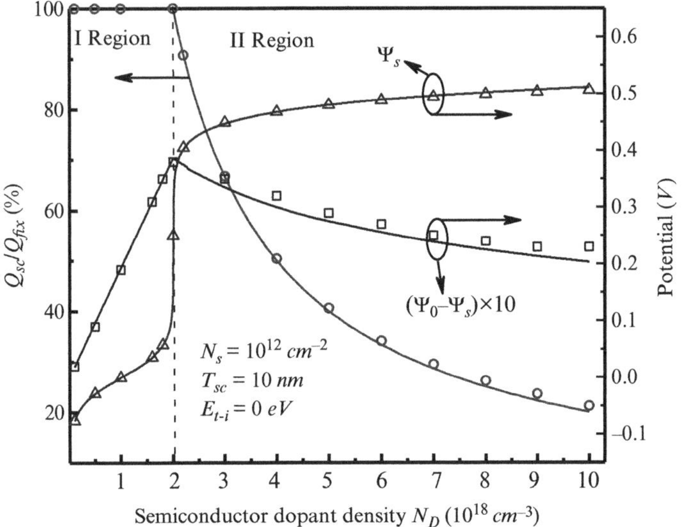

Figure 13.2 Normalized depletion charge in semiconductor film (left axis) and electrostatic potential (right axis) versus doping density (Markers: TCAD simulations; lines: analytical model). Reprinted from [190] with permission.

Model versus Numerical Simulations

Numerical simulations of a long-channel ($1\,\mu m$) junctionless double-gate FET with an oxide thickness of $5\,nm$ hosting surface mid-gap traps used for reference (note that mid-gap means $E_{t-i} = 0$). The trap density is fixed to $10^{12}\ cm^{-2}$. In addition, two silicon thicknesses are considered, 10 and $20\,nm$.

Figure 13.2 shows the semiconductor charge density normalized to the total fixed-charge density ($Q_{fix} = qN_DT_{sc}$) (left axis) for a $10\,nm$ silicon thickness, as a function of the doping density. At low doping levels, below $2 \times 10^{18}\ cm^{-3}$, most of the mobile carriers in the channel are captured by the traps, creating a deep depleted channel. This value of 2×10^{18} corresponds to the threshold at which traps and donors balance each other out. For higher doping densities, the semiconductor is only partially depleted (see Region-II in Figure 13.2) and Q_{sc} remains close to the total trap concentration $2qN_S$ (the factor of two comes from the two gates), explaining the $1/N_D$ trend.

The semiconductor surface potential and the difference between the center and surface potentials (magnified 10 times) are shown (right axis) with respect to the semiconductor-doping density. If each trap is active, the channel will be in full depletion up to a carrier density of $2 \times 10^{18}\ cm^{-3}$. This is where the band bending inside the silicon is maximum. Similarly, a sharp transition of the surface potential occurs for the same critical doping density. In any case, the analytical approach matches the numerical simulations.

The semiconductor-depletion charge normalized to the fixed-charge density for different trap energies E_{t-i} is illustrated in Figure 13.3 as a function of the doping.

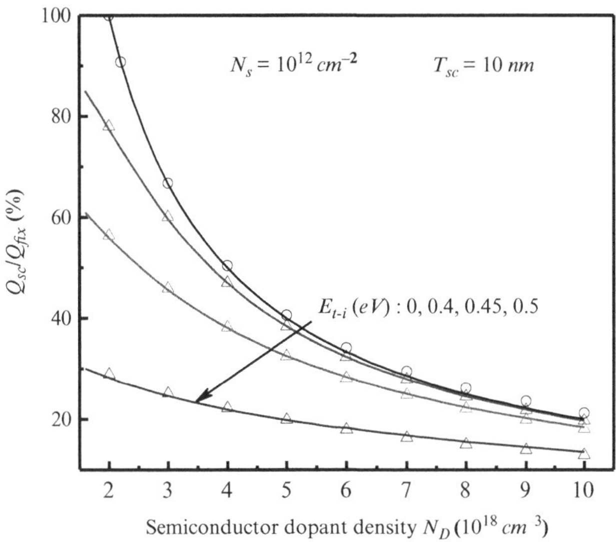

Figure 13.3 Normalized depletion charge in semiconductor-film versus semiconductor-dopant density considering different trap-energy levels. Markers: TCAD simulations; lines: analytical model. Reprinted from [190] with permission.

For a given doping level, increasing the energy of the traps with respect to the intrinsic level decreases the space charge in the channel. This trend is seen in relation (13.1): the farther the traps ($E_{t-i} = 0.45 - 0.52\,eV$), the less active they are and the less their ionization state is dependent on the doping density.

The traps' occupation probability for different trap energies is depicted in Figure 13.4 versus the doping density. As expected, traps with energy close to the conduction band ($E_{t-i} = 0.45 - 0.52\,eV$) are less occupied. Even for high doping densities, the filling factor is around 50 percent, whereas almost full trapping occurs for mid-gap traps. These trends are well captured by the model.

The same analysis carried out for a nanowire geometry using the equivalent model parameters proposed in Chapter 3 is represented in Figure 13.5 (note that the trap density has to be expressed per unit surface of the nanowire). Again, for $E_{t-i} = 0$ the plateau observed below some critical doping density means that all available free carriers (electrons) are captured by the traps. The normalized mobile charge in a $5\,nm$ nanowire radius versus the doping density is illustrated in Figure 13.6 for different interface-trap energies. The mobile-charge density is accurately predicted for a broad range of doping levels and trap energies. In particular, the exponential dependence is well captured. This would not be possible with full-depletion approximation. Obviously, not only mid-gap states are effective at depleting the channel. Higher energy states have to be included as well. In summary, full depletion can take place for in ungated junctionless FETs for common values of interface-trap densities with technological parameters (silicon thickness and doping concentration) compatible with current nanowire technology.

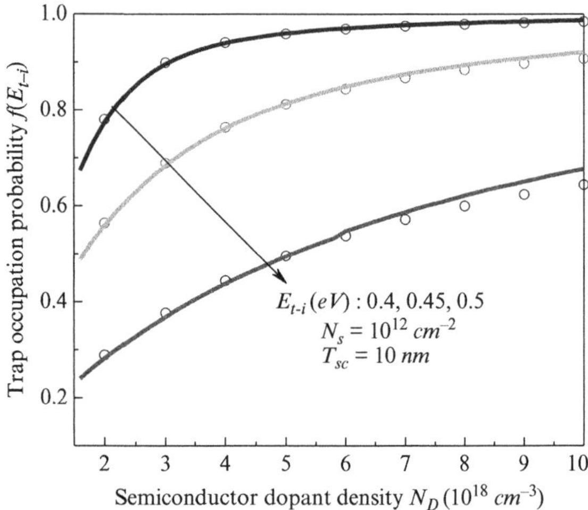

Figure 13.4 The traps' occupation probability at different trap-energy levels versus semiconductor doping for different trap energies. Markers: TCAD simulations; lines: analytical model. Reprinted from [190] with permission.

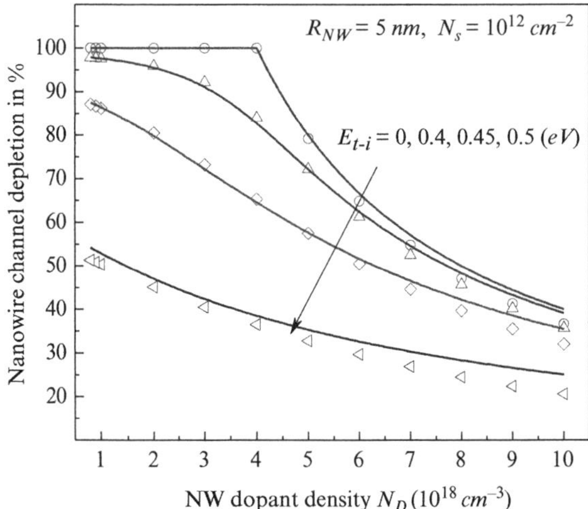

Figure 13.5 Nanowire normalized-depletion charge versus nanowire-dopant density considering different trap-energy levels. Markers: TCAD simulations; lines: analytical model. Reprinted from [190] with permission.

Gated Junctionless FETs with Traps

The effect of the gate and drain potentials was not included in (13.3)–(13.6). To remove this limitation and add a gate electrode to the junctionless FET (double-gate and nanowire), a slightly different analysis is needed since the gate electrode

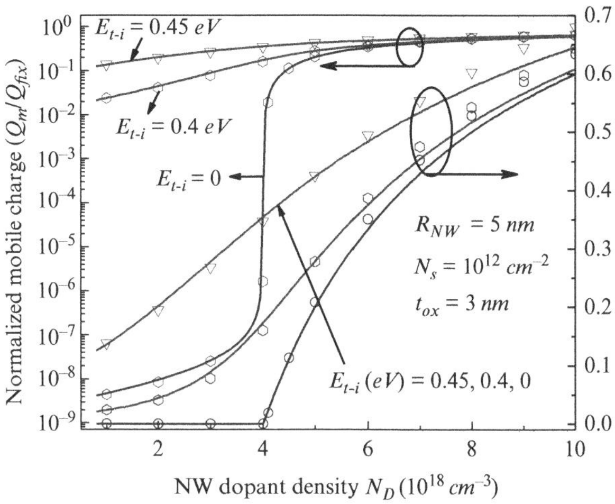

Figure 13.6 Nanowire normalized-mobile charge versus nanowire-dopant density considering different trap-energy levels in log-scale (left axis) and in linear scale (right axis). Markers: TCAD simulations; lines: analytical model. Reprinted from [190] with permission.

must now be part of the charge neutrality condition. Relations (13.3) to (13.5) are rewritten for clarity, while introducing the pseudo-Fermi potential V_{ch}:

$$\Psi_s \approx \Psi_0 + \frac{qn_iT_{sc}^2}{8\varepsilon_{si}}\left[\exp\left(\frac{\Psi_0 - V_{ch}}{U_T}\right) - \frac{N_D}{n_i}\right], \tag{13.7}$$

$$E_s^2 = \left(\frac{Q_{sc}}{2\varepsilon_{si}}\right)^2 = \frac{2qn_iU_T}{\varepsilon_{si}}\left[\exp\left(\frac{\Psi_s - V_{ch}}{U_T}\right) - \exp\left(\frac{\Psi_0 - V_{ch}}{U_T}\right) - \frac{N_D}{n_iU_T}(\Psi_s - \Psi_0)\right], \tag{13.8}$$

$$\bar{Q}_{ss} = \frac{-qN_s}{1 + \exp\left(\dfrac{E_{t-i}}{k_BT}\right)\exp\left(-\dfrac{\Psi_s - V_{ch}}{U_T}\right)}. \tag{13.9}$$

The source Fermi level is set as the reference level. From (13.7), the surface potential at flat-band at the source side is given by $\Psi_{sFB} = U_T \ln(N_D/n_i)$, and does not depend on the charges trapped on the surface. In the absence of surface traps ($\bar{Q}_{ss} = 0$), the applied gate voltage (V_{GS}) at flat-band reverts to the metal-semiconductor work function potential difference (W_{ms}), and thus the effective gate voltage is defined as $V_{GS}^* = V_{GS} - \Delta\phi_{ms} = V_{GS} - W_{ms} - U_T \ln(N_D/n_i)$.

Yet, when surface traps are included, the gate voltage at flat-band should compensate for the band bending generated by the metal semiconductor work function difference and by the charges trapped at the interface. Relation (3.35) becomes:

$$Q_{sc} = -2Q_G = -2C_{ox}\left(V_{GS}^* + \frac{\bar{Q}_{ss}}{C_{ox}} - \Psi_s\right). \tag{13.10}$$

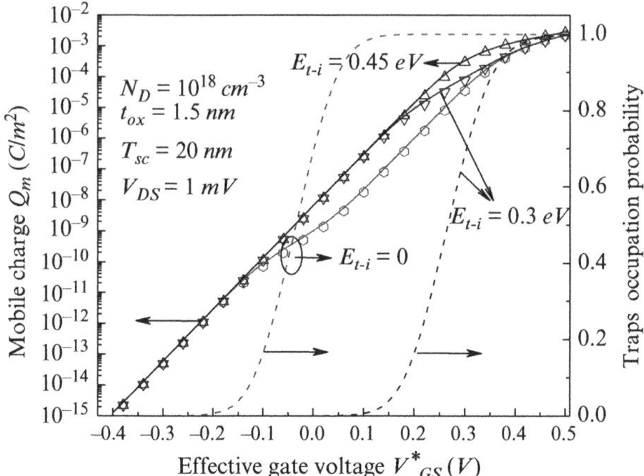

Figure 13.7 The mobile-charge density junctionless double-gate FET calculated from the analytical model (lines) and TCAD simulations (markers) depicted on left axis in logarithmic scale. Reprinted from [190] with permission.

A simple approach to solve the set of relations (13.7)–(13.10) is as follows: assuming a given value for Ψ_0, the surface potential Ψ_s is calculated from relation (13.7). Next, this feeds relations (13.8) and (13.9), where Q_{sc} and $\bar{Q}_{ss}$ are determined, together with Q_G from (13.10). Finally, the Q_{sc} versus V_{GS}^* plots are obtained.

The nanowire geometry follows the same approach: (13.7)–(13.10) are resolved with the equivalent parameters (see Table 3.1) using the gate capacitance for the cylindrical geometry. The mobile-charge density with respect to the effective gate voltage is illustrated in Figure 13.7 and is compared to numerical simulations. Different energy levels for interface-trapped charges are considered as well.

It should be noted that in contrast to the ungated device, for a doping density of $N_D = 1 \times 10^{18}\,cm^{-3}$, high energy levels that were not very active in ungated devices start affecting the mobile-charge density (see Figures 13.3 and 13.4). In fact, increasing the gate voltage lowers the energy of the traps, pulling down the high energies.

Meanwhile, the occupation probability $f(E_t)$ (the dashed lines in the figure) defines the transition regions, and it is clear that any shift in the trap energy turns into a shift in the gate voltage (note that the shift in $E_{t-i}/(-q)$ is not the shift in V_{GS}^*).

In addition, interface traps with mid-gap energies degrade the slope in the deep subthreshold, whereas traps located at 0.3 and $0.45\,eV$ above the intrinsic level tend to modify the charge characteristics near the threshold voltage.

13.2.3   Current Derivation

Estimation of the current in the channel is expected to be different from the expression derived in Chapter 3 since the charge–voltage dependence is now modified by the presence of the traps. However, as can be seen, using the same expressions for the current where the mobile-charge density is evaluated at source and drain (but now including traps) is still accurate. At first sight, relation (13.10) suggests that the effect of the interface-charge density is added potential shift $\delta = \bar{Q}_{ss}/C_{ox}$ to the effective gate voltage. However, strictly speaking this is not correct since the interface-charge density will not keep the same value at $(V^*_{GS} + \delta)$. Nevertheless, this approximation is found to be accurate for Q_{sc}.

Therefore, $\bar{Q}_{ss}$ is calculated first with the real voltages. Next, the potential shift δ is added to V^*_{GS} and the approximated semiconductor charge density Q_{sc} is obtained using the charge–voltage relationships (3.43) and (3.46). These values are used to estimate the drain current with interface traps. In depletion mode, this looks like:

$$I_{Dep}(V^*_{GS} + \delta, V_{DS})$$

$$= \frac{W}{L_G}\mu \left\{ \left(\frac{1}{8C_{si}} - \frac{1}{4C_{ox}} \right) Q^2_{sc} - \frac{Q^3_{sc}}{12 Q_{fix} C_{si}} + \left(\frac{Q_{fix}}{2C_{ox}} + 2U_T \right) Q_{sc} - U_T Q_{fix} \ln\left(1 + \frac{Q_{sc}}{Q_{fix}} \right) \right\} \Bigg|^D_S ,$$

$$(13.11)$$

where $Q_{sc} = Q_{sc}(V^*_{GS} + \delta, V_{DS})$ is calculated either with (3.43) or (3.46) or explicitly by means of relation (13.14) in [92].

Figure 13.8 displays the current density for double-gate (a) and nanowire FETs (b) with a mid-gap trap density of $10^{12}\ cm^{-2}$ and a default mobility of $1,417\ cm^2/V.s$ (the default value in TCAD Sentaurus assuming constant mobility) versus the effective gate voltage at low ($V_{DS} = 1\ mV$, $V_{DS} = 0.1\ V$) and high ($V_{DS} = 1\ V$) drain voltages.

The subthreshold swing degradation due to the presence of interface traps is seen in logarithmic scale (left axis). For comparison, the drain current is also calculated for the case where no traps are introduced ($V_{DS} = 1\ V$).

Figure 13.9 depicts the gate transconductance (g_m) for a double-gate junctionless MOSFET. Low (0.1 V) and high (1 V) drain voltages are considered. The continuity of the model is satisfied.

13.2.4 The Case of Continuous Energy-Trap Distribution

Discrete interface-trap levels are useful to validate the approach but are not representative of real devices where a U-shaped distribution of interface states in the band-gap is common [214, 215]. The model developed for discrete level traps can still be used but an exponential distribution for the trap density with respect to the energy needs to be introduced. Therefore, N_s is replaced with

$$N^*_s = N_{sT} \int\limits_0^{E_{sT}} \exp\left(-\left| \frac{E - E_{sT}}{E_j} \right| \right) dE, \qquad (13.12)$$

and E_{t-i} becomes E_{sT}.

N_{sT} is the maximum value of the interface state density, E_{sT} is the energy level for N_{sT}, and E_j is set to $0.035\ eV$ to be consistent with the default value of the numerical TCAD simulations.

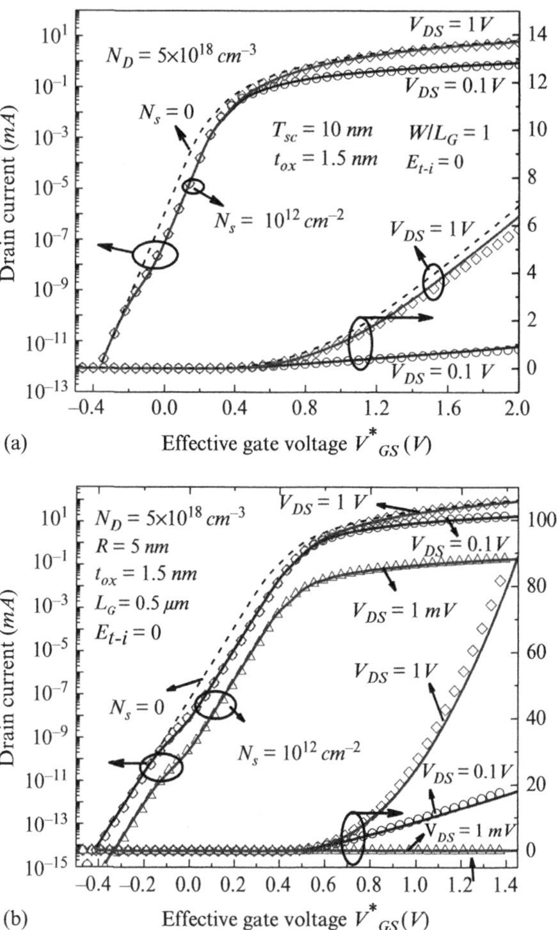

Effective gate voltage $V^*_{GS} (V)$

Figure 13.8 Drain current versus gate voltage in linear scale (right axis) and logarithmic scale (left axis) in (a) junctionless double-gate FET and (b) nanowire junctionless FET. The drain current at $N_s = 0$ and $V_{Ds} = 1\ V$ is shown by the black dashed line in logarithmic scale. Analytical model, lines; TCAD simulations, markers. Reprinted from [190] with permission.

The trapped charge density is

$$\bar{Q}_{ss} = \frac{-qN_s^*}{1 + \exp\left(\dfrac{E_{sT}}{KT}\right) exp\left(-\dfrac{\Psi_s - V_{ch}}{U_T}\right)}. \tag{13.13}$$

Using the data reported in [215] with the parameters $E_{sT} = 0.36\ eV$ i.e., traps $0.92\ eV$ above the valence band, $N_{sT} = 4 \times 10^{13} cm^{-2}\ eV^{-1}$, and assuming $E_j = 0.035\ eV$, the model predictions are as shown in Figure 13.10. Modeling the interface traps with the exponential energy distribution with the tuned parameters (N_s^*, E_{sT}) is consistent with the numerical simulations. Importantly, the continuous energy-trap distribution shifts the $I-V$ transfer characteristics and distorts the subthreshold swing.

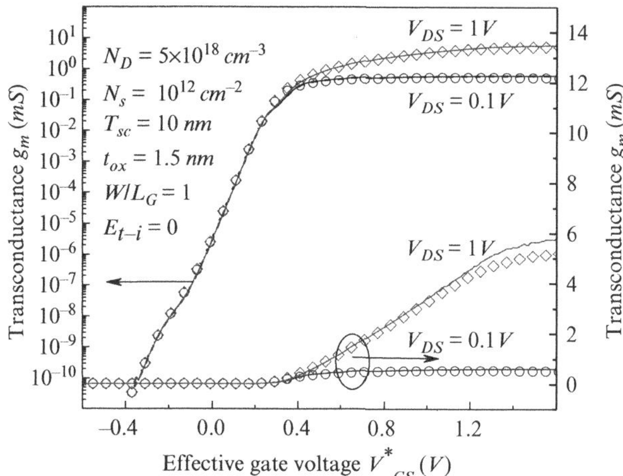

Figure 13.9 Gate transconductance versus effective gate voltage in a junctionless double-gate FET. Analytical model accounting for trapped interface, lines; TCAD simulations, markers. Reprinted from [190] with permission.

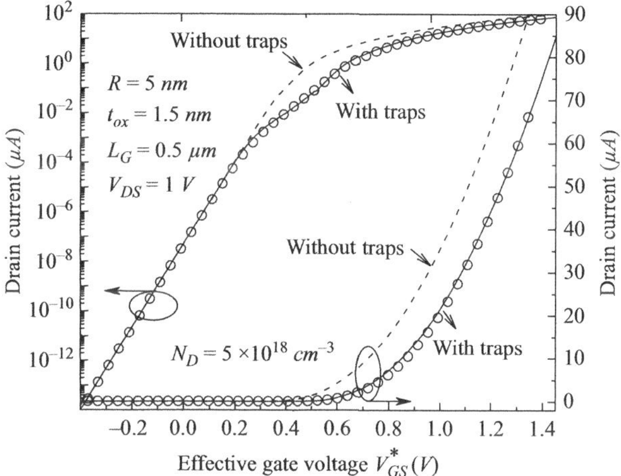

Figure 13.10 Drain current versus effective gate voltage in linear scale (right axis) and logarithmic scale (left axis) in nanowire junctionless FET calculated at energy distribution of interface traps. The drain current calculated at $N_s = 0$ is shown by the black dashed line. Markers, TCAD simulations; lines, analytical model. Reprinted from [190] with permission.

13.2.5 Summary

Inclusion of surface traps in junctionless ungated and gated in planar and nanowire topologies was done on top of a charge-based model, extending the work done in Chapter 3 to simulate the nanowires used in biosensing applications. As such, this approach can be used to predict and characterize transfer characteristics of semiconductor-based nanowire-targeting biosensors in which the bare surface is likely to generate critical artifacts during measurements.

Appendix A   Design-Space of Twin-Gate Junctionless Vertical Slit FETs

A.1   Design-Space of Twin-Gate Junctionless Vertical Slit FETs

As discussed in Chapter 1, among the junctionless devices that can potentially sustain scalability is the junctionless vertical slit field-effect transistor (JL VeSFET), which uses a novel 3D integration scheme [17, 19]. This architecture combined with easier fabrication steps has received attention for its ability to integrate digital circuits with a highly regular layout combined and easier fabrication steps [32, 33, 216, 217].

However, there are still some technological limitations that have to be quantitatively assessed in order to optimize device performance, while scaling down the dimensions below tens of nanometers. The impact of technological parameters such as the doping concentration, the oxide thickness, and the channel width on the intrinsic *off*-current, subthreshold swing, DIBL rail-to-rail voltage, and I_{on}/I_{off} ratio was investigated in [218, 219] from a TCAD aspect.

A.1.1   Device Structure

The 3D geometry and the top view of the VeSFET is shown in Figure A.1. The minimum and maximum values of the silicon thickness along the channel are denoted $T_{sc,min}$ and $T_{sc,max}$, and the geometrical physical gate length as L_G. According to the device geometry, these parameters are linked to the radius r and oxide thickness t_{ox} through the following relations:

$$T_{sc,min} = 2r\left(\sqrt{2} - 1\right) - 2t_{ox}, \tag{A.1}$$

$$T_{sc,max} = \sqrt{2}\left(r - t_{ox}\right), \tag{A.2}$$

$$L_G = \sqrt{2}\,r. \tag{A.3}$$

A.1.2   Electrostatics in Junctionless VeSFET and Design-Space

The electrostatic potential distribution and the channel electron density obtained from TCAD simulations for different values of radius (r) are illustrated in Figure A.2a–d. The gate-oxide thickness t_{ox} is $2\,nm$ and the doping concentration

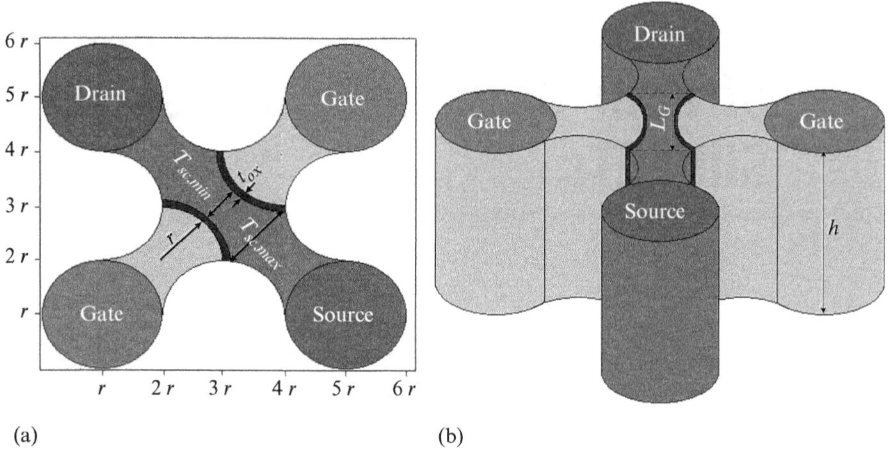

(a)   (b)

Figure A.1 (a) Top view and (b) 3D geometry of the n-type junctionless VeSFET investigated here. Reprinted from [218] with permission.

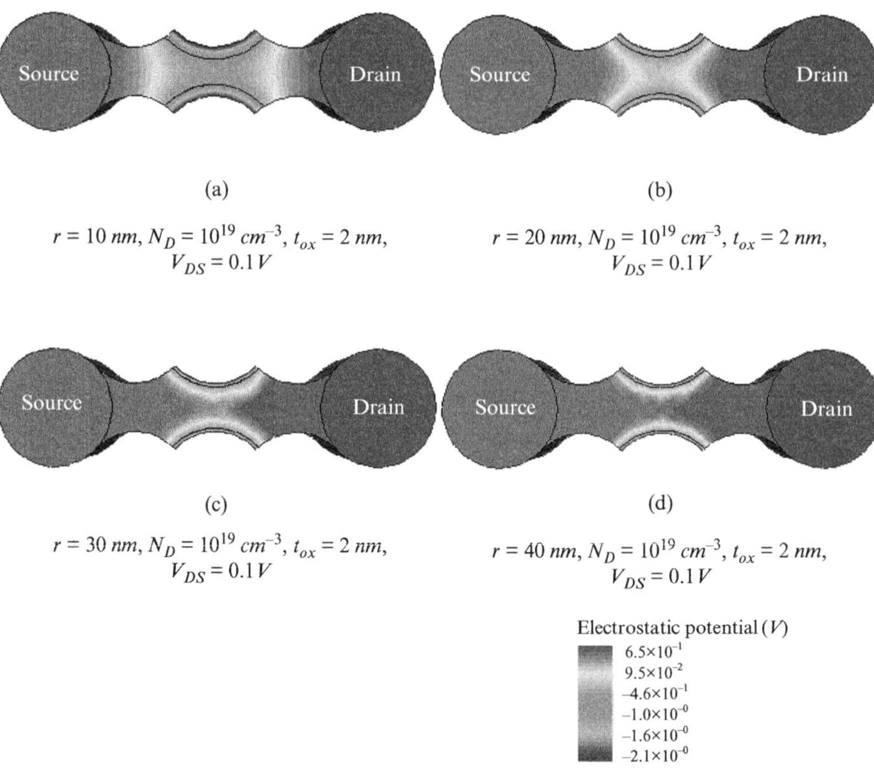

(a)

$r = 10\ nm$, $N_D = 10^{19}\ cm^{-3}$, $t_{ox} = 2\ nm$, $V_{DS} = 0.1\ V$

(b)

$r = 20\ nm$, $N_D = 10^{19}\ cm^{-3}$, $t_{ox} = 2\ nm$, $V_{DS} = 0.1\ V$

(c)

$r = 30\ nm$, $N_D = 10^{19}\ cm^{-3}$, $t_{ox} = 2\ nm$, $V_{DS} = 0.1\ V$

(d)

$r = 40\ nm$, $N_D = 10^{19}\ cm^{-3}$, $t_{ox} = 2\ nm$, $V_{DS} = 0.1\ V$

Figure A.2 Electrostatic potential distribution through the channel for different values of r (a–d) in the n-type junctionless VeSFET investigated. The gate voltage is $-2.5\ V$ and the x and y dimensions are normalized to $r = 10\ nm$. Reprinted from [218] with permission.

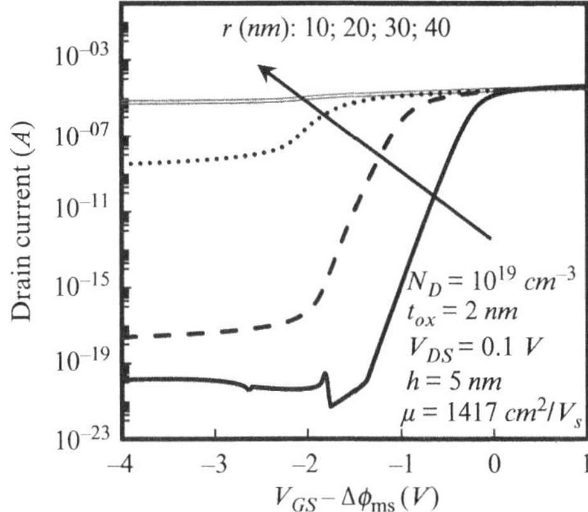

Figure A.3 $I_{DS} - V_{GS}$ characteristics obtained with TCAD simulations for different radius values in an n-type junctionless VeSFET. Note that $\Delta\phi_{ms}$ is the difference between the metal work function and an intrinsic reference semiconductor. Reprinted from [218] with permission.

is considered uniform through the channel and set to $10^{19}\ cm^{-3}$. Next, the drain and gate potential (with respect to the source) are given as 0.1 and $-2.5\ V$, respectively. For the lowest radius considered here i.e., $r = 10\ nm$, we observed that when applying a relatively large negative gate voltage ($-2.5\ V$), the whole channel potential decreased to about $90\ mV$, and still follows the gate voltage (not shown here) as there is full depletion of electrons in the silicon layer. However, increasing the radius up to $r = 40\ nm$, the channel potential is increased and ultimately reverts to its flat-band value (in the center). In other words, the gate voltage no longer affects the center potential for such large silicon thickness i.e., $T_{sc,min} = 29\ nm$; see Figure A.2d. Therefore, the center potential remains almost neutral and the channel cannot be switched off. As for the double-gate junctionless FET, this limitation is not a matter of gate voltage, but rather a phenomenon that occurs when an inversion layer of holes is generated at the channel interface. This layer screens the electric field and impedes any modification of the center potential.

I–V characteristics for a highly doped channel ($1 \times 10^{19}\ cm^{-3}$) (TCAD simulations) for different values of r are shown in Figure A.3. As expected, for relatively thick silicon slits, the *off*-state current becomes of the order of the *on*-current. As for double-gate topology, a design-space for the junctionless VeSFET where the input parameters are the doping density and the slit radius is proposed in [218].

Appendix B   Transient *Off*-Current in Junctionless FETs

Chapter 4 addressed the design space of a double-gate junctionless FET from a technological point of view. It was predicted that an inversion layer at the semiconductor–insulator interface could prevent full depletion of the channel, giving rise to a lower limit for the *off*-state current. In this appendix, the occurrence of such an inversion layer is inferred through transient *off*-state current measurements on a double-gate junctionless FET in a FinFET topology.

The silicon layer is 200 *nm* in height with a phosphorous doping density of about 10^{19} cm^{-3}. The gate stack consists of a nominal 4 *nm* thick layer of SiO_2. Two widths for the silicon fin have been implemented i.e., 15 and 40 *nm* with $L_G = 500$ and $L_G = 100$ *nm*, respectively. Parallel configuration was used to enhance the signal-to-noise ratio i.e., 200 and 100 identical fins for 15 and 40 *nm*, respectively.

Figure B.1 displays the transfer characteristics in linear and logarithmic scales for the devices biased at $V_{DS} = 0.25$ *V* i.e., negligible impact ionization. For the 15 *nm*-wide device, the drain current varies by more than five orders of magnitude when V_{GS} is varied from 1 to −0.8 *V* (subthreshold operation), confirming that such a device can effectively be turned off. However, in the 40 *nm* channel device, the current decreases only by 50 percent when V_{GS} is varied from 0 to −2.5 *V* (see Figure B.1).

Unexpectedly, when the gate voltage is pulsed from 0 volt to negative values V_{GS}^{low}, the current is decreased beyond the steady-state intensity for the 40 *nm*-wide junctionless FET (see Figure B.2) and its DC value (which depends on V_{GS}^{low}) is recovered after some milliseconds. The correlation between poor switch-off capabilities and transient current undershooting can be understood by invoking minority carrier generation (holes in this case), which is a relatively slow process as the channel is *n*-type doped. Assuming there is no inversion layer, the gate-induced electric field is not screened and can achieve deep depletion of the silicon channel i.e., the transistor can be turned off. With time, a hole-inverted layer will emerge gradually at the channel interface and the depletion layer will start shrinking (to maintain charge neutrality). Then the current increases until reaching the steady-state value.

The occurrence of an inversion layer is supported by TCAD simulations carried out on idealized structures with comparable technological parameters i.e., T_{sc}, t_{ox}, and N_D. Figure B.3 shows that for the thicker device ($T_{sc} = 40$ *nm*), an inversion layer i.e., holes is indeed created for low gate voltages (quantum effects have been included), whereas such a situation does not occur for the thinner device i.e., $T_{sc} = 15$ *nm*. Similarly, for gate voltages below −2 *V*, the thick channel cannot be

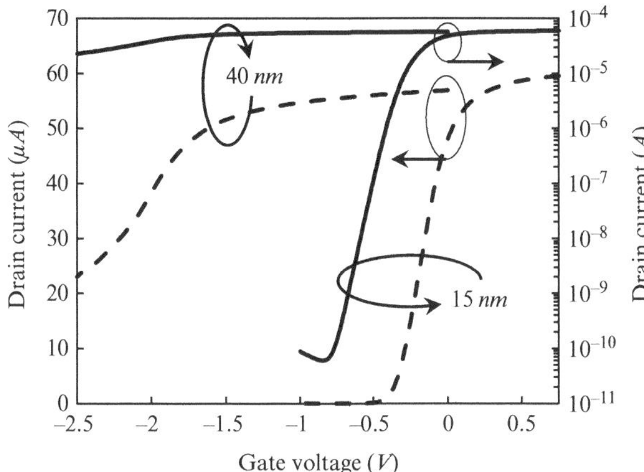

Figure B.1 Transfer characteristics in linear (full line) and logarithmic scales (dashed line) for the 15 *nm* and respectively 40 *nm* silicon thickness junctionless FET. Reprinted from [144] with permission.

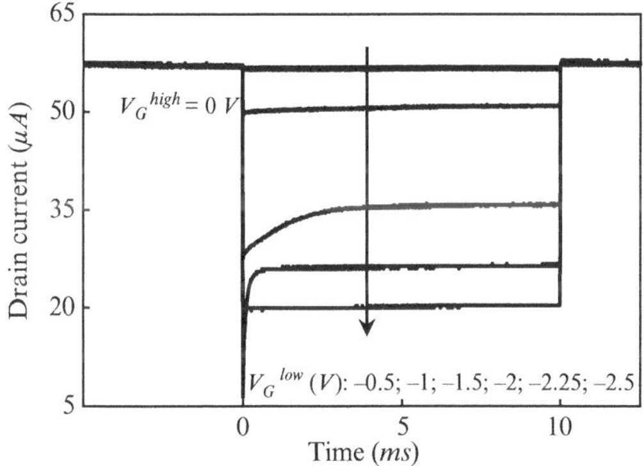

Figure B.2 Transient drain current for the 40 *nm* silicon thickness junctionless FET. Here the gate voltage is pulsed between 0 V and V_{GS}^{low}, and each curve corresponds to a given value of V_{GS}^{low}. Reprinted from [144] with permission.

emptied from electrons. Thus the poor switch-off capabilities of the device can be attributed to the presence of an inversion layer. On the other hand, in thinner devices that we can be switched off there is no inversion layer (it is worth noting that all devices that could not be switched off and have different channel thicknesses exhibited the same transient behavior). Similar effects have been reported in enhancement-mode SOI MOSFETs [220]. Drain-current overshoot related to electrically floating-body partially depleted SOI MOSFETs was used as a novel method to extract recombination lifetimes in thin silicon films [221–223]. Note that the time constants reported in [220–223] are much higher than what was measured in [144], likely due to the nonoptimized fabrication process and the high density of

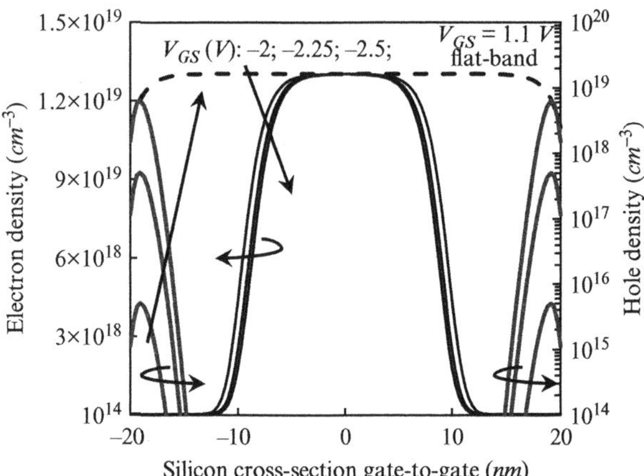

Figure B.3 TCAD simulations of electron (black, linear scale) and hole density (grey, logarithmic scale) through the channel width, for an _n_-type doping level 10^{19} _cm_$^{-3}$. Hole density at the interface increases rapidly as V_{GS} is lowered to more negative values. Reprinted from [144] with permission.

recombination traps. This suggests that using nonswitching-off devices might be an interesting solution to characterize minority carrier generation lifetimes upon transient drain current in junctionless FETs, making it possible to "inspect" channels in nanowires.

Appendix C Derivatives of Mobile Charge Density with Respect to V_{GS} and V_{DS}

Mobile charge density (Q_m) as the mean value of local mobile charge densities at source and drain at **low** V_{DS} is defined by

$$Q_m = \frac{1}{2}\left(Q_{ms} + Q_{md}\right). \tag{C.1}$$

From (C.1), the derivative of Q_m with respect to V_{GS} is given by

$$\frac{\partial Q_m}{\partial V_{GS}} = \frac{1}{2}\left(\frac{\partial Q_{ms}}{\partial V_{GS}} + \frac{\partial Q_{md}}{\partial V_{GS}}\right). \tag{C.2}$$

Note that the charge at the source is not affected by the drain potential.

$$\frac{\partial Q_{ms}}{\partial V_{DS}} = 0. \tag{C.3}$$

The derivative of Q_m with respect to V_{DS} is

$$\frac{\partial Q_m}{\partial V_{DS}} = \frac{1}{2}\left(\frac{\partial Q_{md}}{\partial V_{DS}}\right). \tag{C.4}$$

In addition, at low V_{DS} the derivatives of Q_{ms} and Q_{md} with respect to V_{GS} are also equal, meaning that (C.2) can be written as

$$\frac{\partial Q_m}{\partial V_{GS}} = \frac{1}{2}\left(\frac{\partial Q_{ms}}{\partial V_{GS}} + \frac{\partial Q_{md}}{\partial V_{GS}}\right) = \frac{\partial Q_{md}}{\partial V_{GS}}. \tag{C.5}$$

In addition, the mobile charge densities at source and drain satisfy

$$\frac{\partial Q_{md}}{\partial V_{DS}} = -\frac{\partial Q_{md}}{\partial V_{GS}}. \tag{C.6}$$

Finally, combining (C.4), (C.5), and (C.6) links gate and drain "control" on the mean value of the mobile charge density in the channel:

$$\frac{\partial Q_m}{\partial V_{GS}} = -2\frac{\partial Q_m}{\partial V_{DS}}. \tag{C.7}$$

Appendix D Global Charge Density at Drain in Depletion Mode

$$\overline{Q}_{m,d} = \eta \log\left(Q_m + 2Q_{fix}\right)\left[32AU_TQ_{fix}^5 - 16BU_TQ_{fix}^4 + 8CU_TQ_{fix}^3 + 4U_Tf(Q_{m,s})Q_{fix}^2\right]$$

$$+ Q_m^4\left(\frac{Q_{fix}(I_6 + I_5/I_7)}{2} - \frac{I_4}{16C_{ox}N_D\varepsilon_{si}q}\right) + Q_m^6\left(\frac{AQ_{fix}}{12N_D\varepsilon_{si}q} - \frac{I_8}{24C_{ox}N_D\varepsilon_{si}q}\right)$$

$$- Q_m^5\left(\frac{I_6}{5} + \frac{I_5}{20C_{ox}N_D\varepsilon_{si}q}\right) + Q_m^2\left[Q_{fix}\left(I_3 - \frac{I_1}{I_7}\right) + \frac{I_2}{8C_{ox}N_D\varepsilon_{si}q}\right]$$

$$- Q_m\left\{2Q_{fix}\left[2Q_{fix}\left(I_3 - \frac{I_1}{I_7}\right) + \frac{I_2}{I_7}\right] - 2Q_{fix}U_Tf(Q_{m,s})\right\}$$

$$- Q_m^3\left(\frac{I_3}{3} - \frac{I_1}{12C_{ox}N_D\varepsilon_{si}q}\right) - \frac{AQ_m^7}{28N_D\varepsilon_{si}q}\Bigg|_{Q_{m,s}}^{Q_{m,d}}, \tag{D.1}$$

where

$$I_1 = 3C_{ox}Q_{fix}f(Q_{m,s}) - 2CC_{ox}Q_{fix}^2 + 2N_D\varepsilon_{si}f(Q_{m,s})q - 4CN_DQ_{fix}\varepsilon_{si}q$$

$$+ 8CC_{ox}N_DU_T\varepsilon_{si}q + 8BC_{ox}N_DQ_{fix}U_T\varepsilon_{si}q, \tag{D.2}$$

$$I_2 = 2C_{ox}Q_{fix}^2f(Q_{m,s}) + 4N_DQ_{fix}\varepsilon_{si}f(Q_{m,s})q - 8C_{ox}N_DU_T\varepsilon_{si}f(Q_{m,s})q$$

$$+ 8CC_{ox}N_DQ_{fix}U_T\varepsilon_{si}q, \tag{D.3}$$

$$I_3 = 2Q_{fix}\left(2Q_{fix}\left(I_6 + \frac{I_5}{I_7}\right) - \frac{I_4}{I_7}\right), \tag{D.4}$$

$$I_4 = 3CC_{ox}Q_{fix} - C_{ox}f(Q_{m,s}) + 2BC_{ox}Q_{fix}^2 + 2CN_D\varepsilon_{si}q + 4BN_DQ_{fix}\varepsilon_{si}q$$

$$- 8BC_{ox}N_DU_T\varepsilon_{si}q - 8AC_{ox}N_DQ_{fix}U_T\varepsilon_{si}q, \tag{D.5}$$

$$I_5 = CC_{ox} + 3BC_{ox}Q_{fix} + 2AC_{ox}Q_{fix}^2 + 2BN_D\varepsilon_{si}q$$

$$+ 4AN_DQ_{fix}\varepsilon_{si}q - 8AC_{ox}N_DU_T\varepsilon_{si}q, \tag{D.6}$$

$$I_6 = 2Q_{fix}\left(\frac{AQ_{fix}}{2N_D\varepsilon_{si}q} - \frac{I_8}{I_7}\right), \tag{D.7}$$

$$I_7 = 4C_{ox}N_D\varepsilon_{si}q, \tag{D.8}$$

$$I_8 = BC_{ox} + 3AC_{ox}Q_{fix} + 2AN_D\varepsilon_{si}q, \tag{D.9}$$

$$A = -\frac{1}{12qN_D\varepsilon_{si}}, \tag{D.10}$$

$$B = -\left(\frac{1}{8C_{si}} + \frac{1}{4C_{ox}}\right), \tag{D.11}$$

$$C = 2U_T. \tag{D.12}$$

Appendix E Global Charge Density at Drain in Accumulation Mode

$$\overline{Q}_{m,d} = \eta \int_{Q_{m,s}}^{Q_{m,d}} \left[Q_m F(Q_{m,s}) + \frac{Q_m^3}{4C_{ox}} - 2U_T Q_m^2 \right] \times \left[-\frac{Q_m}{2C_{ox}} + \frac{2U_T Q_m (Q_m + Q_{fix})}{8U_T Q_{fix} C_{si} + \left(Q_m + Q_{fix} \right)^2} \right] dQ_m$$

$$= \eta Q_m^3 \left(\frac{J_3}{192 C_{ox}^4} - \frac{J_4}{24 C_{ox}^2} + \frac{2Q_{fix} J_5}{3} \right)$$

$$- \eta Q_m \left(2Q_{fix} \left(\frac{J_1}{8C_{ox}^2} - 2Q_{fix} J_2 + \frac{J_5 J_3}{8C_{ox}^2} \right) + \frac{J_7 + J_3 J_2}{8C_{ox}^2} \right) - \eta \frac{Q_m^5}{40 C_{ox}^2}$$

$$+ \eta Q_m^2 \left(\frac{J_1}{16 C_{ox}^2} - Q_{fix} J_2 + \frac{J_5 J_3}{16 C_{ox}^2} \right)$$

$$- \eta Q_m^4 \left(\frac{J_6}{32 C_{ox}^2} - \frac{Q_{fix}}{16 C_{ox}^2} \right) - \eta \frac{J_8}{32 C_{ox}^2} \log \left(Q_m^2 + 2Q_m Q_{fix} + Q_{fix}^2 + 8 C_{si} U_T Q_{fix} \right)$$

$$- \eta \frac{4\sqrt{2}}{C_{ox}} \sqrt{C_{si} U_T^3 Q_{fix}^3} \arctan \left(\frac{\sqrt{2} \sqrt{Q_{fix}}}{4\sqrt{C_{si} U_T}} + \frac{\sqrt{2}}{4\sqrt{C_{si} U_T}} \frac{Q_m}{\sqrt{Q_{fix}}} \right)$$

$$\times \left[Q_{fix}^2 + 2C_{ox} F(Q_{m,s}) + 6C_{ox} Q_{fix} U_T - 8 C_{si} Q_{fix} U_T - 16 C_{ox} C_{si} U_T^2 \right] \Big|_{Q_{m,s}}^{Q_{m,d}},$$

$$(\text{E.1})$$

where

$$J_1 = -32 C_{ox}^2 Q_{fix} U_T^2 + 16 F(Q_{m,s}) C_{ox}^2 U_T + 8 C_{ox} Q_{fix}^2 U_T$$
$$+ 64 C_{si} C_{ox} Q_{fix} U_T^2 - 8 F(Q_{m,s}) C_{ox} Q_{fix}, \qquad (\text{E.2})$$

$$J_2 = \frac{J_3}{64 C_{ox}^4} - \frac{J_4}{8 C_{ox}^2} + 2Q_{fix} J_5, \qquad (\text{E.3})$$

$$J_3 = 8 C_{ox}^2 Q_{fix}^2 + 64 C_{si} U_T C_{ox}^2 Q_{fix}, \qquad (\text{E.4})$$

$$J_4 = 32 C_{ox}^2 U_T^2 - 20 C_{ox} Q_{fix} U_T + 4F(Q_{m,s}) C_{ox} + Q_{fix}^2 + 8 C_{si} Q_{fix} U_T, \qquad (\text{E.5})$$

$$J_5 = \frac{J_6}{8 C_{ox}^2} - \frac{Q_{fix}}{4 C_{ox}^2}, \qquad (\text{E.6})$$

$$J_6 = 2Q_{fix} - 12 C_{ox} U_T, \qquad (\text{E.7})$$

$$J_7 = -16F(Q_{m,s})U_T C_{ox}^2 Q_{fix} + 4F(Q_{m,s})C_{ox}Q_{fix}^2 + 32C_{si}F(Q_{m,s})U_T C_{ox}Q_{fix}, \quad (\text{E.8})$$

$$J_8 = -1536C_{ox}^2 C_{si}Q_{fix}^2 U_T^3 - 256F(Q_{m,s})C_{ox}^2 C_{si}Q_{fix}U_T^2 + 64C_{ox}^2 Q_{fix}^3 U_T^2$$

$$+ 32F(Q_{m,s})C_{ox}^2 Q_{fix}^2 U_T + 512C_{ox}C_{si}^2 Q_{fix}^2 U_T^3 - 384C_{ox}C_{si}Q_{fix}^3 U_T^2 + 8C_{ox}Q_{fix}^4 U_T.$$

$$(\text{E.9})$$

Appendix F The EPFL Junctionless MODEL

After having developed complete DC, AC, and noise models, implementation in electrical simulators must be done before the model can be used by designers. In its simplest version, the EPFL-JL model restricted to symmetric double gate and cylindrical architectures covers the whole range of operation without introducing fitting parameters, and covers deep depletion to accumulation and linear to saturation regimes without introducing fitting parameters as it roots in clear and simple physical basis. The model is suitable for digital and analog simulations. Its main features are:

- Simulation of I–V characteristics (constant mobility μ)
- Simulation of gate transconductance g_{ms} and output conductance g_{ds}
- Simulation of complete AC transcapacitance matrix elements
- Implementation of a unique core model to simulate symmetric double-gate and nanowire architectures.

Even though the following features have been developed, this first version does not include them:

- Field dependent mobility
- Asymmetric operation
- Short- and narrow-channel effects including reverse short-channel effect (RSCE)
- Thermal and induced gate noise
- Charge trapping on silicon/insulator interface deep traps
- Channel-length modulation
- Flicker noise
- Gate leakage current
- Gate-induced drain leakage
- Quantum mechanical corrections
- Nonuniform doping profile.

F.1 The EPFL-Junctionless Model Modules

As illustrated in Figure F.1, the EPFL-JL model includes the following modules:

- Charge–voltage module (see Chapter 3)
- Explicit charges and current module (see details in [92])

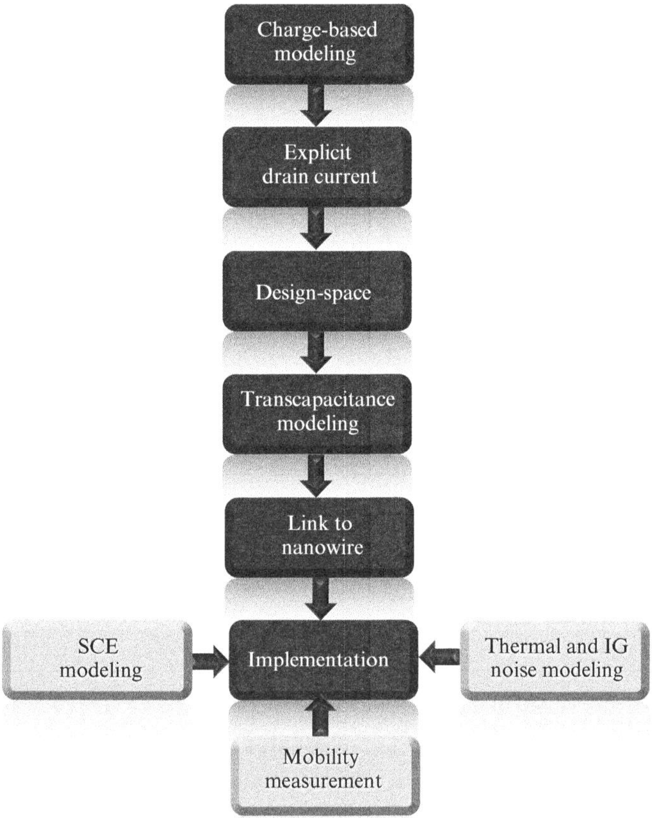

Figure F.1 Modules implemented in the EPFL-JL version.

- Design-space module (see Chapter 4)
- Transcapacitance matrix module (see Chapter 8)
- Nanowire module (see Chapter 3).

In its current status, the model can be used for preliminary simulation of junctionless FET circuits and nanowire sensors made of idealized devices. Moreover, this also includes cylindrical junction FETs geometries.

F.2 Source Code Modules and Library

The EPFL-JL is implemented as **subcircuits**-compatible for Hspice simulations for n-channel and p-channel junctionless double-gate MOSFETs and junctionless nanowire FET structures.

Both transcapacitance matrix elements and DC characteristics are mixed in a single subcircuit. It should be noted that the model implemented as a subcircuit

Table F.1. The transistor parameters assigned to the EPFL-JL model

Physical parameters	Symbol	Unit
Radius	R	m
Gate length	L_G	m
Uniform doping concentration	N_D	cm^{-3}
Channel width	W	m
Silicon thickness	T_{sc}	m
Oxide thickness	t_{ox}	m
Metal–semiconductor work function difference	W_{fms}	V

Table F.2. The value and description of the various physical constants used in the EPFL-JL model

Physical parameters	Symbol	Default value
Temperature scale	T	300
Boltzmann constant	KB	$1.380658 \times 10^{-23}\,J/K$
Elementary unit charge	q	$1.6021918 \times 10^{-19}\,C$
Permittivity of free space (ε_0)	$epzero$	$8.8541878176 \times 10^{-12}\,F/m$
Permittivity of silicon (ε_{si})	$epsi$	$1.035918 \times 10^{-10}\,F/m$
Permittivity of oxide (ε_{ox})	$epox$	$3.45306 \times 10^{-11}\,F/m$

in Hspice can be imported as a library to Cadence Virtuoso or Advanced Design System (ADS). The EPFL-JL 1.0 is made of four subcircuits.

The EPFL-JL model Parameters: As an option, it is possible to modify the transistor parameters given in Table F.1. The value and the description of the physical constants used in the EPFL-JL 1.0 model are given in Table F.2.

F.3   DC Implementation in Junctionless FETs

The drain-current model derived in Chapter 3 is implemented in EPFL-JL for long-channel double-gate junctionless transistors. Since this is three-terminal device (source (S), gate (G), and drain (D)), the source is used as the reference for the potentials. A total of eight switches are used to calculate the drain current and to select the right component for the three domains of operation, namely depletion (I_{dep}), accumulation (I_{acc}), and hybrid (I_{hyb}) (see Figure F.2). For instance, if the junctionless FET is working in a depletion regime, SW_1 should be closed and SW_2 should remain open. At the same time, other current components should be disabled and set to zero ($I_{acc} = I_{hyb} = 0$), which is ensured when SW_3, SW_4, SW_5, and SW_6 are open. Note that to avoid convergence issues, additional switches are introduced in parallel with the independent-current sources to make them short circuited.

The different possibilities, which are merely functions of the gate-to-source voltage V_{GS}, are summarized in Table F.3. This is how the drain current is evaluated in the symmetric double-gate junctionless FETs.

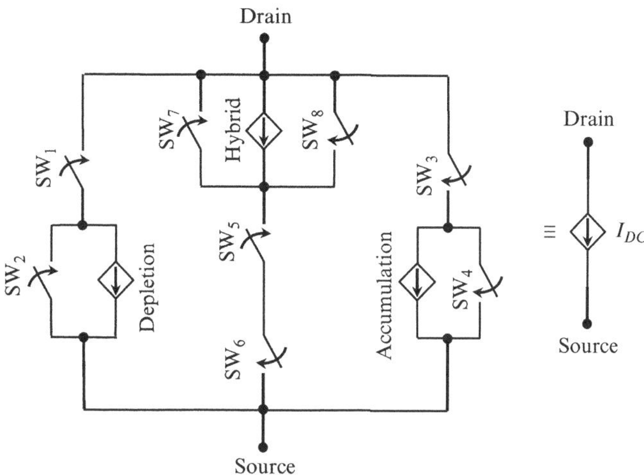

Figure F.2 A total of eight switches are used to calculate the drain current in nanowire and double-gate junctionless FETs.

Table F.3. Combination of the *on*-state and the *off*-state of eight switches depending on V_{GS} as a control signal and compared to the flat-band gate voltage in the EPFL-JL model

	Depletion $V_{GS} < V_{GS,FB}$	Hybrid $V_{GS,FB} < V_{GS} < V_{GS,FB} + V_{DS}$	Accumulation $V_{GS} > V_{GS,FB} + V_{DS}$
SW_1	Close	Open	Open
SW_2	Open	Close	Close
SW_3	Open	Open	Close
SW_4	Close	Close	Open
SW_5	Open	Close	Close
SW_6	Close	Close	Open
SW_7	Close	Open	Open
SW_8	Open	Open	Close

Drain current versus gate-to-source potential in a junctionless double-gate MOS-FET with $N_D = 10^{19}\, cm^{-3}$, $T_{sc} = 10\,nm$, $L_G = W = 1\,\mu m$, and $t_{ox} = 1.5\,nm$ is shown in Figure F.3a and b in linear and logarithmic scales. Similarly, the current versus the drain voltage is shown in Figure F.3c for different values of gate-to-source potential. Note that the continuity of the current is satisfied at the transition between depletion, hybrid, and accumulation even though virtual switches are introduced.

F.4   AC Implementation in Junctionless FETs

Using the same approach proposed in subsection F.3, the transcapacitance matrix elements in all the regions of operation are implemented as illustrated in Figure F.4 and are used to simulate AC operation of junctionless FETs around the

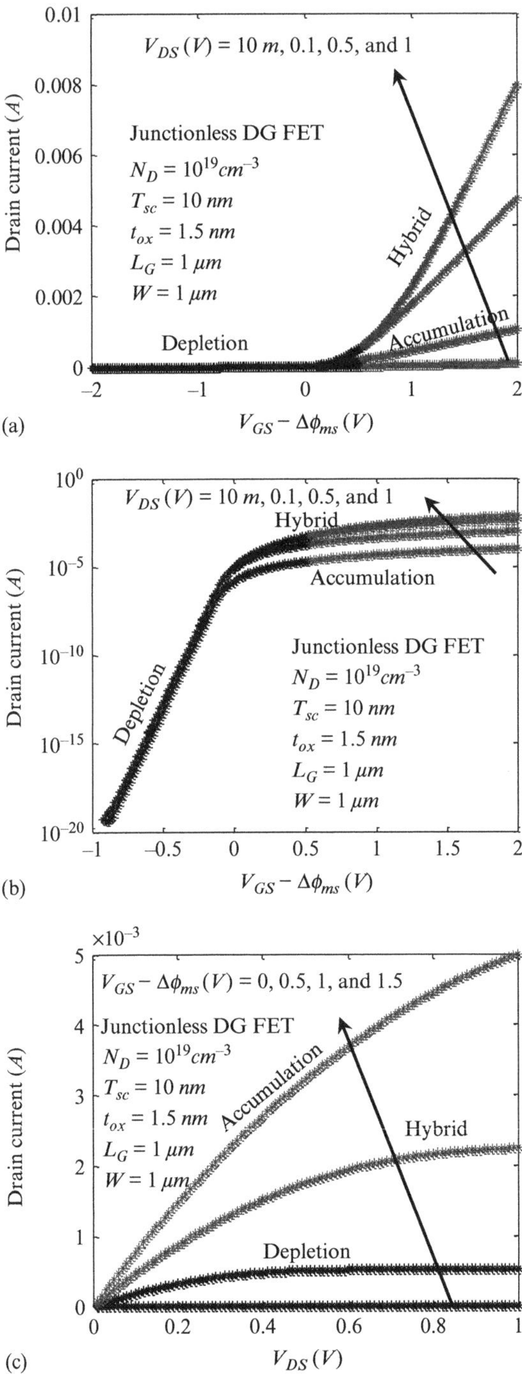

Figure F.3 Drain current versus gate voltage for various drain potentials in $10\,nm$ silicon thickness junctionless double-gate doped at $10^{19}\,cm^{-3}$ in (a) linear scale, (b) logarithmic scale, and (c) drain current versus drain voltage for different values of gate-to-source potential.

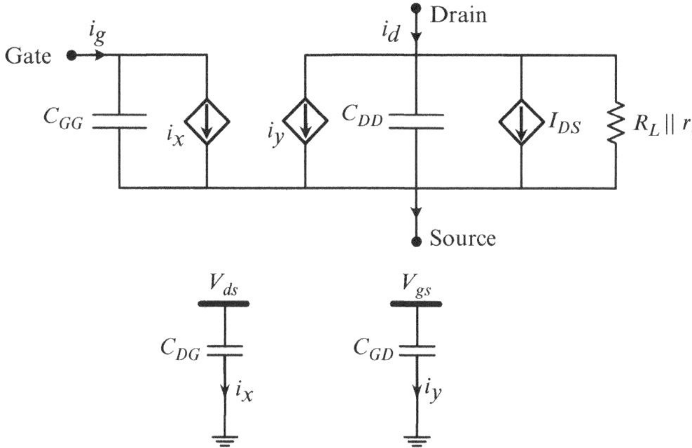

Figure F.4 Small signal-equivalent circuit of double-gate and nanowire junctionless FETs used in the EPFL-JL model.

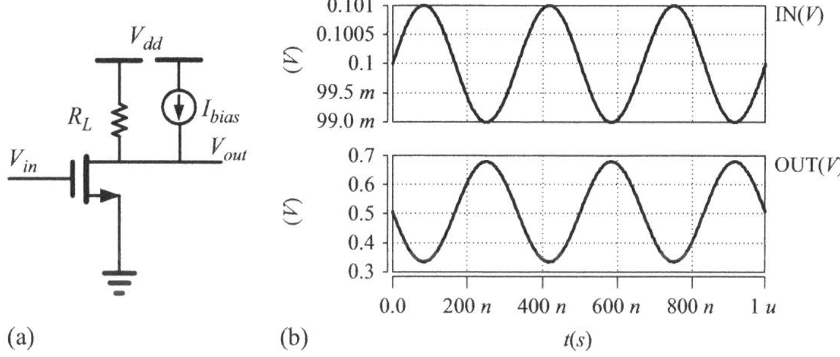

(a) (b)

Figure F.5 (a) A common-source amplifier based on n-type junctionless FETs. (b) Input (V_{in}) and output (V_{out}) signals.

DC bias. C_{GG}, C_{DD}, C_{DG}, and C_{GD} are from the transcapacitance analysis discussed in Chapter 8. Since the source is taken as the reference, the 3×3 matrix transadmittance can be reduced to a 2×2 matrix. The 2×2 capacitance matrix used to simulate small signal characteristics is given by

$$\begin{bmatrix} i_g \\ i_d \end{bmatrix} = \begin{bmatrix} C_{GG} & C_{DG} \\ C_{GD} & C_{DD} \end{bmatrix} \times \begin{bmatrix} v_{gs} \\ v_{ds} \end{bmatrix} \tag{F.1}$$

where i_g, i_d, v_{gs}, and v_{ds} are the small signal components.

F.5 Junctionless Double-Gate and Nanowire FET Amplifier

A common-source junctionless FET (double-gate or nanowire) amplifier with the circuit topology illustrated in Figure F.5a is set at $I_{bias} = 360\,\mu A$ and $V_{GS,bias} = 0.1\,V$. At low frequency, this arrangement would maximize the small-signal gain ($A_V = -180\,V/V$) due to the very large (ideally infinite) resistance associated with a DC

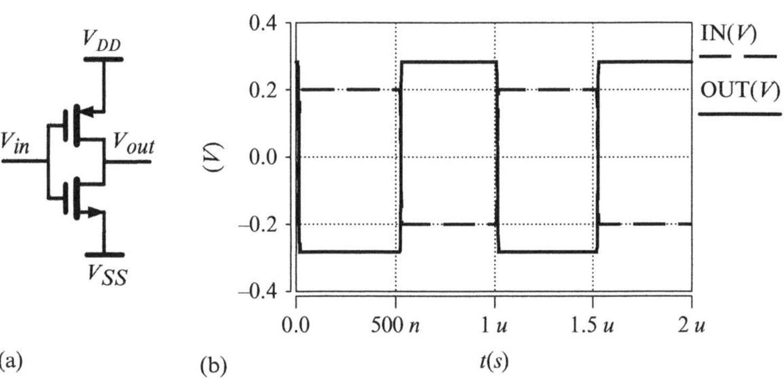

(a)   (b) $t(s)$

Figure F.6 (a) A junctionless logic gate: the junctionless FETs inverter. (b) Input (V_{in}) and output (V_{out}) signals.

current source (frequency $= 3\,MHz$, $V_{DS} = 1\,V$, and $R_L = 40\,K\Omega$). It is worth noting that the simulations plotted in Figure F.5b refer either to a junctionless double-gate MOSFET (with $N_D = 10^{19}\,cm^{-3}$, $T_{sc} = 10\,nm$, $L_G = 100\,nm$, $t_{ox} = 1.5\,nm$, and $W = 15.7\,nm$) or to a junctionless nanowire FET (with $N_D = 2 \times 10^{19}\,cm^{-3}$, $R = 5\,nm$, $L_G = 100\,nm$, $t_{ox} = 1.75\,nm$).

F.6   Junctionless Double-Gate and Nanowire FET Inverter

A junctionless logic gate was simulated and the results are shown in Figure F.6 ($V_{DD} = 0.3\,V$, $V_{SS} = -0.3\,V$). When the input signal is high ($V_{in} = 0.2V$), the n-JLFET on the bottom switches on, which pulls down the output node to $V_{SS} = -0.3$ and the p-J FET on top switches off. When the input voltage is low ($V_{in} = -0.2\,V$), the gate-to-source voltage is below the threshold, so it switches off and the p-JL FET switches on, pushing the output up to $V_{DD} = 0.3$. The simulations plotted in Figure F.6b are for a junctionless double-gate MOSFET (with $N_D = 10^{19}\,cm^{-3}$, $T_{sc} = 10\,nm$, $L_G = 100\,nm$, $t_{ox} = 1.5\,nm$, and $W = 15.7\,nm$), or equivalently for a junctionless nanowire FET (with $N_D = 2 \times 10^{19}\,cm^{-3}$, $R = 5\,nm$, $L_G = 100\,nm$, and $t_{ox} = 1.75\,nm$).

References

[1] J.-P. Colinge, C.-W. Lee, A. Afzalian, et al., "Nanowire transistors without junctions," *Nature Nanotechnology*, vol. 5, pp. 225–9, March 2010.

[2] Y. M. Georgiev, N. Petkov, B. McCarthy, et al., "Fully cmos-compatible top-down fabrication of sub-50 nm silicon nanowire sensing devices," *Microelectronic Engineering*, vol. 118, pp. 47–53, 2014.

[3] H. M. Fahad, H. Shiraki, M. Amani, et al., "Room temperature multiplexed gas sensing using chemical-sensitive 3.5-nm-thin silicon transistors," *Science Advances*, vol. 3, no. 3, pp. 1–8, 2017.

[4] Y. Kamimuta, K. Ikeda, K. Furuse, T. Irisawa, and T. Tezuka, "Short channel polyge junction-less p-type finfets for beol transistors," *2013 International Symposium on VLSI Technology, Systems and Application (VLSI-TSA)*, Hsinchu, Taiwan, pp. 1–2, April 2013.

[5] G. E. Moore, "Cramming more components onto integrated circuits," *Proceedings of the IEEE*, vol. 86, pp. 82–85, January 1998.

[6] E. J. Lilienfeld, "Method and apparatus for controlling electric currents," US patent 1745175, 1935.

[7] W. Shockley, "How we built the transistor," *New Scientist*, 1972.

[8] S. Veeraraghavan and J. Fossum, "Short-channel effects in SOI MOSFETs," *Electron Devices, IEEE Transactions on*, vol. 36, pp. 522–8, March 1989.

[9] T. Ghani, K. Mistry, P. Packan, et al., "Scaling challenges and device design requirements for high performance sub-50 nm gate length planar CMOS transistors," *VLSI Technology, 2000. Digest of Technical Papers. 2000 Symposium on*, Honolulu, HI, pp. 174–5, June 2000.

[10] M. Lundstrom, "Elementary scattering theory of the Si MOSFET," *Electron Device Letters, IEEE*, vol. 18, pp. 361–3, July 1997.

[11] W. Tsai, L.-A. Ragnarsson, L. Pantisano, et al., "Performance comparison of sub 1nm sputtered TiN/HfO2 nMOS and pMOSFETs," *Electron Devices Meeting, 2003. IEDM '03 Technical Digest. IEEE International*, Washington, DC, pp. 13.2.1–13.2.4, December 2003.

[12] Y. Mii, S. Wind, Y. Taur, et al., "An ultra-low power 0.1 μm CMOS," *VLSI Technology, 1994. Digest of Technical Papers. 1994 Symposium on*, Honolulu, HI, pp. 9–10, June 1994.

[13] H. Wong and H. Iwai, "On the scaling of subnanometer EOT gate dielectrics for ultimate nano CMOS technology," *Microelectronic Engineering*, vol. 138, pp. 57–76, 2015.

[14] A. Chatterjee, R. Chapman, K. Joyner, et al., "CMOS metal replacement gate transistors using tantalum pentoxide gate insulator," *Electron Devices Meeting, 1998. IEDM '98. Technical Digest., International*, San Francisco, CA, pp. 777–80, December 1998.

[15] D. Frank, S. Laux, and M. Fischetti, "Monte Carlo simulation of a 30 nm dual-gate MOSFET: how short can Si go?," *Electron Devices Meeting, 1992. IEDM '92. Technical Digest., International*, San Francisco, CA, pp. 553–6, December 1992.

[16] K. Ismail, "Si/SiGe high-speed field-effect transistors," in *Electron Devices Meeting, 1995. IEDM '95, International*, pp. 509–12, December 1995.

[17] W. P. Maly, "Integrated circuit device, system, and method of fabrication," US patent WO2007133775A2, May 2012.

[18] W. Maly, "Vertical slit transistor based integrated circuits (VeSTICs) paradigm," *ISPD*, pp. 63–4, 2009.

[19] W. Maly, N. Singh, Z. Chen, and N. Shen, "Twin gate, vertical slit FET (VeSFET) for highly periodic layout and 3D integration," *Proceedings of the 18th International Conference Mixed Design of Integrated Circuits and Systems – MIXDES 2011*, Gliwice, pp. 145–50, 2011.

[20] H. Lu and A. Seabaugh, "Tunnel field-effect transistors: State-of-the-art," *IEEE Journal of the Electron Devices Society*, vol. 2, pp. 44–9, July 2014.

[21] J. Welser, S. Tiwari, S. Rishton, K. Lee, and Y. Lee, "Room temperature operation of a quantum-dot flash memory," *Electron Device Letters, IEEE*, vol. 18, pp. 278–80, June 1997.

[22] A. Pfitzner, M. Staniewski, and M. Strzyga, "DC characteristics of junction vertical slit field-effect transistor (JVeSFET)," *2009 MIXDES-16th International Conference Mixed Design of Integrated Circuits & Systems*, Lodz, pp. 420–3, June 2009.

[23] X. Qiu, M. Marek-Sadowska, and W. Maly, "Vertical slit field effect transistor in ultra-low power applications," *Quality Electronic Design (ISQED), 2012 13th International Symposium on*, Santa Clara, CA, pp. 384–90, March 2012.

[24] N. Sugii, "Low-power-consumption fully depleted silicon-on-insulator technology," *Microelectronic Engineering*, vol. 132, pp. 226–35, 2015.

[25] S. Zhu, J. Chen, M.-F. Li, et al., "N-type Schottky barrier source/drain MOSFET using Ytterbium silicide," *IEEE Electron Devices Letters*, vol. 25, pp. 565–7, April 2004.

[26] M. K. Ieong, P. M. Solomon, S. E. Laux, H.-S. Philip Wong, and D. Chidambarrao, "Comparison of raised and Schottky source/drain MOSFETs using a novel tunneling contact model," *Electron Devices Meeting, 1998. IEDM '98. Technical Digest., International*, San Francisco, CA, pp. 733–8, December 1998.

[27] C.-W. Lee, I. Ferain, A. Afzalian, et al., "Performance estimation of junctionless multigate transistors," *Solid-State Electronics*, vol. 54, pp. 97–103, February 2010.

[28] C.-W. Lee, A. Afzalian, N. D. Akhavan, et al., "Junctionless multigate field effect transistor," *Applied Physical Letters*, vol. 94, pp. 053511-1–053511-2, February 2009.

[29] J. P. Colinge, C. W. Lee, A. Afzalian, et al., "SOI gated resistor: CMOS without junctions," *IEEE International SOI Conference*, Foster City, CA, pp. 1–2, October 2009.

[30] M.-H. Han, C.-Y. Chang, H.-B. Chen, Y.-C. Cheng, and Y.-C. Wu, "Device and circuit performance estimation of junctionless bulk FinFETs," *Electron Devices, IEEE Transactions on*, vol. 60, pp. 1807–13, June 2013.

[31] M.-H. Han, C.-Y. Chang, H.-B. Chen, et al., "Performance comparison between bulk and SOI junctionless transistors," *Electron Device Letters, IEEE*, vol. 34, pp. 169–71, February 2013.

[32] Z. Chen, A. Kamath, N. Singh, et al., "*N*-channel junctionless vertical slit field effect transistor (VeSFET): fabrication-based feasibility assessment," *International Conference on Solid-State and Integrated Circuit (ICSIC 2012)*, Singapore, vol. 32, pp. 152–4, 2012.

[33] A. Kamath, Z. Chen, N. Shen, et al., "Realizing *and* and *or* functions with single vertical-slit field effect transistor," *IEEE Electron Device Letters*, vol. 33, pp. 152–4, February 2012.

[34] V. Nandakumar and M. Marek-Sadowska, "A low energy network-on-chip fabric for 3-D multi-core architectures," *Emerging and Selected Topics in Circuits and Systems, IEEE Journal on*, vol. 2, pp. 266–77, June 2012.

[35] M. Pastre, F. Krummenacher, L. Barbut, J.-M. Sallese, and M. Kayal, "Towards circuit design using VeSFETs," *Mixed Design of Integrated Circuits and Systems (MIXDES), 2011 Proceedings of the 18th International Conference*, Gliwice, Poland, pp. 139–44, June 2011.

[36] M. Weis, A. Pfitzner, K. Kasprowicz, et al., "Low power SRAM cell using vertical slit field effect transistor (VeSFET)," *European Solid-State Circuits Conference ESSCIRC Fringe*, Edinburgh, UK, 2008.

[37] E. Gnani, A. Gnudi, S. Reggiani, et al., "Numerical investigation on the junctionless nanowire FET," *IEEE Ultimate Integration on Silicon (ULIS) Conference*, Cork, Ireland, pp. 1–4, March 2011.

[38] R. Rios, A. Cappellani, M. Armstrong, et al., "Comparison of junctionless and conventional trigate transistors with Lg down to 26 nm," *IEEE Transaction Electron Devices*, vol. 32, pp. 1170–2, September 2011.

[39] W. Vitale, M. Mohamed, and U. Ravaioli, "Monte Carlo study of transport properties in junctionless transistors," *2010 14th International Workshop on Computational Electronics*, Pisa, pp. 1–3, October 2010.

[40] H.-C. Lin, C.-I. Lin, and T.-Y. Huang, "Characteristics of *n*-type junctionless poly-Si thin-film transistors with an ultrathin channel," *IEEE Electron Device letters*, vol. 33, pp. 53–5, January 2012.

[41] P. Razavi, G. Fagas, I. Ferain, et al., "Performance investigation of short-channel junctionless multigate transistors," *IEEE Ultimate Integration on Silicon (ULIS)*, pp. 1–4, March 2011.

[42] P. Francis, A. Terao, D. Flandre, and F. Van de Wiele, "Moderate inversion model of ultrathin double-gate nMOS/SOI transistors," *Solid State Electronics*, vol. 38, pp. 171–6, March 1995.

[43] P. Francis, A. Terao, D. Flandre, and F. van de Wiele, "Modeling of ultra-thin double-gate nMOS/SOI transistors," *IEEE Transaction Electron Devices*, vol. 41, pp. 715–20, May 1994.

[44] G. Baccarani and S. Reggiani, "A compact double-gate MOSFET model comprising quantum-mechanical and nonstatic effects," *IEEE Transaction Electron Devices*, vol. 46, pp. 1656–66, August 1999.

[45] A. Bansal and K. Roy, "Analytical subthreshold potential distribution model for gate underlap double-gate MOS Transistors," *IEEE Transaction Electron Devices*, vol. 54, pp. 1793–8, July 2007.

[46] K. Suzuki, Y. Tosaka, and T. Sugii, "Analytical threshold voltage model for short-channel double-gate SOI MOSFETs," *IEEE Transaction Electron Devices*, vol. 43, pp. 1166–8, July 1996.

[47] Q. Chen, E. M. Harrell, and J. D. Meindl, "A physical short-channel threshold voltage model for undoped symmetric double-gate MOSFETs," *IEEE Transaction Electron Devices*, vol. 50, pp. 1631–7, July 2003.

[48] Q. Chen, B. Agrawal, and J. D. Meindl, "A comprehensive analytical subthreshold swing model for double-gate MOSFETs," *IEEE Transaction Electron Devices*, vol. 49, pp. 1086–90, June 2002.

[49] T. Holtij, M. Schwarz, A. Kloes, and B. Iniguez, "Analytical 2D modeling of junctionless and junction-based short-channel multigate MOSFETs," *ESSDERC Fringe*, Bordeaux, France, 2012.

[50] J.P. Colinge, A. Kranti, R. Yan, et al., "Junctionless nanowire transistor (JNT): properties and design guidelines," *Solid-State Device Research Conference (ESSDERC)*, Sevilla, Spain, pp. 33–37, September 2010.

[51] J.-P. Colinge, "Junctionless transistors," *Future of Electron Devices, Kansai (IMFEDK), 2012 IEEE International Meeting for*, Suita, Japan, pp. 1–2, 2012.

[52] A. Koukab, F. Jazaeri and J.-M. Sallese, "On performance scaling and speed of junctionless transistors," *Solid-State Electronics*, vol. 79, pp. 18–21, January 2013.

[53] K.-I. Goto, T.-H. Yu, J. Wu, C. H. Diaz, and J. P. Colinge, "Mobility and screening effect in heavily doped accumulation-mode metal-oxide-semiconductor field-effect transistors," *Applied Physics Letters*, vol. 101, no. 7, pp. 073503-1–073503-2, 2012.

[54] B. Sorée, W. Magnus, and W. Vandenberghe, "Low-field mobility in ultrathin silicon nanowire junctionless transistors," *Applied Physics Letters*, vol. 99, no. 23, pp. 233509-1–233509-3, 2011.

[55] J.-P. Colinge, A. Kranti, R. Yan, et al., "A simulation comparison between junctionless and inversion-mode mugfets," *ECS Transactions*, vol. 35, pp. 63–72, 2011.

[56] G. Leung and C .O. Chui, "Variability impact of random dopant fluctuation on nanoscale junctionless FinFETs," *IEEE Electron Device Letters*, vol. 33, pp. 767–9, June 2012.

[57] A. Asenov, G. Slavcheva, A. R. Brown, J. H. Davies, and S. Saini, "Increase in the random dopant induced threshold fluctuations and lowering in sub-100 nm MOSFETs due to quantum effects: a 3-D density-gradient simulation study," *IEEE Transactions on Electron Devices*, vol. 48, pp. 722–9, April 2001.

[58] A. Asenov, "Random dopant induced threshold voltage lowering and fluctuations in sub-0.1 mu MOSFETs: a 3-D atomistic simulation study," *IEEE Transactions on Electron Devices*, vol. 45, pp. 2505–13, December 1998.

[59] A. Asenov, A. R. Brown, J. H. Davies, S. Kaya, and G. Slavcheva, "Simulation of intrinsic parameter fluctuations in decananometer and nanometer-scale mosfets," *IEEE Transactions on Electron Devices*, vol. 50, pp. 1837–52, September 2003.

[60] T. Hagiwara, K. Yamaguchi, and S. Asai, "Threshold voltage variation in very small MOS transistors due to local impurity fluctuations," in *VLSI Technology, 1982. Digest of Technical Papers. Symposium on*, Oiso, Japan, pp. 46–7, September 1982.

[61] S. Washburn, A. B. Fowler, H. Schmid, and D. Kern, "Possible observation of transmission resonances in GaAs–AlGaAs transistors," *Physical Review B*, vol. 38, pp. 1554–7, July 1988.

[62] J. H. Davies and J. A. Nixon, "Fluctuations in submicrometer semiconducting devices caused by the random positions of dopants," *Physical Review B*, vol. 39, pp. 3423–6, February 1989.

[63] R. Wang, J. Zhuge, R. Huang, et al., "Investigation on variability in metal-gate Si nanowire MOSFETs: analysis of variation sources and experimental characterization," *IEEE Transactions on Electron Devices*, vol. 58, pp. 2317–25, August 2011.

[64] A. Asenov, "Random dopant induced threshold voltage lowering and fluctuations in sub 50 nm mosfets: a statistical 3D 'atomistic' simulation study," *Nanotechnology*, vol. 10, no. 2, p. 153, 1999.

[65] P. A. Stolk and D. B. M. Klaassen, "The effect of statistical dopant fluctuations on mos device performance," *Electron Devices Meeting, IEDM '96, International*, San Francisco, CA, pp. 627–30, December 1996.

[66] P. A. Stolk, F. P. Widdershoven, and D. B. M. Klaassen, "Modeling statistical dopant fluctuations in mos transistors," *IEEE Transactions on Electron Devices*, vol. 45, pp. 1960–71, September 1998.

[67] H.-S. Wong and Y. Taur, "Three-dimensional atomistic simulation of discrete random dopant distribution effects in sub-0.1 m MOSFET's," *Electron Devices Meeting, 1993. IEDM '93. Technical Digest, International*, Washington, DC, pp. 705–8, December 1993.

[68] N. Seoane, A. Martinez, A. R. Brown, J. R. Barker, and A. Asenov, "Current variability in Si nanowire MOSFETs due to random dopants in the source/drain regions: a fully 3-D NEGF simulation study," *IEEE Transactions on Electron Devices*, vol. 56, pp. 1388–95, July 2009.

[69] A. Martinez, A. R. Brown, S. Roy, and A. Asenov, "NEGF simulations of a junctionless Si gate-all-around nanowire transistor with discrete dopants," *Ultimate Integration on Silicon (ULIS), 2011 12th International Conference on*, Cork, Ireland, pp. 1–4, March 2011.

[70] A. Gnudi, S. Reggiani, E. Gnani, and G. Baccarani, "Analysis of threshold voltage variability due to random dopant fluctuations in junctionless FETs," *IEEE Electron Device Letters*, vol. 33, pp. 336–8, March 2012.

[71] A. Ueda, M. Luisier, and N. Sano, "Enhanced impurity-limited mobility in ultra-scaled Si nanowire junctionless field-effect transistors," *Applied Physics Letters*, vol. 107, no. 25, 2015.

[72] T. Rudenko, A. Nazarov, I. Ferain, et al., "Mobility enhancement effect in heavily doped junctionless nanowire silicon-on-insulator metal-oxide-semiconductor field-effect transistors," *Applied Physics Letters*, vol. 101, no. 21, pp. 213502-1–213502-4, 2012.

[73] S.-J. Choi, D.-I. Moon, J. Duarte, S. Kim, and Y.-K. Choi, "A novel junctionless all-around-gate SONOS device with a quantum nanowire on a bulk substrate for 3D stack NAND flash memory," *VLSI Technology (VLSIT), 2011 Symposium on*, Honolulu, HI, pp. 74–5, June 2011.

[74] S.-J. Choi, D.-I. Moon, S. Kim, et al., "Nonvolatile memory by all-around-gate junctionless transistor composed of silicon nanowire on bulk substrate," *Electron Device Letters, IEEE*, vol. 32, pp. 602–4, May 2011.

[75] M. Parihar, D. Ghosh, and A. Kranti, "Ultra low power junctionless MOSFETs for subthreshold logic applications," *Electron Devices, IEEE Transactions on*, vol. 60, pp. 1540–6, May 2013.

[76] J.-P. Colinge and S. Dhong, "Prospective for nanowire transistors," *Custom Integrated Circuits Conference (CICC), 2013 IEEE*, San Jose, CA, pp. 1–8, September 2013.

[77] J. S. T. Huang and G. W. Taylor, "Modeling of an ion-implanted silicon-gate depletion-mode IGFET," *IEEE Transactions on Electron Devices*, vol. 22, pp. 995–1001, November 1975.

[78] J. S. T. Huang, "Characteristics of a depletion-type IGFET," *IEEE Transactions on Electron Devices*, vol. 20, pp. 513–14, May 1973.

[79] J. R. Edwards and G. Marr, "Depletion-mode IGFET made by deep ion implantation," *IEEE Transactions on Electron Devices*, vol. 20, pp. 283–9, March 1973.

[80] J. P. Duarte, S.-J. Choi, D.-I. Moon, and Y.-K. Choi, "Simple analytical bulk current model for long-channel double-gate junctionless transistors," *IEEE Transactions on Electron Devices*, vol. 32, pp. 704–6, June 2011.

[81] J. P. Colinge, "Conduction mechanisms in thin-film accumulation-mode SOI p-channel MOSFETs," *IEEE Transactions on Electron Devices*, vol. 37, pp. 718–23, March 1990.

[82] E. Gnani, A. Gnudi, S. Reggiani, and G. Baccarani, "Physical model of the junctionless UTB SOI-FET," *IEEE Transactions on Electron Devices*, vol. 59, pp. 941–8, April 2012.

[83] Z. Chen, Y. Xiao, M. Tang, et al., "Surface-potential-based drain current model for long-channel junctionless double-gate MOSFETs," *Electron Devices, IEEE Transactions on*, vol. 59, pp. 3292–98, December 2012.

[84] C. Sahu, P. Swami, S. Sharma, and J. Singh, "Simplified drain current model for pinch-off double gate junctionless transistor," *IEEE Electronics Letters*, vol. 50, pp. 116–18, January 2014.

[85] A. Cerdeira, M. Estrada, R. Trevisoli, et al., "Analytical model for potential in double-gate junctionless transistors," *Microelectronics Technology and Devices (SBMicro), 2013 Symposium on*, Curitiba, Brazil, pp. 1–3, September 2013.

[86] J.-M. Sallese, N. Chevillon, C. Lallement, B. Iñiguez, and F. Prégaldiny, "Charge-based modeling of junctionless double-gate field effect transistors," *IEEE Transactions on Electron Devices*, vol. 58, pp. 2628–37, August 2011.

[87] X. Jin, X. Liu, M. Wu, et al., "A unified analytical continuous current model applicable to accumulation mode (junctionless) and inversion mode MOSFETs with symmetric and asymmetric double-gate structures," *Solid-State Electronics*, vol. 79 (suppl. C), pp. 206–9, 2013.

[88] X. Jin, X. Liu, J.-H. Lee, and J.-H. Lee, "A continuous current model of fully-depleted symmetric double-gate MOSFETs considering a wide range of body doping concentrations," *Semiconductor Science Technology*, vol. 25, p. 055018, May 2010.

[89] Y. Taur, X. Liang, W. Wang, and H. Lu, "A continuous, analytic drain-current model for DG MOSFETs," *Electron Device Letters, IEEE*, vol. 25, pp. 107–9, February 2004.

[90] J. P. Duarte, S.-J. Choi, and Y.-K. Choi, "A full-range drain current model for double-gate junctionless transistors," *IEEE Transactions on Electron Devices*, vol. 58, pp. 4219–25, December 2011.

[91] F. Lime, E. Santana, and B. Iñiguez, "A simple compact model for long-channel junctionless double gate MOSFETs," *Solid-State Electronics*, vol. 80 (suppl. C), pp. 28–32, 2013.

[92] A. Yesayan, F. Prégaldiny, and J.-M. Sallese, "Explicit drain current model of junctionless double-gate field-effect transistors," *Solid-State Electronics*, vol. 89, pp. 134–8, November 2013.

[93] "The international technology roadmap for semiconductors," 2011. www .semiconductors.org/clientuploads/Research_Technology/ITRS/2011/2011ExecSum .pdf

[94] J. Duarte, M.-S. Kim, S.-J. Choi, and Y.-K. Choi, "A compact model of quantum electron density at the subthreshold region for double-gate junctionless transistors," *Electron Devices, IEEE Transactions on*, vol. 59, pp. 1008–12, April 2012.

[95] Y. Taur and T. H. Ning, *Fundamentals of modern VLSI devices*. New York: Cambridge University Press, p. 496, October 1998.

[96] J. Huang, W. Chew, J. Peng, et al., "Model order reduction for multiband quantum transport simulations and its application to *p*-type junctionless transistors," *Electron Devices, IEEE Transactions on*, vol. 60, pp. 2111–19, July 2013.

[97] S. Datta, *Quantum transport: atom to transistor*. New York: Cambridge University Press, p. 417, May 2013.

[98] T.-K. Chiang, "A quasi-two-dimensional threshold voltage model for short-channel junctionless double-gate MOSFETS," *IEEE Transaction Electron Devices*, vol. 59, pp. 2284–9, September 2012.

[99] T.-K. Chiang, "A quasi-2D threshold voltage model for short-channel junctionless (JL) double-gate MOSFETs," *Electron Devices and Solid State Circuit (EDSSC), 2012 IEEE International Conference on*, Bangkok, Thailand, pp. 1–4, December 2012.

[100] P. Gupta, D. Burman, J. Das, et al., "Modeling the channel potential and threshold voltage of a fully depleted double gate junctionless FET," in *Communications, Devices and Intelligent Systems (CODIS), 2012 International Conference on*, Kolkata, India, pp. 149–52, December 2012.

[101] X. Jin, X. Liu, R. Chuai, J.-H. Lee, and J.-H. Lee, "A compact model of subthreshold characteristics for short channel double-gate junctionless field effect transistors," *The European Physical Journal Applied Physics*, vol. 65, no. 3, p. 30101, 2014.

[102] X. Jin, X. Liu, H.-I. Kwon, J.-H. Lee, and J.-H. Lee, "A subthreshold current model for nanoscale short channel junctionless MOSFETs applicable to symmetric and asymmetric double-gate structure," *Solid-State Electronics*, vol. 82, pp. 77–81, April 2013.

[103] X. Jin, X. Liu, M. Wu, et al., "Modeling of the nanoscale channel length effect on the subthreshold characteristics of junctionless field-effect transistors with a symmetric double-gate structure," *Journal of Physics D:applied Physics*, vol. 45, pp. 375102-1–375102-5, July 2012.

[104] B. Ray and S. Mahapatra, "Modeling of channel potential and subthreshold slope of symmetric double-gate transistor," *Electron Devices, IEEE Transactions on*, vol. 56, pp. 260–6, February 2009.

[105] T. Holtij, M. Graef, F. Hain, A. Kloes, and B. Iniguez, "Compact model for short-channel junctionless accumulation mode double gate MOSFETs," *Electron Devices, IEEE Transactions on*, vol. 61, pp. 288–99, February 2014.

[106] T. Holtij, M. Graef, F. Hain, A. Kloes, and B. Iñíguez, "Unified charge model for short-channel junctionless double gate MOSFETs," *Mixed Design of Integrated Circuits and Systems (MIXDES), 2013 Proceedings of the 20th International Conference*, Gdynia, Poland, pp. 75–80, June 2013.

[107] T. Holtij, M. Schwarz, M. Graef, et al., "Model for investigation of I_{on}/I_{off} ratios in short-channel junctionless double gate MOSFETs," *Ultimate Integration on Silicon*

(ULIS), 2013 14th International Conference on, Warwick University Coventry, UK, pp. 85–8, March 2013.

[108] T. Holtij, M. Schwarz, A. Kloes, and B. Iñíguez, "Threshold voltage, and 2D potential modeling within short-channel junctionless DG MOSFETs in subthreshold region," *Solid-State Electronics*, vol. 90 (suppl. C), pp. 107–15, 2013.

[109] K. Young, "Short-channel effect in fully depleted SOI MOSFETs," *Electron Devices, IEEE Transactions on*, vol. 36, pp. 399–402, February 1989.

[110] A. Gnudi, S. Reggiani, E. Gnani, and G. Baccarani, "Semianalytical model of the subthreshold current in short-channel junctionless symmetric double-gate field-effect transistors," *IEEE Transactions on Electron Devices*, vol. 60, pp. 1342–8, April 2013.

[111] F. Jazaeri, L. Barbut, A. Koukab, and J.-M. Sallese, "Analytical model for ultra-thin body junctionless symmetric double gate MOSFETs in subthreshold regime," *Solid-State Electronics*, vol. 82, pp. 103–10, April 2013.

[112] T. Holtij, M. Schwarz, A. Kloes, and B. Iniguez, "2D analytical potential modeling of junctionless DG MOSFETs in subthreshold region including proposal for calculating the threshold voltage," *Ultimate Integration on Silicon (ULIS), 2012 13th International Conference on*, Grenoble, France, pp. 81–4, March 2012.

[113] G. Mariniello, R. Doria, M. de Souza, M. Pavanello, and R. Trevisoli, "Analysis of gate capacitance of *n*-type junctionless transistors using three-dimensional device simulations," *8th International Caribbean Conference on Devices, Circuits and Systems (ICCDCS)*, Playa del Carmen, Mexico, pp. 1–4, September 2012.

[114] S.-W. Hsu, J.-T. Lin, Y.-C. Eng, et al., "Simulation study of junctionless vertical MOSFETs for analog applications," *Junction Technology (IWJT), 2012 12th International Workshop on*, Shanghai, China, pp. 226–9, May 2012.

[115] J. Rahul, A. Srivastava, S. Yadav, and K. Kishor Jha, "Performance evaluation of Junctionless Vertical Double Gate MOSFET," *Devices, Circuits and Systems (ICDCS), 2012 International Conference on*, Coimbatore, India, pp. 440–2, March 2012.

[116] A. Dehzangi, F. Larki, B. Majlis, et al., "Numerical investigation of channel width variation in junctionless transistors performance," in *Micro and Nanoelectronics (RSM), 2013 IEEE Regional Symposium on*, Langkawi, Malaysia, pp. 101–4, September 2013.

[117] R. Baruah and R. Paily, "Double-gate junctionless transistor for low power digital applications," *Emerging Trends and Applications in Computer Science (ICETACS), 2013 1st International Conference on*, Shillong, India, pp. 23–6, September 2013.

[118] F. Jazaeri, L. Barbut, and J.-M. Sallese, "Trans-capacitance modeling in junctionless symmetric double-gate MOSFETs," *IEEE Transactions on Electron Devices*, vol. 60, pp. 4034–40, October 2013.

[119] F. Jazaeri, L. Barbut, and J.-M. Sallese, "Modeling asymmetric operation in double-gate junctionless FETs by means of symmetric devices," *IEEE Transactions on Electron Devices*, vol. 61, pp. 3962–70, December 2014.

[120] S. M. Lee and J. T. Park, "The impact of substrate bias on the steep subthreshold slope in junctionless MuGFETs," *Electron Devices, IEEE Transactions on*, vol. 60, pp. 3856–61, November 2013.

[121] J. Duarte, S.-J. Choi, D.-I. Moon, and Y.-K. Choi, "A nonpiecewise model for long-channel junctionless cylindrical nanowire FETs," *Electron Device Letters, IEEE*, vol. 33, pp. 155–7, February 2012.

[122] E. Gnani, A. Gnudi, S. Reggiani, and G Baccarani, "Theory of the junctionless nanowire FET," *IEEE Transactions on Electron Devices*, vol. 58, pp. 2903–10, September 2011.

[123] J. Xiao-Shi, L. Xi, K. Hyuck-In, and L. Jong-Ho, "A continuous current model of accumulation mode (junctionless) cylindrical surrounding-gate nanowire MOSFETs," *Chinese Physics Letters*, vol. 30, no. 3, p. 38502, 2013.

[124] J.-M. Sallese, F. Jazaeri, L. Barbut, N. Chevillon, and C. Lallement, "A common core model for junctionless nanowires and symmetric double gate FETs," *IEEE Transactions on Electron Devices*, vol. 60, pp. 4277–80, December 2013.

[125] F. Lime, O. Moldovan, and B. Iniguez, "A compact explicit model for long-channel gate-all-around junctionless MOSFETs. Part I: DC characteristics," *Electron Devices, IEEE Transactions on*, vol. 61, pp. 3036–41, September 2014.

[126] R. Trevisoli, R. Doria, M. de Souza, and M. Pavanello, "Drain current model for junctionless nanowire transistors," *Devices, Circuits and Systems (ICCDCS), 2012 8th International Caribbean Conference on*, Playa del Carmen, Mexico, pp. 1–4, March 2012.

[127] C. Li, Y. Zhuang, S. Di, and R. Han, "Subthreshold behavior models for nanoscale short-channel junctionless cylindrical surrounding-gate MOSFETs," *Electron Devices, IEEE Transactions on*, vol. 60, pp. 3655–62, November 2013.

[128] T.-K. Chiang, "A new quasi-2-D threshold voltage model for short-channel junctionless cylindrical surrounding gate (JLCSG) MOSFETs," *Electron Devices, IEEE Transactions on*, vol. 59, pp. 3127–9, November 2012.

[129] S. Shin, I. M. Kang, and K. R. Kim, "Extraction method for substrate-related components of vertical junctionless silicon nanowire field-effect transistors and its verification on radio frequency characteristics," *Japanese Journal of Applied Physics*, vol. 51, p. 06FE20, 2012.

[130] S. Cho, K. R. Kim, B.-G. Park, and I. M. Kang, "RF performance and small-signal parameter extraction of junctionless silicon nanowire MOSFETs," *Electron Devices, IEEE Transactions on*, vol. 58, pp. 1388–96, May 2011.

[131] J. Wang, G. Du, K. Wei, et al., "Mixed-mode analysis of different mode silicon nanowire transistors-based inverter," *Nanotechnology, IEEE Transactions on*, vol. 13, pp. 362–7, March 2014.

[132] P. Razavi, I. Ferain, S. Das, et al., "Intrinsic gate delay and energy-delay product in junctionless nanowire transistors," *Ultimate Integration on Silicon (ULIS), 2012 13th International Conference on*, Grenoble, France, pp. 125–8, March 2012.

[133] S. Singh, P. Kondekar, and A. Dixit, "Digital and analog performance of gate inside p-type junctionless transistor (GI-JLT)," *Computational Intelligence, Modelling and Simulation (CIMSim), 2013 Fifth International Conference on*, Seoul, South Korea, pp. 394–7, September 2013.

[134] F. Jazaeri, L. Barbut, and J.-M. Sallese, "Trans-capacitance modeling in junctionless gate-all-around nanowire FETs," *Solid-State Electronics*, vol. 96, pp. 34–7, June 2014.

[135] O. Moldovan, F. Lime, and B. Iniguez, "A compact explicit model for long-channel gate-all-around junctionless MOSFETs. Part II: total charges and intrinsic capacitance characteristics," *Electron Devices, IEEE Transactions on*, vol. 61, pp. 3042–6, September 2014.

[136] J. Huang, W. C. Chew, L. J. Jiang, et al., "Full-quantum simulation of p-type junctionless transistors with multi-band $k.p$ model," *Electron Devices and Solid-State Circuits*

(EDSSC), 2013 IEEE International Conference of, Hong Kong, China, pp. 1–2, June 2013.

[137] A. T. Pham, B. Soree, W. Magnus, et al., "Quantum simulations of electrostatics in Si cylindrical nanowire pinch-off nFETs and pFETs with a homogeneous channel including strain and arbitrary crystallographic orientations," *Ultimate Integration on Silicon (ULIS), 2011 12th International Conference on*, Cork, Ireland, pp. 1–4, March 2011.

[138] G. Baccarani, E. Gnani, A. Gnudi, S. Reggiani, and M. Rudan, "Theoretical foundations of the quantum drift-diffusion and density-gradient models," *Solid-State Electronics*, vol. 52, no. 4, pp. 526–32, 2008.

[139] J. Peng, Q. Chen, N. Wong, et al., "A multi-scale framework for nano-electronic devices modeling with application to the junctionless transistor," *2013 IEEE International Conference of Electron Devices and Solid-State Circuits*, Hong Kong, China, pp. 1–2, June 2013.

[140] L. Meng, Z. Yin, C. Yam, et al. "Frequency-domain multiscale quantum mechanics/electromagnetics simulation method," *The Journal of Chemical Physics*, vol. 139, no. 24, pp. 244111-1–244111-6, 2013.

[141] C. Yam, J. Peng, Q. Chen, et al., "A multi-scale modeling of junctionless field-effect transistors," *Applied Physics Letters*, vol. 103, no. 6, pp. 062109-1–062109-5, 2013.

[142] Y. Taur, "Analytic solutions of charge and capacitance in symmetric and asymmetric double-gate MOSFETs," *IEEE Transactions on Electron Devices*, vol. 48, pp. 2861–9, December 2001.

[143] C.-W. Lee, A. Borne, I. Ferain, et al., "High temperature performance of silicon junctionless MOSFETs," *IEEE Transactions on Electron Devices*, vol. 57, pp. 620–5, March 2010.

[144] L. Barbut, F. Jazaeri, D. Bouvet, and J.-M. Sallese, "Transient off current in junctionless FETs," *IEEE Transactions on Electron Devices*, vol. 60, pp. 2080–3, June 2013.

[145] R. Yu, S. Das, I. Ferain, et al., "Device design and estimated performance for p-type junctionless transistors on bulk germanium substrates," *IEEE Transactions on Electron Devices*, vol. 59, pp. 2308–13, September 2012.

[146] F. Jazaeri, L. Barbut, and J.-M. Sallese, "Modeling and design space of junctionless symmetric double gate MOSFETs with long channel," *IEEE Transactions on Electron Devices*, vol. 60, pp. 2120–7, July 2013.

[147] F. Jazaeri, L. Barbut, and J.-M. Sallese, "Generalized charge based model of double gate junctionless FETs including inversion," *IEEE Transactions on Electron Devices*, vol. 61, pp. 3553–7, October 2014.

[148] S. J. Choi, D. I. Moon, S. Kim, J. P. Duarte, and Y. K. Choi, "Sensitivity of threshold voltage to nanowire width variation in junctionless transistors," *IEEE Electron Device Letters*, vol. 32, pp. 125–7, February 2011.

[149] R. Chau, S. Datta, M. Doczy, et al., "Benchmarking nanotechnology for high-performance and low-power logic transistor applications," *IEEE Transactions on Nanotechnology*, vol. 4, no. 2, pp. 153–8, 2005.

[150] G. Masetti, M. Severi, and S. Solmi, "Modeling of carrier mobility against carrier concentration in arsenic-, phosphorus-, and boron-doped silicon," *IEEE Transactions on Electron Devices*, vol. 30, pp. 764–9, July 1983.

[151] Z.-M. Lin, H.-C. Lin, K.-M. Liu, and T.-Y. Huang, "Analytical model of subthreshold current and threshold voltage for fully depleted double-gated junctionless transistor," *Japanese Journal of Applied Physics*, vol. 51, pp. 02BC14–02BC14-7, 2012.

[152] T. Holtij, M. Schwarz, A. Kloes, and B. Iniguez, "2D analytical potential modeling of junctionless DG MOSFETs in subthreshold region including proposal for calculating the threshold voltage," *IEEE Ultimate Integration on Silicon (ULIS) Conference*, Grenoble, France, pp. 81–4, June 2012.

[153] K. K. Young, "Analysis of conduction in fully depleted SOI MOSFETs," *IEEE Transaction Electron Devices*, vol. 36, pp. 504–6, February 1989.

[154] Y. Omura, S. Horiguchi, M. Tabe, and K. Kishi, "Quantum-mechanical effects on the threshold voltage of ultrathin-SOI nMOSFETs," *IEEE Electron Device Letters*, vol. 14, pp. 569–71, December 1993.

[155] J.-P. Colinge, "Multiple-gate SOI MOSFETs," *Solid State Electronics*, vol. 48, pp. 897–905, 2004.

[156] S.-Y. Oh, D. E. Ward, and R. W. Dutton, "Transient analysis of MOS transistors," *IEEE Journal of Solid-State Circuits*, vol. 15, pp. 165–73, August 1980.

[157] D. E. Ward and R. W. Dutton, "A charge-oriented model for mos transistor capacitances," *IEEE Journal of Solid-State Circuits*, vol. 13, pp. 703–8, October 1978.

[158] D. Munteanu, J. L. Autran, X. Loussier, et al., "Quantum short-channel compact modeling of drain-current in double-gate MOSFET," *Solid State Electronics*, vol. 50, pp. 680–6, December 2006.

[159] A. Yesayan, F. Prégaldiny, N. Chevillon, C. Lallement, and J.-M. Sallese, "Physics-based compact model for ultra-scaled FinFETs," *Solid-State Electronics*, vol. 62, pp. 165–73, September 2011.

[160] Z.-M. Lin, H.-C. Lin, K.-M. Liu, and T.-Y. Huang, "Analytical model of subthreshold current and threshold voltage for fully depleted double-gated junctionless transistor," *Japanese Journal of Applied Physics*, vol. 51, pp. 02BC14-1–02BC14-7, February 2012.

[161] J.-H. Woo, J.-M. Choi, and Y.-K. Choi, "Analytical threshold voltage model of junctionless double-gate MOSFETs with localized charges," *IEEE Transactions on Electron Devices*, vol. 60, pp. 2951–5, September 2013.

[162] D. Jang, J. W. Lee, C.-W. Lee, et al., "Low-frequency noise in junctionless multigate transistors," *Applied Physics Letters*, vol. 98, pp. 133502-1–133502-3, 2011.

[163] F. Jazaeri and J.-M. Sallese, "Modelling channel thermal noise and induced gate noise in junctionless FETs," *IEEE Transactions on Electron Devices*, vol. 62, no. 8, pp. 2593–7, August 2015.

[164] C. H. Chen, M. J. Deen, Y. Cheng, and M. Matloubian, "Extraction of the induced gate noise, channel thermal noise and their correlation in sub-micron MOSFETs from RF noise measurements," *IEEE Transactions on Electron Devices*, vol. 48, pp. 2884-892, December 2001.

[165] A. Van Der Ziel, "Gate noise in field effect transistors at moderately high frequencies," *Proceedings of the IEEE*, vol. 51, pp. 461–7, March 1963.

[166] D.P. Triantis, A.N. Birbas, and S.E. Plevridis, "Induced gate noise in MOSFETs revisited: the submicron case," *Solid-State Electronics*, vol. 41, pp. 1937–42, December 1997.

[167] R. van Langevelde, J. C. J. Paasschens, A. J. Scholten, et al., "New compact model for induced gate current noise MOSFET," *IEEE International Electron Devices Meeting*, pp. 36.2.1–36.2.4, 2003.

[168] F. H. De La Moneda, H. N. Kotecha, and M. Shatzkfs, "Measurements of MOSFET constants," *IEEE Electron Device Letters*, vol. 3, pp. 10–12, January 1982.

[169] L. Risch, "Electron mobility in short channel MOSFETs with series resistance," *IEEE Transactions on Electron Devices*, pp. 959–61, January 1983.

[170] T. Grotiohn, and B. Hoefflinger, "A parametric short channel MOS transistor model for subthreshold and strong inversion currents," *IEEE Journal of Solid-State Circuits*, vol. 19, pp. 234–41, February 1984.

[171] S. Severi, L. Pantisano, E. Augendre, et al., "A reliable metric for mobility extraction of short-channel MOSFETs," *IEEE Transactions on Electron Devices*, vol. 54, pp. 2690–8, October 2007.

[172] P. I. Suciu and R. L. Johnston, "Experimental derivation of the source and drain resistance of MOS transistors," *IEEE Transactions on Electron Devices*, vol. 27, pp. 1846–8, September 1980.

[173] F. C. J. Kong , Y. T. Yeow and Z. Q. Yao, "Extraction of MOSFET threshold voltage, series resistance, effective channel length, and inversion layer mobility from small-signal channel conductance measurement," *IEEE Transactions on Electron Devices*, vol. 48, pp. 2870–4, December 2001.

[174] G. Ghibaudo, "New method for the extraction of MOSFET parameters," *Electronics Letters*, vol. 24, pp. 543–5, April 1988.

[175] M. Najmzadeh, M. Berthomé, J.-M. Sallese, W. Grabinski, and A. M. Ionescu, "Electron mobility extraction in triangular gate-all-around Si nanowire junctionless nMOSFETs with cross-section down to 5 nm," *Solid-State Electronics*, vol. 98, pp. 55–62, August 2014.

[176] X. Li, W. Han, H. Wang, et al., "Low-temperature electron mobility in heavily n-doped junctionless nanowire transistor," *Applied Physics Letters*, vol. 102, pp. 223507-1–223507-3, June 2013.

[177] L. Barbut, F. Jazaeri, D. Bouvet, and J.-M. Sallese, "Mobility measurement in nanowires based on magnetic field-induced current splitting method in H-shape devices," *IEEE Transactions on Electron Devices*, vol. 61, pp. 2486–94, July 2014.

[178] O. Gunawan, L. Sekaric, A. Majumdar, et al., "Measurement of carrier mobility in silicon nanowires," *Nano Letter*, vol. 8, pp. 1566–71, June 2008.

[179] J. Chen, T. Saraya, and T. Hiramoto, "Electron mobility in multiple silicon nanowires GAA nMOSFETs on (110) and (100) SOI at room and low temperature," *Electron Devices Meeting, 2008. IEDM 2008. IEEE International*, San Francisco, CA, pp. 1–4, December 2008.

[180] E. C. Garnett, Y.-C. Tseng, D. R. Khanal, et al., "Dopant profiling and surface analysis of silicon nanowires using capacitance-voltage measurements," *Nature Nanotechnology*, vol. 4, pp. 311–14, March 2009.

[181] C. G. Sodini, T. W. Ekstedt, and J. L. Moll, "Charge accumulation and mobility in thin dielectric MOS transistors," *Solid-State Electronics*, vol. 25, pp. 833–41, September 1982.

[182] F. Jazaeri and J.-M. Sallese, "Carrier mobility extraction methodology in junctionless and inversion-mode FETs," *IEEE Transactions on Electron Devices*, vol. 62, pp. 3373–8, October 2015.

[183] P. Lauritzen and O. Leistiko, "Field-effect transistors as low-noise amplifiers," *Solid-State Circuits Conference. Digest of Technical Papers. 1962 IEEE International*, vol. V, pp. 62–3, February 1962.

[184] W. M. C. Sansen and C. J. M. Das, "A simple model of ion-implanted JFETs valid in both the quadratic and the subthreshold regions," *IEEE Journal of Solid-State Circuits*, vol. 17, pp. 658–66, August 1982.

[185] W. W. Wong, J. J. Liou, and J. Prentice, "An improved junction field-effect transistor static model for integrated circuit simulation," *IEEE Transactions on Electron Devices*, vol. 37, pp. 1773–5, July 1990.

[186] W. W. Wong and J. J. Liou, "JFET circuit simulation using SPICE implemented with an improved model," *IEEE Transactions on Computer-Aided Design of Integrated Circuits and Systems*, vol. 13, pp. 105–9, January 1994.

[187] H. Ding, J. J. Liou, K. Green, and C. R. Cirba, "A new model for four-terminal junction field-effect transistors," *Solid-State Electronics*, vol. 50, no. 3, pp. 422– 8, 2006.

[188] W. Wu, S. Banerjee, and K. Joardar, "A four-terminal JFET compact model for high-voltage power applications," *Proceedings of the 2015 International Conference on Microelectronic Test Structures*, Tempe, AZ, pp. 37–41, March 2015.

[189] Y. Cui, Q. Wei, H. Park, and C. M. Lieber, "Nanowire nanosensors for highly sensitive and selective detection of biological and chemical species," *Science*, vol. 293, no. 5533, pp. 1289–92, 2001.

[190] A. Yesayan, F. Jazaeri, and J.-M. Sallese, "Charge-based modeling of double-gate and nanowire junctionless FETs including interface-trapped charges," *IEEE Transactions on Electron Devices*, vol. 63, pp. 1368–74, March 2016.

[191] P. Bergveld, "Development of an ion-sensitive solid-state device for neurophysiological measurements," *IEEE Transactions on Biomedical Engineering*, vol. BME-17, pp. 70–1, January 1970.

[192] P. Bergveld, "Thirty years of isfetology: what happened in the past 30 years and what may happen in the next 30 years," *Sensors and Actuators B: Chemical*, vol. 88, no. 1, pp. 1–20, 2003.

[193] M. W. Shinwari, M. J. Deen, and D. Landheer, "Study of the electrolyte-insulator-semiconductor field-effect transistor (eisfet) with applications in biosensor design," *Microelectronics Reliability*, vol. 47, no. 12, pp. 2025–57, 2007.

[194] M. Schöning and A. Poghossian, "Bio feds (field-effect devices): state-of-the-art and new directions," *Electroanalysis*, vol. 18, no. 19–20, pp. 1893–900, 2006.

[195] S. D. Moss, J. Janata, and C. C. Johnson, "Potassium ion-sensitive field effect transistor," *Analytical Chemistry*, vol. 47, no. 13, pp. 2238–43, 1975.

[196] D. C. Grahame, "The electrical double layer and the theory of electrocapillarity," *Chemical Reviews*, vol. 41, no. 3, pp. 441–501, 1947.

[197] R. van Hal, J. Eijkel, and P. Bergveld, "A general model to describe the electrostatic potential at electrolyte oxide interfaces," *Advances in Colloid and Interface Science*, vol. 69, no. 1, pp. 31–62, 1996.

[198] F. Patolsky, G. Zheng, and C. M. Lieber, "Nanowire-based biosensors," *Analytical Chemistry*, vol. 78, no. 13, pp. 4260–9, 2006.

[199] Y. Cui, Q. Wei, H. Park, and C. M. Lieber, "Nanowire nanosensors for highly sensitive and selective detection of biological and chemical species," *Science*, vol. 293, no. 5533, pp. 1289–92, 2001.

[200] J.-I. Hahm and C. M. Lieber, "Direct ultrasensitive electrical detection of DNA and DNA sequence variations using nanowire nanosensors," *Nano Letters*, vol. 4, no. 1, pp. 51–4, 2004.

[201] Z. Li, Y. Chen, X. Li, et al., "Sequence-specific Label-free DNA sensors based on silicon nanowires," *Nano Letters*, vol. 4, no. 2, pp. 245–7, 2004.

[202] J. A. Streifer, H. Kim, B. M. Nichols, and R. J. Hamers, "Covalent functionalization and biomolecular recognition properties of DNA-modified silicon nanowires," *Nanotechnology*, vol. 16, no. 9, p. 1868, 2005.

[203] E. Buitrago, M. Fernández-Bolaños, and A. Ionescu, "Vertically stacked Si nanostructures for biosensing applications," *Microelectronic Engineering*, vol. 97, pp. 345–8, 2012.

[204] L. Chen, F. Cai, U. Otuonye, and W. D. Lu, "Vertical Ge/Si core/shell nanowire junctionless transistor," *Nano Letters*, vol. 16, no. 1, pp. 420–6, 2016.

[205] G. Zheng, F. Patolsky, Y. Cui, W. U. Wang, and C. M. Lieber, "Multiplexed electrical detection of cancer markers with nanowire sensor arrays," *Nature Biotechnology*, vol. 23, no. 10, pp. 1294–301, 2005.

[206] F. Patolsky and C. M. Lieber, "Nanowire nanosensors," *Materials Today*, vol. 8, no. 4, pp. 20–8, 2005.

[207] F. Najam, S. Kim, and Y. S. Yu, "Gate all around metal oxide field transistor: surface potential calculation method including doping and interface trap charge and the effect of interface trap charge on subthreshold slope," *Journal of Semiconductor Technology and Science*, vol. 13, no. 5, pp. 530–7, 2013.

[208] Z. Chen, X. Zhou, G. Zhu, and S. Lin, "Interface-trap modeling for silicon-nanowire MOSFETs," *Reliability Physics Symposium (IRPS), 2010 IEEE International*, Anaheim, CA, pp. 977–80, May 2010.

[209] B. H. Hong, N. Cho, S. J. Lee, et al., "Subthreshold degradation of gate-all-around silicon nanowire field-effect transistors: effect of interface trap charge," *IEEE Electron Device Letters*, vol. 32, pp. 1179–81, September 2011.

[210] A. C. E. Chia and R. R. LaPierre, "Analytical model of surface depletion in gaas nanowires," *Journal of Applied Physics*, vol. 112, no. 6, pp. 063705-1–063705-7, 2012.

[211] S. M. Sze and K. K. Ng, "Metal–insulator–semiconductor capacitors," *Physics of semiconductor devices*, 3rd ed. Hoboken, NJ: Wiley, pp. 214–15, 2007.

[212] J. G. Simmons and G. W. Taylor, "Nonequilibrium steady-state statistics and associated effects for insulators and semiconductors containing an arbitrary distribution of traps," *Physical Reviews B*, vol. 4, pp. 502–11, July 1971.

[213] V. Schmidt, S. Senz, and U. Gosele, "Influence of the Si/SiO2 interface on the charge carrier density of Si nanowires," *Applied Physics A*, vol. 86, no. 2, pp. 187–91, 2006.

[214] Z. Chen, B. J. Bin, and C.-T. Sah, "Effects of energy distribution of interface traps on recombination dc current–voltage line shape," *Journal of Applied Physics*, vol. 100, no. 11, p. 114511, 2006.

[215] M. S. Kim, H. T. Kim, S. S. Chi, et al., "Distribution of interface states in MOS systems extracted by the subthreshold current in MOSFETs under optical illumination," *Journal of the Korean Physical Society*, vol. 43, pp. 873–8, November 2003.

[216] M. Marek-Sadowska and X. Qiu, "A study on cell-level routing for VeSFET circuits," *Proceedings of the 18th International Conference Mixed Design of Integrated Circuits and Systems – MIXDES 2011*, Gliwice, Poland, pp. 127–32, June 2011.

[217] X. Qiu and M. Marek-Sadowska, "Can pin access limit the footprint scaling?," *Proceedings of the 49th Annual Design Automation Conference*, San Francisco, CA, pp. 1100–6, 2012.

[218] L. Barbut, F. Jazaeri, D. Bouvet, and J.-M. Sallese, "Design space of twin gate junctionless vertical slit field effect transistors," *Mixed Design of Integrated Circuits and Systems (MIXDES), 2013 Proceedings of the 20th International Conference*, Gdynia, Poland, pp. 393–6, June 2013.

[219] A. Pfitzner, "Vertical-slit field-effect transistor (VeSFET) – design space exploration and DC model," *Proceedings of the 18th International Conference Mixed Design of Integrated Circuits and Systems – MIXDES 2011*, Gliwice, pp. 151–6, June 2011.

[220] D. E. Ioannou, S. Cristoloveanu, M. Mukherjee, and B. Mazhari, "Characterization of carrier generation in enhancement-mode SOI MOSFET's," *IEEE Electron Device Letters*, vol. 11, pp. 409–11, September 1990.

[221] S. P. Sinha, A. Zaleski, D. E. Ioannou, "Investigation of carrier generation in fully depleted enhancement and accumulation mode SOI MOSFET's," *IEEE Transactions on Electron Devices*, vol. 41, pp. 2413–16, December 1994.

[222] D. Munteanu, D. A. Weiser, S. Cristoloveanu, et al., "Generation–recombination transient effects in partially depleted SOI transistors: systematic experiments and simulations," *IEEE Transactions on Electron Devices*, vol. 45, pp. 1678–83, August 1998.

[223] D. Munteanu and A. M. Ionescu, "Modeling of drain current overshoot and recombination lifetime extraction in floating-body submicron SOI MOSFETs," *IEEE Transactions on Electron Devices*, vol. 49, pp. 1198–205, July 2002.

Index